Molecular and Ionic Recognition with Imprinted Polymers

ACS SYMPOSIUM SERIES **703**

Molecular and Ionic Recognition with Imprinted Polymers

Richard A. Bartsch, EDITOR
Texas Tech University

Mizuo Maeda, EDITOR
Kyushu University

Developed from a symposium sponsored by the Division
of Industrial and Engineering Chemistry
at the 213th National Meeting
of the American Chemical Society,
San Francisco, California,
April 13–17, 1997

American Chemical Society, Washington, DC

Library of Congress Cataloging-in-Publication Data

Molecular and ionic recognition with imprinted polymers / Richard A. Bartsch, editor, Mizuo Maeda, editor.

p. cm.—(ACS symposium series, ISSN 0097–6156; 703)

"Developed from a symposium sponsored by the Division of Industrial and Engineering Chemistry at the 213th National Meeting of the American Chemical Society, San Francisco, California, April 13–17, 1997."

Includes bibliographical references and indexes.

ISBN 0–8412–3574–0

1. Imprinted polymers—Congresses. 2. Molecular imprinting—Congresses. 3. Molecular recognition—Congressses.

I. Bartsch, Richard A. II. Maeda, Mizuo. III. American Chemical Society. Division of Industrial and Engineering Chemistry. IV. American Chemical Society. Meeting (213th : 1997 : San Francisco, Calif.) V. Series.

QD382.I43M65   1998
543'.1'08—dc21   98–6979
                                                                                  CIP

The paper used in this publication meets the minimum requirements of American National Standard for Information Sciences—Permanence of Paper for Printed Library Materials, ANSI Z39.48–1984.

PRINTED IN THE UNITED STATES OF AMERICA

Foreword

THE ACS SYMPOSIUM SERIES was first published in 1974 to provide a mechanism for publishing symposia quickly in book form. The purpose of the series is to publish timely, comprehensive books developed from ACS sponsored symposia based on current scientific research. Occasionally, books are developed from symposia sponsored by other organizations when the topic is of keen interest to the chemistry audience.

Before agreeing to publish a book, the proposed table of contents is reviewed for appropriate and comprehensive coverage and for interest to the audience. Some papers may be excluded in order to better focus the book; others may be added to provide comprehensiveness. When appropriate, overview or introductory chapters are added. Drafts of chapters are peer-reviewed prior to final acceptance or rejection, and manuscripts are prepared in camera-ready format.

As a rule, only original research papers and original review papers are included in the volumes. Verbatim reproductions of previously published papers are not accepted.

ACS BOOKS DEPARTMENT

Contents

METAL ION RECOGNITION
WITH ORGANIC-BASED POLYMERS

RECOGNITION WITH INORGANIC-BASED POLYMERS

Preface

MOLECULAR RECOGNITION is a central feature of life in which biomolecules, such as proteins and nucleic acids, control biological activities. Because such biological recognition involves simple interactions between chemical units, synthetic molecules that are capable of molecular recognition are the object of intensive research investigations.

A new approach to the preparation of "host" molecules that can recognize "guest" species is a template polymerization technique called "molecular imprinting". In this method, a template molecule or ionic species associates with one or more functional monomers to form a complex. This complex is then polymerized with a matrix-forming monomer (cross-linker) to produce a resin. Upon removal of the template species, cavities that recognize the spatial features and bonding preference of the template are produced. These cavities may then be used to selectively rebind the template from a mixture of chemical species.

Molecular imprinting has now been successfully applied to the creation of recognition sites in synthetic polymers for a variety of small organic molecules, inorganic ions, and even biological macromolecules. In recent reports, imprinted polymers have been utilized as stationary phases for chromatographic separations of molecules, including some with very minor structural differences (e.g. optical isomers), for molecular and ionic separations by selective sorption, as recognition entities in sensors and immunoassays, and as catalysts.

At the Spring 1997 National Meeting of the American Chemical Society, the world's first international symposium on imprinted polymers was held. The three-day Symposium on Recognition with Imprinted Polymers included plenary lectures by Günter Wulf (University of Düsseldorf, Germany) and Klaus Mosbach (Lund University, Sweden), pioneers in the field of polymer imprinting. Presentations were also made by most of the other established investigators in this rapidly developing field, as well as by several relative newcomers. This world's first book on imprinted polymers is based upon papers that were presented in the symposium.

Within the 22 chapters of this book, the most recent information regarding the preparation of imprinted polymers and their applications in recognition of a variety of molecular and ionic species is summarized. The overview chapter provides an introduction to the two important types of polymer imprinting and brief discussions of reported applications for imprinted polymeric materials.

This is followed by three survey chapters describing the overall status of the field, as well as general principles for the preparation of imprinted polymers, and summaries of applications of these novel materials. The remainder of the book is divided into sections on molecular recognition with organic-based polymers (10 chapters), metal ion recognition with organic-based polymers (6 chapters), and recognition with inorganic-based polymers (2 chapters). These chapters are primarily focused upon the applications of imprinted polymers in the determination and separation of such diverse species as amino acids, drugs, herbicides, metal ions, and steroids. Also included is the use of imprinted polymers as catalysts for organic and inorganic reactions.

These descriptions of imprinted polymer technology provide state-of-the-art information for both the novice and the practitioner. A wide variety of scientists and engineers, including analytical chemists, biochemists, chemical engineers, clinical chemists, pharmaceutical scientists, polymer scientists, and separation scientists, will find this book to be a valuable resource.

Acknowledgments

For the Symposium on Recognition with Imprinted Polymers held during the 213th ACS National Meeting in San Francisco on April 13–17, 1997, we gratefully acknowledge financial support from the Separation Science and Technology Subdivision of the ACS Division of Industrial and Engineering Chemistry; The Donors of the Petroleum Research Fund, administered by the American Chemical Society; IGEN, Inc. (Gaithersburg, Maryland); and Eichrom Industries, Inc. (Darien, Illinois). We also acknowledge the members of the ACS Books Department for their efforts in assembling this volume. Finally, we thank the authors for preparing their manuscripts and the referees who reviewed each chapter.

RICHARD A. BARTSCH
Department of Chemistry and Biochemistry
Texas Tech University
Lubbock, TX 79409–1061

MIZUO MAEDA
Department of Chemical Science and Technology
Faculty of Engineering
Kysushu University
Hakozaki, Fukuoka 812-81, Japan

Chapter 1

Molecular and Ionic Recognition with Imprinted Polymers: A Brief Overview

Mizuo Maeda[1] and Richard A. Bartsch[2]

[1]Department of Chemical Science and Technology, Faculty of Engineering, Kyushu University, Hakozaki, Fukuoka 812-81, Japan
[2]Department of Chemistry and Biochemistry, Texas Tech University, Lubbock, TX 70409–1061

The concepts of chemical imprinting of polymeric materials and recognition of molecular and ionic species with imprinted polymers are introduced. Following a brief description of the two general methodologies for the preparation of imprinted polymers, applications are outlined for their use in chromatographic separations, as artificial antibodies in drug and pesticide analysis, in selective sorption of molecular and ionic species, in sensors and membranes, and as artificial enzymes. Recent advances in molecular imprinting as described in the twenty-one contributed chapters are highlighted.

Molecular recognition is crucial for the functioning of living systems, where biological macromolecules including proteins (which may serve as enzymes, antibodies, and receptors), nucleic acids, and saccharides play decisive roles in biological activities. Since the recognition should be based on simple interactions between chemical units, the challenge of synthesizing artificial molecules which are capable of molecular recognition has drawn special attention to this field of chemistry. In the effort to obtain host molecules with precise recognition of guest species, the design, synthesis, and evaluation of supramolecules are being intensely investigated in laboratories throughout the world (*1*). However, the requisite multi-step preparative routes to the molecular receptors are frequently tedious and often provide only a low overall yield of the final product.

An alternative approach to the synthesis of host molecules which can recognize targeted guest species is a much simpler template polymerization technique called "molecular imprinting" (*2-4*). In this method, precise molecular design is not necessary since a crosslinked polymeric resin is prepared in the presence of a template (Figure 1). The template molecule or ionic species associates with the functional monomer(s) to form a covalent or non-covalent-bonded complex. This complex is then polymerized with a matrix-forming monomer (crosslinker) to give a resin. Upon removal of the template species, cavities are formed in the polymer matrix. These cavities have memorized the spatial features and bonding preference(s) of the template, so the imprinted polymer will selectively rebind the template from a mixture of chemical species.

1

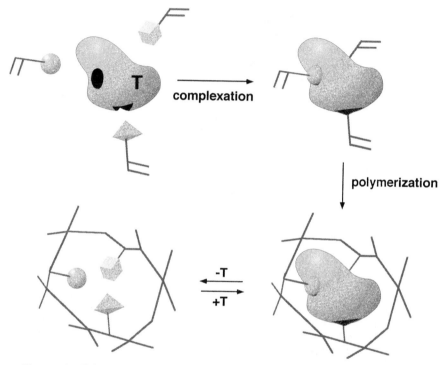

Figure 1. Schematic representation of the molecular imprinting technique. Specific cavities are formed in a crosslinked polymer by a template (T) with different complementary binding monomers.

Molecular imprinting has been successfully applied to the creation of recognition sites in synthetic polymers for various kinds of small organic molecules, inorganic ions, and even biological macromolecules. The imprinted polymers have been used as stationary phases for chromatographic separation of molecules, including those which have minor structural differences (*e.g.*, optical isomers), for molecular and ionic separations by selective sorption, as recognition entities in sensors and immunoassays, and as catalysts.

The original concept of molecular imprinting was first inspired by Linus Pauling in the 1940's with his postulated mechanism for the formation of antibodies (*5*). However, the current interest in imprinted materials as artificial antibodies is attributed to a landmark paper by Wulff and Sarhan in 1972 (*6*). By template polymerization, these researchers succeeded in preparing polymers with chiral cavities for the separation of racemic mixtures. Research with imprinted materials accelerated when Mosbach *et al.* introduced a simpler and synthetic approach which utilized non-covalent interactions of the template species with the functional monomer(s) (*7*).

The number of articles on molecular imprinting published each year from 1931 to 1997 is shown in Figure 2 (adapted from (*8*)). As can be seen, the annual number of publications on polymer imprinting remained at a low level until the mid-1980's. This was followed by a gradual increase in the late 1980's and into the early 1990's. During the period of 1994 to 1997, a dramatic increase in the frequency of reports on imprinted polymers is readily evident. At present, research groups throughout the world are focusing attention on the preparation of novel imprinted polymers and new analytical applications of molecular-imprinted materials in HPLC, capillary electrophoresis, selective sorption, various types of sensors, and membrane separations.

Imprinting Methodologies

In general, molecular imprinting processes may be classified in one of two groups depending upon the type of interactions which exist between the template guest and the host functional monomer(s): a) covalent bonding; and b) non-covalent bonding (Figures 3a and 3b, respectively). The covalent bonding system (Figure 3a) employs a template-monomer complex which is formed by covalent, but reversible, bonding. The first example of an imprinted polymer reported by Wulff *et al.* utilized the reversible formation of ester linkages between a sugar and phenylboronic acid which was derivatized with a vinyl group (*6*). Covalent bonding involving Schiff bases and ketals was employed subsequently in the covalent imprinting of amino acid derivatives and ketones, respectively (*2*).

An important feature of imprinting with covalent interactions is that it has allowed the structures of imprinted cavities to be probed in detail. In addition to Wulff's pioneering investigations (*2*), Shea *et al.* (*3*) have made leading contributions in this area. However, the covalent system is not very flexible in terms of the choices of functional monomer(s) and template species.

For non-covalent imprinting systems, the interactions include hydrogen bonding, electrostatic interactions, and metal ion coordination. A typical functional monomer for non-covalent imprinting is methacrylic acid, which was first proposed by Mosbach *et al.* (*4*). Methacrylic acid has been utilized as the functional monomer for a variety of template species, which include peptides, nucleotides, saccharides, drugs, herbicides, and inorganic ions. Although the non-covalent imprinting system is simple and convenient, each interaction is weak so that a large excess (at least four times) of the functional monomer should be used. This may result in formation of a variety of binding sites with different affinities (*i.e.* polyclonality).

Another important factor in molecular imprinting is a crosslinker whose type and quantity are critical factors for attaining high affinity. Ethyleneglycol dimethacrylate has been used frequently. Divinylbenzene is an alternative, but less popular,

4

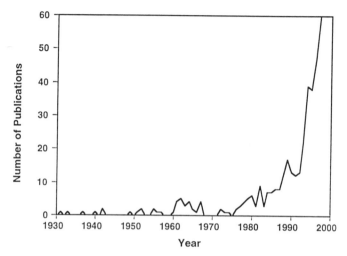

Figure 2. The number of articles on molecular imprinting published from 1931 to 1977. (Adapted from (8)).

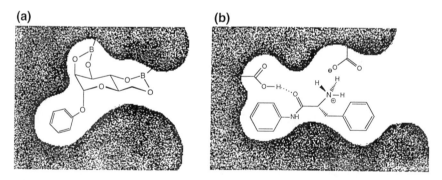

Figure 3. Examples for the two categories of molecular imprinting based upon the interaction between the template guest and the functional host monomer: a) Covalent bonding, (4-vinylphenyl)boronic acid imprinting sugar derivatives; b) Non-covalent bonding, methacrylic acid imprinting of amino acid derivatives. In both cases, the crosslinker was ethyleneglycol dimethacrylate.

crosslinker. Utilization of a porogen or solvent to give a macroporous structure is important to allow guest molecules ready access to the imprinted sites.

Factors to be considered in the choice of functional monomers, crosslinkers, porogens, and reaction conditions for the preparation of imprinted polymers have been discussed in several reviews (2-4,8,9). In this monograph, excellent summaries concerning the selection of these experimental variables is provided in chapters by Wulff et al., Mosbach et al., and Sellergren (Chapters 1-3, respectively).

Applications of Imprinted Polymers

Chromatographic Separations of Molecular Species with Imprinted Polymers. Molecular-imprinted polymeric materials have been used extensively as stationary phases for separations by high performance liquid chromatography (HPLC). Molecules with closely related structures can be separated (e.g., different steroids, various herbicides, and underivatized amino acids) and chiral separations of peptides and of drugs have been achieved. It is now possible to prepare tailor-made HPLC column materials for separations of a variety of molecules of interest (9). In this volume, applications of imprinted polymers as stationary phases for HPLC are described by Sellergren, by Ramström et al., by Karube and coworkers, by Arnold et al., by Takeuchi and Matsui, and by Haupt in Chapters 4-9, respectively.

In many instances, the stationary phase is derived from an imprinted polymer which has been prepared by bulk polymerization. In Chapter 10, Hosoya and Tanaka describe the development of a method for synthesizing uniform-sized, molecular-imprinted polymer particles to enhance column efficiency; whereas in Chapter 7, Arnold et al. coat an imprinted polymer on silica gel particles. Hosoya and Tanaka also explore the in situ surface modification of imprinted polymers to produce 'multi-channel selectors'. In Chapter 8, Takeuchi and Matsui describe an in situ method for the preparation of the molecular-imprinted stationary phase directly in an HPLC column.

A post-polymerization technique to enhance the selectivity of HPLC stationary phases by reducing the 'polyclonality' of the imprinted sites is reported by Karube and coworkers in Chapter 6. This is the first successful attempt to reduce the polyclonality which diminishes selectivity for the template molecule.

In addition to HPLC, the use of molecular-imprinted materials is being explored in capillary electrophoresis (10,11) for which very high numbers of theoretical plates may be achieved.

Thin layer chromatography is another frequently encountered chromatographic technique. Preliminary results for the application of imprinted materials in TLC have been published (12).

Selective Sorption of Molecular Species on Imprinted Polymers. Separations of template molecules from other molecular species can also be accomplished by solid phase extraction with imprinted polymers. Such batch sorption experiments are utilized in this monograph for the separation of underivatized amino acids by Arnold et al. (Chapter 7) and Yoshikawa (Chapter 12) and of agrochemicals by Takeuchi and Matsui (Chapter 8) and by Haupt (Chapter 9).

Selective Sorption of Ionic Species on Imprinted Polymers. Compared with the application of molecular imprinting in the construction of selective polymers for relatively large molecules, imprinted polymeric adsorbents for ionic species have received much less attention. The combination of high levels of crosslinking with macromolecular ligating monomers produced a selective polymer for Ca(II), a metal ion which does not exhibit directional bonding (13). In this monograph, several chapters are devoted to the preparation and evaluation of polymers for transition and heavy metal ions which exhibit directional bonding. Templated polymers with high selectivity for Pb(II) are described by Murray *et al.* in Chapter 15. High selectivities for sorption of Cu(II) and of Hg(II) with imprinted polymeric materials are reported by Fish in Chapter 16. In these investigations, the imprinted polymers were prepared by bulk polymerization.

The use of 'Surface Imprinting' to prepare microspheres with transition metal-templated surfaces and the metal ion-sorption properties of these material are described in Chapters 17-20 by Tsukagoshi *et al.*, Koide *et al.*, Uezu *et al.*, and Miyajima *et al.*, respectively. Uezu *et al.* also report results for post irradiation of the imprinted polymer particles with γ rays to make the polymeric matrix more rigid.

In Chapter 22, Sasaki *et al.* utilize molecular imprinted receptors in sol-gel materials for the aqueous phase recognition of phosphates and phosphonates.

In this monograph, two chapters describe the utilization of metal complex imprinted polymers for applications in which the metal ion species are retained in the resin. In Chapter 7, Arnold *et al.* utilized metal complexes to imprint polymers with amines and underivatized amino acids. In another novel application, Krebs and Borovik (Chapter 11) employed the microenvironments of metal chelate complexes to design surrounding network polymers for complexation of CO. In this case, the imprinted polymers are mimics of metalloproteins.

Selective Transport of Molecular Species Through Imprinted Polymer Membranes. Separation of chemical species by selective permeation through imprinted polymeric membranes is in its infancy. In this volume, Yoshikawa (Chapter 12) describes the preparation of imprinted polymer membranes which under electrodialysis provide chiral separations of amino acids. Kobayashi *et al.* (Chapter 13) show the preferential adsorption of the template to an acrylonitrile-acrylic acid copolymer membrane which was prepared by a phase inversion method in the presence of the template. However, only sorption of the template by the membrane, not permeation, is reported at this time.

Imprinted Polymers as Artificial Antibodies in Drug and Pesticide Analysis. Imprinted polymers were originally designed as antibody mimics. Mosbach *et al.* have employed imprinted polymers as recognition elements in immunoassay-type analyses instead of antibodies (*14*). Trace analysis of drugs is another landmark in the field of molecular imprinting. Such artificial antibodies are especially important for compounds against which it is difficult or impossible to prepare antibodies, *e.g.* non-immunogenic molecules, immunosuppressive drugs, *etc.* In Chapter 9, Haupt provides information about this application of imprinted polymers.

Sensors Involving Imprinted Polymers. As described by Mosbach *et al.* in Chapter 3, applications of imprinted polymers in sensors are receiving ever increasing attention. Arnold *et al.* have reported recently a glucose sensor which uses a ligand exchange reaction on an imprinted metal complex (*15*). In this monograph, Karube and coworkers (Chapter 6) describe a FIA (flow injection analysis)-based sensor for the antibiotic chloramphenicol in which a dye conjugate is displaced from a molecular-imprinted polymer by the substrate. Murray *et al.* (Chapter 15) imprint polymers with Pb(II) and, after their incorporation in polyvinyl chloride membranes, obtain ion-selective electrodes with high selectivity for Pb(II). The use of imprinted polymers in uranyl ion-selective electrodes is also described. An optical sensor for Pb(II) based upon imprinted polymers is presented by Murray *et al.* in Chapter 15. In other work, Murray and coworkers are developing sensors for the nerve gasses sarin and soman (*16*).

Imprinted Polymers as Artificial Enzymes. For application as artificial enzymes, imprinted polymers are imprinted with transition state analogues. The first example of an artificial catalytic antibody was reported by Robinson and Mosbach (*17*). Recent advances in this area are reported in this volume by Wulff (Chapter 1) and by Mosbach *et al.* (Chapter 2). Shea and coworkers have been important contributors in this area. In this monograph, Morihara *et al.* describe the preparation of 'footprints' on an alumina-doped silica surface which are molecular-imprinted sites with complementary cavities for a template species and a Lewis acid site at the bottom. In Chapter 5, Ramström *et al.* employ an imprinted polymer to shift the equilibrium for a chemical reaction.

Future Trends

Molecular imprinting is in the adolescent stage (*i.e.* between infancy and maturity). Although our understanding of the experimental variables which control molecular imprinting is increasing rapidly, much remains to be learned for the method to realize its full potential.

One important issue is that of 'polyclonality' (*i.e.* selectivity differences between individual cavities) which reduces the selectivity attainable with non-covalently imprinted materials. The first successful attempt to reduce the 'polyclonality' of imprinted sites is reported by Karube and coworkers in Chapter 6. For an imprinted polymer prepared with acrylic acid as the functional monomer, the template species is readsorbed to block sites at which the template molecule is strongly bound. Subsequent methylation is then restricted to the non-specific or weak binding sites. Removal of the template provides a modified imprinted polymer with enhanced selectivity (monoclonality) for the print species.

Another factor is the effect of aqueous media in the imprinting process. In most of the reported studies, the polymerization which forms the templated site is performed in an organic solvent to maximize the weak, non-covalent interactions between the template and the functional monomer(s). Even for metal ions, organic media have been employed for the preparation of the imprinted polymers. In this volume, Tsukagoshi *et al.*, Koide *et al.*, and Uezu *et al.* (Chapters 17-19, respectively) use water-in-oil emulsions to imprint the surfaces of microspheres with transition metal ions. Koide *et al.* (Chapter 18) employed a functional polymerizable surfactant for formation of surface imprinted microspheres at the oil-water interface. Ferrocyanide, an inorganic ion, has been examined as a template in an aqueous system by a similar surface imprinting method using a cationic surfactant (*18*). However, aqueous conditions are inevitable when biopolymers, such as proteins and polysaccharides, are the targets. Arnold suggested surface imprinting on a vesicle membrane, constituted of polymerizable lipids (*19*). A different approach was made by Hjerten *et al.* (*20*) for enzyme imprinting. These authors imprinted enzymes, such as ribonuclease, in crosslinked polyacrylamide gel without the use of a functional monomer. However, the degree of crosslinking was relatively low. On the other hand, imprinting with bacteria on the surface of polymer particles was reported recently (*21*).

An absolutely different approach is a so-called 'bioimprinting'. Klibanov *et al.* freeze-dried a protein which was dissolved in water with a template. The dried protein was found to adsorb the template selectively in organic media (*22*). Mosbach *et al.* succeeded in reversing the substrate specificity of α-chymotrypsin from L- to D-, by treating it with the D-substrate (*23*). Thus bioimprinting appears to be a promising alternative to polymer imprinting with biomolecules. In this volume, Umeno *et al.* (Chapter 14) report the polymer modification of DNA double helix in which imprinting by a protein was quite effective. Template polymerization of DNA-binding monomers on DNA had been attempted by Bunemann *et al.* (*24*).

In this overview, we have attempted to provide the reader with a brief introduction to the field of imprinted polymers. More detailed summaries are available in recent reviews [*2-4,8,9*] and in survey Chapters 2-4.

8

A very recent development is organization of the Society for Molecular Imprinting (SMI) in 1997. The goal of this non-profit organization is to serve as a platform for researchers in molecular imprinting to contact one another and exchange ideas leading to accelerated development of this exciting technology. SMI is a truly international organization with worldwide membership and a Board consisting of distinguished scientists from Europe, Japan, and the United States. Address: Society for Molecular Imprinting, c/o Pure and Applied Biochemistry, Center for Chemistry and Chemical Engineering, Lund University, POB 124, SE-221 00 Lund, Sweden. Fax: +46 46 2224611 Email: SMI@ng.hik.se URL: http://www.ng.hik.se/SMI

Acknowledgment

RAB wishes to express his appreciation to the Texas Higher Education Coordinating Board Advanced Research Program for support of his initial research efforts in the field of molecular imprinting.

Literature Cited

1. Lehn, J. M. *Angew. Chem., Int. Ed. Engl.* **1988**, *27*, 89.
2. Wulff, G. *Angew. Chem., Int. Ed. Engl.* **1995**, *34*, 1812.
3. Shea, K. J. *Trends Polym. Sci.* **1994**, *2*, 166.
4. Mosbach, K. *Trends Biochem.* **1994**, *19*, 9.
5. Pauling, L. *J. Am. Chem. Soc.* **1940**, *60*, 2643.
6. Wulff, G.; A. Sarhan, A. *Angew. Chem., Int. Ed. Engl.* **1972**, *11*, 341.
7. Andersson, L.; Sellergren, B.; Mosbach, K. *Tetrahedron Lett.* **1984**, *25*, 5211.
8. Anderson, H. S.; Nicholls, I. A, *Recent Res. Develop. Pure Appl. Chem.* **1997**, *1*, 133.
9. Takeuchi, T.; Matsui, J. *Acta Polymer* **1996**, *47*, 471.
10. Nilsson, K.; Lindell, J.; Norrlow, O.; Sellergren, B. *J. Chromatogr. A* **1994**, *680*, 57.
11. Lin, J. M.; Nakagawa, K.; Uchiyama, K.; Hobo, T. J. *Liq. Chromatogr.* **1997**, *20*, 1489.
12. Kriz, D.; Kriz, C. B.; Andersson, L. I.; Mosbach, K. *Anal. Chem.* **1994**, *66*, 2636.
13. Rosatzin, T.; Andersson, L. I.; Simon, W.; Mosbach, K. *J. Chem. Soc. Perkin Trans. 2* **1991**, 1261.
14. Vlatakis, G.; Andersson, L. I.; Muller, R.; Mosbach, K. *Nature (London)* **1993**, *361*, 645.
15. Chen, G.; Guan, Z.; Chen, C.-T.; Sundaresan, V.; Fu, L.; Arnold, R. H. *Nature Biotech.* **1997**, *15*, 354.
16. Murray, G.; Fish, R. H. *New Scientist*, September 13, 1997, p. 34.
17. Robinson, D. K.; Mosbach, K. *J. Chem. Soc., Chem. Commun.* **1989**, 969.
18. Fujiwara, I.; Maeda, M.; Takagi, M. *Anal. Sci.* **1996**, *12*, 545.
19. Shnek, D. R.; Pack, D. W.; Sasaki, D. Y.; Arnold, F. H. *Langmuir* **1994**, *10*, 2382.
20. Hjerten, S.; Liao, J.-L.; Nakazato, K.; Wang, Y.; Zamaratskaia, G.; Zhang, H.-X. *Chromatographia* **1997**, *44*, 227.
21. Aherne, A.; Alexander, C.; Payne, M. J.; Perez, N.; Vulfson, E. N.; *J. Am. Chem. Soc.* **1996**, *118*, 8771.
22. Braco, L.; Dabulis, K.; Klibanov, A. M. *Proc. Natl. Acad. Sci. USA* **1990**, *87*, 274.
23. Stahl, M.; Mansson, M.-O.; Mosbach, K. J. *Am. Chem. Soc.* **1991**, *113*, 9366.
24. Bunemann, H.; Dattagupta, N.; Schuetz, H. J.; Muller, W. *Biochemistry* **1981**, *20*, 2864.

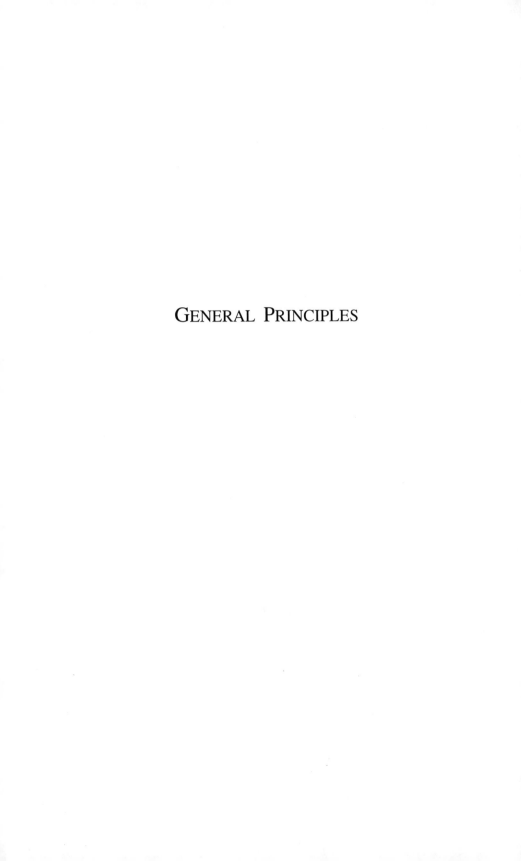

GENERAL PRINCIPLES

Chapter 2

Molecular Imprinting for the Preparation of Enzyme-Analogous Polymers

Günter Wulff, Thomas Gross, Rainer Schönfeld, Thomas Schrader, and Christian Kirsten

Institute of Organic Chemistry and Macromolecular Chemistry, Heinrich-Heine-University Duesseldorf, Universitätsstrasse 1, 40225 Duesseldorf, Germany

The principle of molecular imprinting is based on the crosslinking of a polymer in the presence of interacting monomers around a molecule that acts as a template. After removal of the template, an imprint of specific shape and with functional groups capable of chemical interactions remains in the polymer. We give an introduction into the basic principles of this method, and present new developments in our institute regarding this procedure. Recently we embarked on the development of new types of binding-site interactions which are noncovalent and stoichiometric (due to high binding constants), but do not show the disadvantages of other types of noncovalent binding. Furthermore, new catalytic systems have been designed which exhibit high esterolytic activity and Michaelis-Menten kinetics.

The initial goal of our work on imprinting was to synthesize artificial enzyme mimics which possess the high chemo-, substrate-, and stereoselectivities of enzymes but which, we hope, are at the same time better accessible, more stable, and catalyze a larger variety of reactions (*1*). This was intended to be accomplished by transferring the principles of enzyme action to synthetic polymers.

As is generally known, enzymes catalyze in such a way that the substrate is first bound in the active center in a very defined orientation. Then the bound substrate reacts with other partners from the solution or with a coenzyme under catalysis of specific groups inside the active center. The final product is released in the last step.

The synthesis of the active center therefore remains the crucial problem for the construction of enzyme models (*2*). The following conditions have to be fulfilled in order to approach enzyme similarity (*3*):

a)   A cavity or a cleft must be prepared with a defined shape corresponding to the shape of the substrate of the reaction. For catalysis, it is preferable if the shape of the cavity resembles the shape of the transition state of the reaction, which needs to be more strongly bound than the substrate.

b)   A procedure must be developed in which the functional groups acting as coenzyme analogs, binding sites, or catalytic sites are introduced in the cavity in a predetermined arrangement.

c)   Since binding of the substrate or the transition state of the reaction in the active center is a rather complex overlap of different interactions (which are not known in detail), simplified binding mechanisms have to be developed.

d)   Many enzymes contain coenzymes or prosthetic groups that play an important role in enzyme catalysis. Analogs of such groups should also become useful.

e)   An important part of the preparation of an enzyme model is the choice of a suitable, catalytically active environment. Again simplified systems have to be used in enzyme models without omitting the essential characteristics.

In the past, remarkable results have been obtained using macrocyclic compounds as models for the active sites of enzyme models. For example, crown ethers (4), cryptates (5), cyclophanes (6), cyclodextrins (7), and concave molecules (8) have been used as binding sites to which catalytically active groups have been attached in the correct orientation.

We wanted to use synthetic polymers as carriers of the active site since enzymes are also macromolecules. In this way it should be possible to even imitate dynamic effects, such as *induced fit*, *steric strain*, and *allosteric effects*, more exactly than with low-molecular-weight compounds. On the other hand, generating a model of an active center in polymers becomes much more complicated.

Some years ago, an extremely interesting approach towards enzyme models used monoclonal antibodies (9). These antibodies were generated against the transition state analog of a reaction, and they showed, in some cases, good catalytic activity for that reaction.

We reported an approach that resembles antibody generation in 1972 (10,11). In this case, polymerizable binding site groups are bound by covalent or non-covalent interaction to a suitable template molecule (see Scheme I). The resulting template monomer or template monomer aggregate is copolymerized in the presence of a large amount of crosslinking agent. The templates are then removed to produce a microcavity with a shape and an arrangement of functional groups that are complementary to the template used.

This review gives a short discussion of the principle, (12-14) and outlines, first, the role of the binding sites and new possibilities for binding and, second, the preparation of catalytically active polymers obtained by imprinting with transition-state analogs.

Some Basic Considerations on Imprinting in Crosslinked Polymers

We now describe, as an example for the imprinting procedure, the polymerization of template monomer **1** (Scheme II) (15). This monomer was used to optimize the imprinting procedure and to elucidate the factors that influence the correct copying of

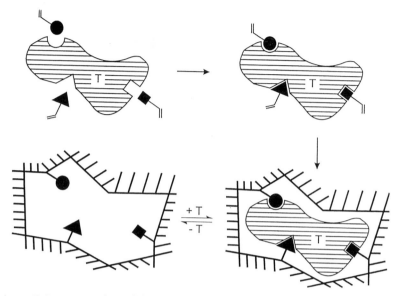

Scheme I. Representation of the Imprinting of Specific Cavities in a Crosslinked Polymer by a Template (T) with Three Different Binding Groups.

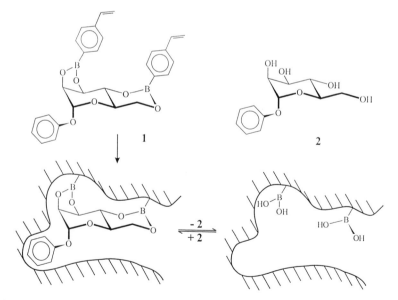

Scheme II. Representation of the Polymerization of **1** to Obtain a Specific Cavity. The template **2** can be removed with water or methanol to give the free cavity.

the template in the imprint. In this case, phenyl-α-D-mannopyranoside **2** is the actual template. Two molecules of 4-vinylphenyl boronic acid are bound to the template by boronic ester formation. This monomer was copolymerized by free radical polymerization with a large amount of crosslinking agent and in the presence of an inert solvent as a porogen. The template could be split off from the resultant macroporous polymer by water or water/methanol to an extent of up to 95%. The accuracy of the steric arrangement of the binding sites in the cavity and the shape of the cavity can be tested by the ability of the polymer to resolve the racemate of the template. This is achieved by equilibrating the polymer with the racemate of the template under thermodynamic control in a batch procedure. Enrichment of the optical antipodes in solution and on the polymer is determined and, from this, the separation factor α (K_D/K_L) can be calculated. After optimization of the polymer structure and the equilibration conditions, α-values of up to 6.0 were obtained (*16*). This is, in fact, an extremely high selectivity for racemic resolution that cannot be attained with other synthetic polymers. Polymers of this type can also be used in a chromatographic mode (see Figure 1). Baseline separation with a resolution $R_s = 4.2$ was achieved (*17*).

Figures 2 and 3 show computer-graphical representations of the cavity both with and without the template (*12*). Optimization of the polymer structure was rather complicated. On the one hand, the polymers should be rather rigid to preserve the structure of the cavity after splitting off the template. On the other hand, a high flexibility of the polymers should be present to facilitate a fast equilibrium between release and reuptake of the template in the cavity. These two properties are contradictory to each other, and a careful optimization has to be performed in these cases. Furthermore, good accessibility of as many cavities as possible is required as well as high thermal and mechanical stability. Since the initial experiments nearly all cases until now were based on macroporous polymers with a high inner surface area (100-600 m^2/g) which show, after optimization, good accessibility as well as good thermal and mechanical stability.

The selectivity is mainly influenced by the kind and amount of crosslinking agent used. Figure 4 shows the dependence of the selectivity for racemic resolution of the racemate of **2** on polymers prepared from **1** (*18*). Below a certain amount of crosslinking in the polymer (around 10%), no selectivity can be observed because the cavities are not sufficiently stabilized. Above 10% crosslinking, selectivity increases steadily. Between 50 and 66%, a surprisingly high increase in selectivity takes place especially in the case of ethylene dimethacrylate as a crosslinker. This crosslinking agent is now preferred by groups all over the world.

After our first detailed investigations, many other groups have entered the field (*12-14*). The group of K. Mosbach (*14*) worked especially with noncovalent binding. For example, they used L-phenylalanine-anilide as template and acrylic or methacrylic acid served as noncovalent binding sites (*19*). Scheme III shows the binding through electrostatic interaction and hydrogen bonding. The first symposium on imprinting and also this book show the broad acceptance of the new procedure for preparing binding sites with highly selective molecular recognition.

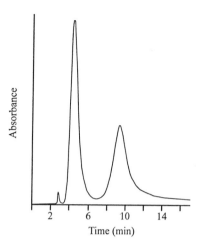

Figure 1. Chromatographic Resolution of D,L-**2** on a Polymer Imprinted with **1**. Elution with a solvent gradient at 90°C (*16*).

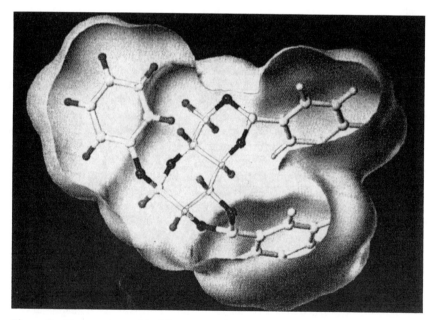

Figure 2. Computer-graphical Representation of an Occupied Cavity. The cavity corresponds to the electrostatic potential surface of **1** as computed by the Molcad 1.0 program (*12*).

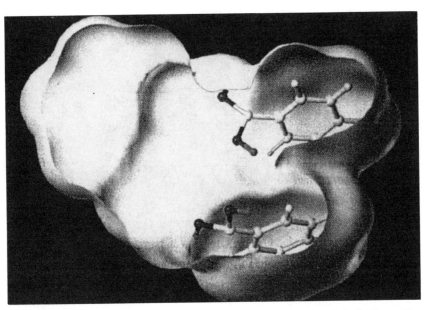

Figure 3. Computer-graphical Representation of an Unoccupied Cavity. The cavity corresponds to the electrostatic potential surface of **1** as computed by the Molcad 1.0 program (*12*).

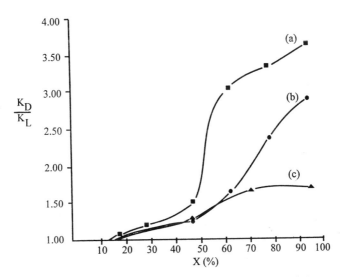

Figure 4. Selectivity of Polymers as a Function of the Type and Amount (X) of Crosslinking Agent (*18*). The crosslinking agents were: a) ethylene dimethacrylate; b) tetramethylene dimethacrylate; and c) divinylbenzene.

16

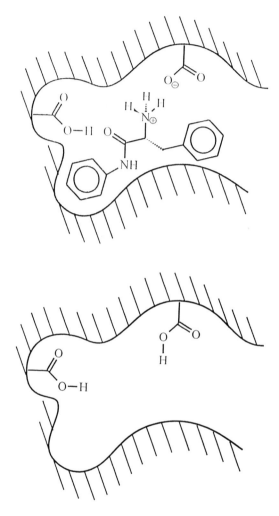

Scheme III. Representation of a Cavity Produced in the Presence of L-Phenylalanine-anilide. The polymerization takes place in the presence of acrylic acid. Noncovalent electrostatic interactions and hydrogen bond formation occur (*14,19*).

The Role of the Binding Sites and Their Improvement

Detailed investigations have shown that the selectivity depends both on the orientation of the functional groups inside the cavities and the shape of the cavities (20-22). The dominant factor, however, is the orientation of the functional groups inside the cavity (22). If two binding sites per template exist, several single-point bindings can occur but only one two-point binding. It is the two-point binding that provides high selectivity. Therefore, this portion may be increased by raising the temperature (23).

Another problem of binding is the reuptake of template in the cavity. In the case of covalent binding and all other types of stoichiometric binding, binding sites are only situated in the cavity. After removal of the template, this usually leads to a swelling of the cavities (3) which guarantees a high proportion (90-95%) of reuptake after the first removal. At the same time, it facilitates a quick mass transfer during equilibration of the template with the polymer (see Scheme IV). On reuptake of the template, the cavity shrinks to its original volume (induced fit). Noncovalent imprinting, e.g. with acrylic acid, requires a four fold excess of binding sites to ensure good selectivity. The binding sites are distributed throughout the polymer (see Scheme V). As was found recently (24), only 15% of the cavities can reuptake a template under these conditions, the remaining 85% are lost irreversibly for use in separation. This might be due to a shrinking of a majority of the cavities. Therefore such imprinted polymers are not well suited for preparative separations or for investigations of catalysis, and new and better binding sites need to be explored.

The binding sites have several functions in the imprinting procedure:
a)   During polymerization, the interaction between binding site and template needs to be stable. Therefore, stoichiometric binding with stable covalent bonds or with noncovalent interactions and high association constants are desirable.
b)   The template should be split off under mild conditions and as completely as possible.
c)   Equilibration with substrates has to be rapid and reversible. The binding site equilibrium should be adjustable to a favorable position.

These conditions are not easily met, and it requires careful optimization to come to a satisfactory interaction. Different types of interactions will be discussed briefly:
a)   Covalent interaction through boronic acids fulfills the first two points easily. A complete and stable conversion of diols to the boron diester is possible (see Equation 1).
    Furthermore, the removal of the diols from the polymer proceeds in very high yields (90-95%) (Equation 2). Binding of diols to the polymeric boronic esters (the reverse reaction in Equation 2) is, however, a relatively slow process. It can be made extremely fast by converting the boronic acid binding site from a trigonal to a tetragonal boron species by addition of ammonia (Equation 3). Under these condition, a rapid equilibration takes place (17,25).
b)   Noncovalent interaction by electrostatic interaction or hydrogen bonding is problematic during the polymerization. To shift the equilibrium of binding completely to the associated complex, a four fold excess of the binding site monomer is usually employed (see Scheme III) (19). As discussed before, this

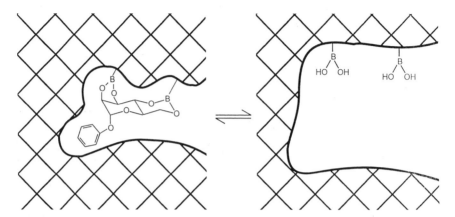

Scheme IV. Representation of the Removal of the Template **2** Bound Covalently from a Polymer and the Swelling of the Cavity. Afterwards, 90-95% of the cavities can be reoccupied.

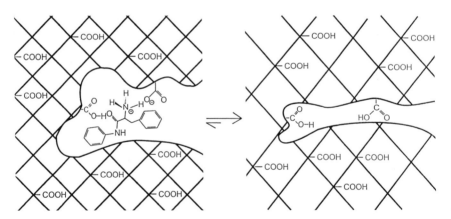

Scheme V. Representation of the Removal of the Template L-Phenylalanine-anilide Bound by Noncovalent Interaction. Owing to the large excess of carboxyl groups, a substantial part of the cavities may shrink. Only around 15% of the cavities can be reoccupied (*24*).

introduces nonspecific binding sites into the polymer. Template extraction and equilibration on the other hand are very favorable.

$$\text{(1)}$$

$$\text{(2)}$$

$$\text{(3)}$$

c) Coordinative binding is an interesting type of interaction as demonstrated by F. Arnold *et al.* (*26*) who introduced binding *via* copper complexes (see Equation 4).

$$\text{(4)}$$

These complexes bind imidazole-containing compounds. Higher association constants for the binding of the imidazole would be desirable. During equilibration, this type of interaction has very favorable properties as was already known from similar interactions in ligand exchange chromatography. In some cases chromatography in presence of Zn^{2+} ions provides more rapid interactions. Quite recently, very similar binding sites have also been used for the binding of glucose (*27*) and amino acids (*28*).

d) The amidine binding site possesses very promising properties for the binding of carboxylic acids or phosphonic acid monoesters. Since these complexes tend to be insoluble, a number of derivatives have been prepared (*29*). The *N,N'*-diethyl-4-vinyl benzamidine is especially suitable (see Equation 5).

$$\text{(5)}$$

$$\text{(6)}$$

Association constants are typically well above 10^6 L mol^{-1} (in CHCl$_3$). Templates are easily split off and the equilibrations are rapid. Even in aqueous solutions sufficient binding occurs. A reverse interaction of an amidine-containing drug (pentamidine) and acrylic acid has been reported previously (*30*).

e)   Multiple interactions under ring formation may be used to increase binding constants through hydrogen bonding. Thus, in our institute Schrader and Kirsten (*31*) performed an investigation on the binding of amido-pyrazoles with dipeptides. It is expected that one amido-pyrazole is bound to the top face *via* three-point binding, whereas a second amido-pyrazole interacts with the bottom face *via* two-point binding. By such complexation, the β-sheet conformation in the dipeptide should be stabilized (see Figure 5). NMR titrations of Ac-L-Val-L-Val-OMe with 3-methacryl amido-pyrazole show large downfield shifts for the peptide amide proton on the top face (see Figure 6). Addition of one equivalent of 3-amido-pyrazole results in complexation from the top by three-point binding with an association constant of 80.0 L mol^{-1}. In contrast, further addition of 3-amido-pyrazole induces complexation from the bottom by two-point binding with an association constant of only 2.0 L mol^{-1}. The association constant can be increased substantially by varying the acid portion in the amide (up to 890 L mol^{-1} for the CF$_3$-CO derivative). The dipeptide possesses a β-sheet conformation in both complexes as shown by NH-α-CH coupling constants and careful intermolecular NOE measurements. Thus, the binding site monomer 3 has been successfully used for imprinting with dipeptides and for racemic resolution of the racemate of the template dipeptide (*32*).

Catalytically Active Imprinted Polymers

Imprinting should also be an excellent method to prepare active sites of enzyme analogs. It has already been reported that antibodies prepared against the transition state of a reaction show considerable catalytic activity (*9*). Thus, antibodies prepared against a phosphonic ester (as a transition state analog for alkaline ester hydrolysis) enhance the rate of ester hydrolysis by 10^3-10^4 fold. First attempts of different groups (*33-35*) to use this concept for the preparation of catalytically active imprinted polymers for ester hydrolysis were rather disappointing. Enhancements of 1.6, 2, and,

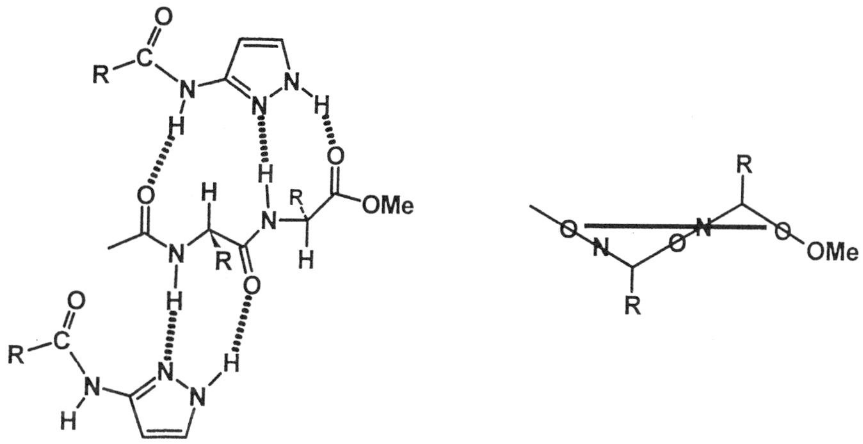

Figure 5. Side and Top Views of the Computer-calculated Dipeptide-amidopyrazole 2:1-complex (*31*). When seen from above, the heterocycle on the top-side is symbolised by a horizontal bar (for a better overview the second heterocycle is omitted).

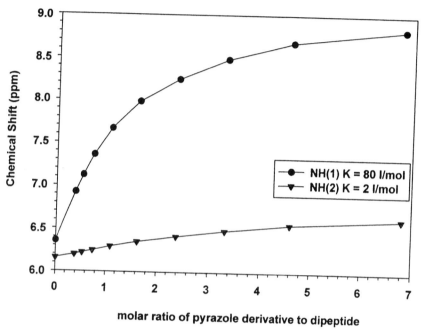

NMR Titration
Ac-L-Val-L-Val-OMe / 3-Methacrylamidopyrazole

Chemical Shift (ppm)

- NH(1) K = 80 l/mol
- NH(2) K = 2 l/mol

molar ratio of pyrazole derivative to dipeptide

Figure 6. ^{1}H NMR Titration Curves for the Divaline Complex with Amidopyrazole. The upper curve gives the downfield shift of the NH of the upper face and the lower curve that of the NH of the bottom face (*31*).

3

in one case, 6.7 are quite low and should be improved. Other reactions were investigated, *e.g.* by K. J. Shea *et al.* (*36*), K. Mosbach *et al.* (*37*), and K. Morihara *et al.* (*38*), and had somewhat higher enhancements. It appears that the shape of the transition state alone does not lead to sufficient catalysis. In addition, catalytically active groups have to be introduced. This is also true for catalytic antibodies since S. J. Benkovic *et al.* (*39*) showed that a guanidinium group of the amino acid L-arginin plays an important role in catalyzing the basic hydrolysis of esters by a catalytic antibody.

We have applied amidine groups both for binding and catalysis in investigation of the alkaline hydrolysis of ester **4** (Scheme VI). Phosphonic acid monoester **5** was used as the transition state analog (*40*). Addition of two equivalents of the new binding site monomer **6** furnished the bisamidinium salt. By the usual polymerization, work-up, and removal of the template, active sites were obtained with two amidine groups each. Owing to the stoichiometric interaction of the binding sites, the amidine groups are only located in the active sites.

At pH 7.6, the imprinted polymer accelerated the rate of hydrolysis of ester **4** by more than 100 fold compared to reaction in solution at the same pH. Addition of an equivalent amount of the monomeric amidine to the solution only slightly increased the rate. The same is true for the addition of polymers with statistically distributed amidine groups. Polymerizing the amidinium-benzoate gave somewhat stronger enhancement in rate.

Table I. Relative Rates of Hydrolysis of Ester 4 with Different Catalysts in Buffer Solution at pH = 7.6

blank	with **6b**	with a polymer imprinted with 6a-benzoate	with a polymer imprinted with **5 and 6a**
relative rate (1.0)	2.4	20.5	102.2

These examples show the strongest catalytic effects obtained to date by the imprinting method. This level of catalysis is only 1-2 orders of magnitude less than those achieved with antibodies. This is especially remarkable since we used "polyclonal" active sites and rigid, insoluble polymers. It should also be mentioned that these hydrolyses occur with non-activated phenol esters and not, as in nearly all other cases, with activated 4-nitrophenyl esters.

To see if these polymers show typical enzyme-analog properties, we investigated the kinetics of the catalyzed reaction in the presence of various excesses of substrate with respect to the catalyst. Figure 7 shows that typical Michaelis-Menten kinetics were observed. Saturation phenomena occur at the higher substrate concentrations. When all active sites are occupied the reaction becomes independent of substrate concentration. In contrast, in solution and in solution with the addition of amidine much slower reactions with a linear relationship is observed. The amidinium benzoate also shows some type of Michaelis-Menten kinetics. Therefore the benzoate acts as a less effective template.

Kinetic data can be obtained from a double-reciprocal Lineweaver-Burke plot (Figure 8). This plot provides evidence of strong binding with a Michaelis constant of $K_m = 0.60$ mM. Turnover is relatively low ($k_{cat} = 0.8 \times 10^{-4}$ min^{-1}), but is definitely

24

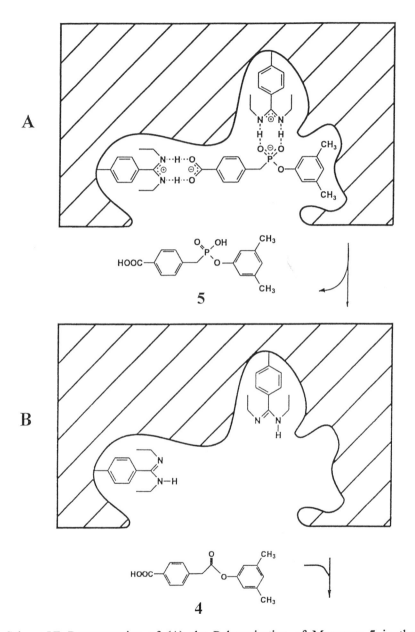

Scheme VI. Representation of (A) the Polymerization of Monomer **5** in the Presence of Two Equivalents of **6a**, (B) the splitting off of **5**, and (C) catalysis causing alkaline hydrolysis of **4** through a tetrahedral transition state.

Continued on next page.

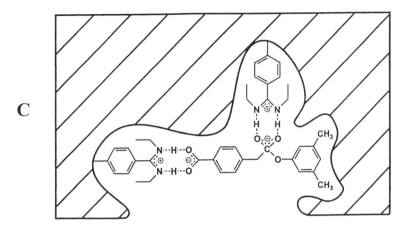

Scheme VI. *Continued.*

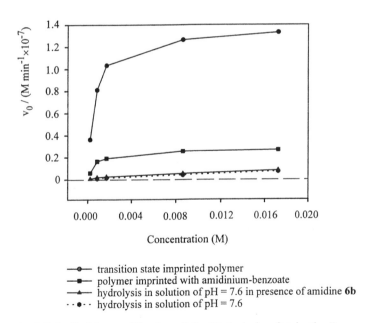

- ─●─ transition state imprinted polymer
- ─■─ polymer imprinted with amidinium-benzoate
- ─▲─ hydrolysis in solution of pH = 7.6 in presence of amidine **6b**
- ··●·· hydrolysis in solution of pH = 7.6

Figure 7. Michaelis-Menten Kinetics of the Hydrolysis of **4** in the Presence of Different Catalysts. The initial rates are plotted *versus* the substrate concentration.

26

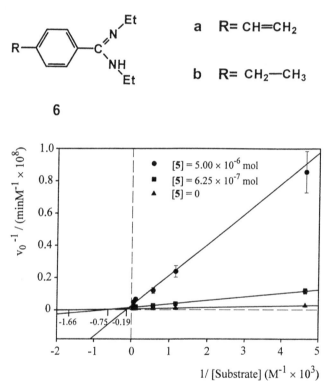

Figure 8. Double-reciprocal (Lineweaver-Burke) Plot of the Initial Rate *versus* the Substrate Concentration. In addition, the rates for the addition of the competitive inhibitor **5** at two different concentrations are plotted.

present. Furthermore, we found that the template molecule **5** is a powerful competitive inhibitor with $K_i = 0.025$ mM, *i.e.* it is bound more strongly than the substrate by a factor of 20. It is remarkable that such a strong binding of the substrate and template occurs in water-acetonitrile (1:1). Binding in aqueous solution by electrostatic interaction or hydrogen bonding is much weaker.

If imprinted polymers are prepared from different transition state analog templates and cross-selectivity with corresponding substrates is investigated, the corresponding "own" substrate is hydrolyzed at the highest rate. Thus, these polymers also show substrate selectivity.

Future Prospects

The generation of imprints in polymers and other materials (including surfaces) is now reaching a high level of sophistication. Applications of these materials are becoming more and more interesting. The first industrial applications of imprinted materials are approaching, especially as stationary phases in chromatography for the resolution of racemates. Other interesting applications are in membranes and in sensors. The use of imprinted polymers for radioimmunoassays has also been reported. Reactions in imprinted cavities are another interesting area. Regio- and stereoselective reactions inside these microreactors have been described. Of great importance to the field is catalysis with imprinted polymers and imprinted silicas. For broader application, further improvement of the method will be necessary. The following problems are at the forefront of present-day investigations:

a) Direct preparation of microparticles by suspension or emulsion polymerisation;
b) Imprinting procedures in aqueous solutions;
c) Imprinting with high-molecular-weight biopolymers or even with bacteria by surface imprinting;
d) Development of new and better binding sites in imprinting;
e) Improvement of mass transfer in imprinted polymers;
f) Reduction of the "polyclonality" of the cavities;
g) Increase in capacity of chromatographic columns, especially those with noncovalent interactions;
h) Development of extremely sensitive detection methods for use in chemosensors; and
i) Development of suitable groupings for catalysis.

Acknowledgment

These investigations were supported by grants from the *Deutsche Forschungsgemeinschaft* and *Fonds der Chemischen Industrie*.

Literature Cited

1. Wulff, G.; Sarhan, A.; Zabrocki, K. *Tetrahedron Lett.* **1973,** 4329.
2. see e.g. Kirby, A. J. *Angew. Chem.* **1996,** *108,* 770; *Angew. Chem., Int. Ed. Engl.* **1996,** *35,* 705.

28

3. Wulff, G.; Sarhan, A. In *Chemical Approaches to Understanding Enzyme Catalysis;* Green, B. S.; Ashani, Y.; Chipman, D., Eds.; Elsevier: Amsterdam, Netherlands, 1982, pp 106-118.
4. Cram, D. J. *Angew. Chem.* **1988**, *100*, 1041; *Angew. Chem., Int. Ed. Engl.* **1988**, *27*, 1009.
5. Lehn, J. M. *Angew. Chem.* **1988**, *100*, 91; *Angew. Chem., Int. Ed. Engl.* **1988**, *27*, 89.
6. Wenz, G. *Angew. Chem.* **1994**, *106*, 851; *Angew. Chem., Int. Ed. Engl.* **1994**, *33*, 803.
7. Schneider, H.-J. *Angew. Chem.* **1991**, *103*, 1419; *Angew. Chem., Int. Ed. Engl.* **1991**, *30*, 1417.
8. Rebek, J. *Angew. Chem.* **1990**, *102*, 261; *Angew. Chem., Int. Ed. Engl.* **1990**, *29*, 245.
9. Lerner, R. A.; Benkovic, S. J.; Schulz, R. G. *Science* **1991**, *252*, 659.
10. Wulff, G.; Sarhan, A. *Angew. Chem.* **1972**, *84*, 364; *Angew. Chem., Int. Ed. Engl.* **1972**, *11*, 341.
11. Wulff, G.; Sarhan, A. *DOS 2242796* **1974**, [*Chem. Abstr.* **1975**, *83*, P60300w].
12. Wulff, G. *Angew. Chem.* **1995**, *107*, 1958; *Angew. Chem., Int. Ed. Engl.* **1995**, *34*, 1812.
13. Shea, K. J. *Trends Polym. Sci.* **1994**, *2*, 166.
14. Mosbach, K. *Trends Biochem. Sci.* **1994**, *19*, 9.
15. Wulff, G.; Vesper, W.; Grobe-Einsler, R.; Sarhan, A. *Makromol. Chem.* **1977**, *178*, 2799.
16. Wulff, G.; Vietmeier, J.; Poll, H. G. *Makromol. Chem.* **1987**, *188*, 731.
17. Wulff, G.; Minarik, M. *J. Liquid Chromatogr.* **1990**, *13*, 2987.
18. Wulff, G.; Kemmerer, R.; Vietmeier, J.; Poll, H.-G. *Nouv. J. Chim.* **1982**, *6*, 681.
19. Sellergren, B.; Lepistö, M.; Mosbach, K. *J. Am. Chem. Soc.* **1988**, *110*, 5853.
20. O`Shannessy, D. J.; Andersson, L. I.; Mosbach, K. *J. Mol. Recognit.* **1989**, *2*, 1.
21. Shea, K. J.; Sasaki, D. Y.; Stoddard, G. J. *Macromolecules* **1989**, *22*, 1722.
22. Wulff, G.; Schauhoff, S. *J. Org. Chem.* **1991**, *56*, 395.
23. Wulff, G.; Kirstein, G. *Angew. Chem.* **1990**, *102*, 706; *Angew. Chem. Int. Ed. Engl.* **1990**, *29*, 684.
24. Sellergren, B.; Shea, K. J. *J. Chromatogr.* **1993**, *635*, 31.
25. Wulff, G.; Dederichs, W.; Grotstollen, R.; Jupe, C. In *Affinity Chromatography and Related Techniques;* Gribnau, T. C. J.; Visser, J.; Nivard, R. J. F., Eds.; Elsevier: Amsterdam, Netherlands, 1982, pp 207-216.
26. Dhal, P. K.; Arnold, F. H. *J. Am. Chem. Soc.* **1991**, *113*, 7417.
27. Chen, G.; Guan, Z.; Chen, C.-T.; Fu, L.; Sundaresan, V.; Arnold, F. H. *Nature Biotechnology* **1997**, *15*, 354.
28. Sundaresan, V.; Ru, M.; Arnold, F. H. *J. Chromatogr.* in press.
29. Wulff, G.; Schönfeld, R.; Grün, M. *unpublished results.*
30. Sellergren, B. *Anal. Chem.* **1994**, *66*, 1578.
31. Schrader, T.; Kirsten, C. *Chem. Commun.* **1996**, 2089.
32. Schrader, T.; Kirsten, C. *J. Am. Chem. Soc.* in press.
33. Robinson, D. K.; Mosbach, K. *Chem. Commun.* **1989**, 969.
34. Sellergren, B.; Shea, K. J. *Tetrahedron: Asymmetry* **1994**, *5*, 1403.
35. Ohkubo, K.; Funa Koshi, Y.; Sagawa, T. *Polymer* **1996**, *37*, 3993 and earlier papers.
36. Beach, J. V.; Shea, K. J. *J. Am. Chem. Soc.* **1994**, *116*, 379.
37. Müller, R.; Andersson, L. I.; Mosbach, K. *Makromol. Chem., Rapid Commun.* **1993**, *14*, 637.
38. For example: Shimada, T.; Nakanishi, K.; Morihara, K. *Bull. Chem. Soc. Jpn.* **1992**, *65*, 954.
39. Stewart, J. D.; Liotta, L. J.; Benkovic, S. J. *Acc. Chem. Res.* **1993**, *26*, 396.
40. Wulff, G.; Groß, T.; Schönfeld, R. *Angew. Chem.* **1997**, *109*, 2049; *Angew. Chem., Int. Ed. Engl.* **1997**, *36*, 1962.

Chapter 3

Molecular Imprinting: *Status Artis et Quo Vadere?*

Klaus Mosbach, Karsten Haupt, Xiao-Chuan Liu, Peter A. G. Cormack, and Olof Ramström

Pure and Applied Biochemistry, Chemical Center, Lund University, P.O. Box 124, SE-221 00 Lund, Sweden

The technique of molecular imprinting allows the formation of specific recognition and catalytic sites in macromolecules via the use of templates. Molecularly imprinted polymers have been applied in an increasing number of applications where molecular binding events are of interest. These include: i) the use of molecularly imprinted polymers as tailor-made separation materials; ii) antibody and receptor binding site mimics in recognition and assay systems; iii) enzyme mimics for catalytic applications; and iv) recognition elements in biosensors. The stability and low cost of molecularly imprinted polymers make them advantageous for use in analysis as well as in industrial scale production and application.

Although so many thoughts and so much data fills our minds, practically urging us to write a book on molecular imprinting, a symposium contribution, by definition, has to be restricted, and should more or less reflect what has been presented at such a meeting. Thus we follow the proverb formulated by Goethe "in der Beschränkung zeigt sich der Meister" (Restriction reveals the master).

A definition of the topic of molecular imprinting is, we feel, not required in the context of this symposium. We shall therefore just refer to some recent review articles in the field (*1-7*).

It is striking just how much the interest in the area has intensified over the last few years, with a present publication rate of around 50 communications per year. Our group alone has published a total of about 75 articles, of which 50 or so have appeared or were submitted within only the last four years. In addition, more than 10 patents/applications from our group were filed, including a more general patent covering imprints mimicking antibodies and enzymes which is of particular note (*8*). We would like to ascribe a considerable part of this upsurge of interest in the area to the general acceptance of the conceptually and practically simple non-covalent (self-assembly) approach, which we developed in our laboratory, and probably also to some degree on the publication of an article on a more "catchy" subject, where we described the use of imprints as artificial antibodies for drug assays (*9*). Finally, a contributing factor to the rapid development could be the increased use of acrylic monomers in polymer formation, as they offer a plethora of building blocks with different functionalities, which considerably extends the scope of the chemistry.

While on the topic of acrylate based polymers for imprints, it might be worth mentioning that one of us (K. M.) used this material many years ago in the first reported entrapment of cells and enzymes (10), and it was in that context that for him the concept, or at that time the dream, of imprinting was perceived, based on the idea that after the removal of an enzyme or a smaller molecule from the polymer network, cavities in the network expressing memory in terms of shape for the original entrapped biomaterial would be revealed. It did not work well at all at that time, but it is interesting to note that very recently this concept now seems to have been realized for enzymes (11).

Molecularly Imprinted Polymers

In the following, we would like to describe the current state-of-the-art of the non-covalent approach (self-assembly approach), as developed in our laboratory, and present some recent developments in this area. Furthermore, we would like to point out some of the potential developments of the technology, and focus upon some expected future directions in the area. An alternative approach (the pre-organized approach), which was pioneered by Wulff and involves covalent interactions, is covered elsewhere in this volume.

Several polymer systems have been developed for use in molecular imprinting technology, in general, and self-assembly systems, in particular. At present, primarily acrylate-based, styrene-based or silane-based polymeric materials have been used in imprinting protocols. Of these, polyacrylate-based matrices have been the most extensively studied for a number of reasons. For example, the polymerization of acrylates is normally very rapid, and may be initiated in several ways (thermo- or photo-initiation are most commonly used). Furthermore, the polymers may be prepared successfully in a wide range of solvents, and at ambient temperatures and pressures. Finally, the polymerization process does not interfere severely with most print molecules, although in certain cases radical quenching can occur.

A range of functional monomers has been assessed for use in self-assembly imprinting protocols (c.f. Table I). Typical examples include: carboxylic acids, sulfonic acids, amides, and heteroaromatic (weak) bases. For metal chelating interactions, an iminodiacetic acid derivative has been commonly used. In the case of polysiloxanes (12), various silanes are used as monomers, although these have not been widely adopted.

Several different crosslinking monomers have been utilized, although with varying degrees of success. Of these, ethylene glycol dimethacrylate (EDMA) and trimethylolpropane trimethacrylate (TRIM) are presently the most widely employed. These crosslinkers have the advantage of being relatively stable and also compatible with the most frequently used functional monomers in respect of their reactivity. For styrene-type functional monomers, divinylbenzene offers an alternative. This crosslinker may be particularly advantageous under harsher conditions where ester-type crosslinkers may undergo hydrolysis.

Recognition Mechanism

The recognition properties displayed by molecularly imprinted materials are highly attractive. In several cases, the selectivities and affinities acquired from the molecular imprinting process have been shown to be on a par with natural binding entities such as antibodies and receptors (see below). A putative self-assembly imprinting mechanism can be outlined as follows (c.f. Figure 1): In the first step(s), the functional monomers form spontaneous interactions with the print molecule. This gives rise to a somewhat complicated situation involving several different complexes between one, or several, of the monomers, print molecules, solvent molecules, etc. Once polymerization has been initiated, the polymer chains begin to grow, leading tothe formation of new self-assembly complexes. In the early stages of polymerization the polymer chains can "unfold" and "refold" so that new recognition

Table I. Examples of interaction types and the corresponding monomers used in self-assembly imprinting system

Functional monomer	Type of interaction	Reference example
Acrylic acids (R = H, CH₃, CF₃, CH₂COOH)	ionic charges hydrogen bonds	*(13-16)*
Vinylbenzoic acids	ionic charges hydrogen bonds	*(13)*
Acrylamido-sulfonic acids	ionic charges	*(17)*
Amino-methacrylamides (R = H, C₂H₅)	ionic charges	*(18,19)*
Vinylpyridines	ionic charges hydrogen bonds charge transfer	*(20,21)*, (Haupt, this volume)
Vinylimidazoles	ionic charges hydrogen bonds metal coordination	*(21,22)*
Acrylamides	hydrogen bonds	*(23)*, (Ramström *et al.*, this volume)
Vinyl-iminodiacetic acids	metal coordination	*(24,25)*

32

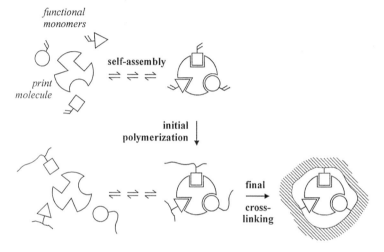

Figure 1. Schematic Model of a Self-assembly Imprinting Mechanism. In the first step, functional monomers are allowed to form spontaneous interactions with the print molecule. Following initiation, polymer chains begin to grow, leading to new assemblies between the template and the functional moieties of the newly formed polymer. Finally, the polymer is sufficiently crosslinked and the complexes are "locked" in position, inhibited from further, major chain motions.

complexes are generated continuously. This situation is likely to persist to some extent until the polymer chains are sufficiently crosslinked so as to inhibit further (major) chain motions.

The principle means for exerting specificity in non-covalent systems is through ionic interactions and hydrogen bonding between the ligands and the polymer functional groups (*c.f.* Table I) (*14*), much akin to the situation in natural systems. These interactions are of utmost importance for the initial self-assembly step, and also for maintaining the integrity of these complexes during polymerization. In addition to these strong polar interactions, other types of interactions may also come into play, such as van der Waals interactions, charge-transfer interactions, and π-π-interactions. Non-polar interactions between the polymer backbone and the print molecule are likely to occur during polymerization. These interactions give rise to the three-dimensional geometry of the sites, and contribute to the overall site quality.

Specificity

The specificity of a molecularly imprinted material is a measure of how well it can distinguish a particular compound from other related compounds. Several factors influence the specificity of molecularly imprinted materials. First and foremost is the quality of the recognition site, which is related to the stability of the complex in the self-assembly and polymerization steps. Secondly, the solvent used for the polymerization is of great importance, since this may either enhance or destabilize the specific interactions which lie at the heart of the imprinting effect. Thirdly, factors such as the temperature and the pressure may also influence the specificity.

The specificity is primarily governed by the three-dimensional arrangement of the various interactions between the functional monomers and the print molecule. Since the self-assembly process is a complicated one involving an equilibrium between free species and complexed species, strong specific interactions favor complex formation and so improve the quality of the recognition site. In addition, the greater the number of interactions available, the better the outcome. Although the spatial arrangement of interactions in the binding site is responsible for most of the selectivity, the shape of the site also plays an important role. The configuration of the surrounding polymer backbone, as molded in place around the print molecule, contributes to the overall ligand specificity. This effect is sometimes less pronounced, and substantial freedom in ligand structure can be observed (*26,27*). However, often the effect is considerable, and minute differences in sizes between analytes can be distinguished. In several cases, the presence or absence of a single methyl, or methylene, group can be crucial for recognition (*9,15,26,28*). In this respect, the shape effect of the site is particularly pronounced for those parts of an imprint molecule in close proximity to ionic- or hydrogen bonds between host and guest. For structural features of the analyte which are more distant from such bonds, the shape is apparently less important.

Analogous to features observed in affinity chromatography, the specificity is dependent on two separate interaction events: i) binding of the ligands to the matrix; and ii) release of the ligands from the matrix. These two events may be controlled concertedly, as is the case in normal chromatographic analyses. They may also be controlled individually, so that the specificity of binding is optimized in a first step and the release conditions optimized in a subsequent step. Thus, fine tuning of the solvent effect on the recognition may lead to a high control of specificity. This latter method has been successfully used in solid phase extraction protocols. In this context, it should be pointed out that when these binding events are compared in the same solvent, the association kinetics are considerably faster than the dissociation kinetics (unpublished data from our laboratory). Therefore, the binding and the release of the ligands may be optimized separately.

Another factor which influences the recognition properties of molecularly imprinted materials is the entropy penalty that has to be paid in the complexation process. This effect may be considerable, and a high degree of rigidity of the print molecules and the functional monomers serves to enhance the recognition. It is likely that complex formation during polymerization is assisted by the fact that the polymer chains become progressively less mobile as the polymerization proceeds, and thus induce an enhanced recognition quality. Generally, a high degree of crosslinking results in a higher recognition quality, thus demonstrating the importance of a rigid polymer backbone structure. However, although not as yet completely assessed, chain motions in the final polymer may be of sufficient magnitude to facilitate mass transfer into the pores, and may also result in phenomena similar to the "induced fit" situation so often encountered in natural systems. By controlling the degree of crosslinking, sufficient flexibility of the polymer chains may be provided, thus giving rise to recognition sites with the ability to accommodate a broader spectrum of structurally related ligands.

The solvent plays a crucial role in the outcome of the molecular imprinting process. Besides its function as a porogen to control the polymer morphology, it governs the strength of non-covalent interactions. Generally, the more polar the porogen, the greater the disruption of the non-covalent interactions, and the weaker the resulting recognition effect. In order to optimize the binding strengths, the best imprinting porogens are solvents of very low dielectric constant, such as toluene and chloroform. The use of more polar solvents weakens the interactions between the print species and the functional monomers, resulting in potentially poorer recognition. On the other hand, hydrophobic effects, of paramount importance in many biological systems, may be employed to overcome this obstacle. Although hydrophobic forces are potentially more difficult to master, being less specific and less directional, combinations of polar and hydrophobic interactions may be used to generate strong and selective binding. In this way even better biomimetic matrices may be produced. This was recently demonstrated in a study where the use of the polar solvents methanol and water in the imprinting as opposed to non-polar solvents, such as toluene, actually resulted in improved recognition matrices for certain compound classes (Haupt, this volume).

In the recognition step, similar questions about the choice of solvent arise. Since all non-covalent forces are influenced by the properties of the solvent, non-polar solvents generally lead to the best recognition. Upon exposing the polymers to gradually more polar solvents, the recognition is progressively diminished. Nevertheless, molecularly imprinted materials prepared in organic solvents often retain a high degree of recognition in aqueous solvents, as we have been able to show in a number of instances. Recently, we have also been able to show that high water concentrations in the recognition solvent also leads to an increased recognition quality of the sites (Ramström, et al., this volume). It is believed that this increase is in part assisted by the hydrophobic effect.

In nature, chirality is a specific source of information and distinction that is inherent in most binding events. In biomimetic chemistry, chiral recognition by artificial binders is therefore a prerequisite. Originally, the use of chirality in our studies was intended more as a means for monitoring and validating the imprinting effect, than a goal in itself, since the geometrical structure of enantiomeric pairs is their principle differentiating quality. Amino acid derivatives were often chosen as model compounds because they are inexpensive, stable, and possess a range of functional groups. Furthermore they are soluble in convenient imprinting solvents. Through the use of a single enantiomer as the print molecule, the imprinting effect exhibited by a polymer is easily monitored, since chiral discrimination would not be observed were the separation based only on ion-exchange or size-exclusion processes.

By use of molecular imprinting, chirality may be introduced in macromolecular matrices starting from, in most cases, achiral building blocks. The induced chirality is thus a consequence of the asymmetry carried by the print molecule. If the enantiomer to the print molecule is added, much lower binding strengths are found compared to the original species. On the other hand, the use of achiral print molecules will not lead to any detectable chiral discrimination by the molecularly imprinted material. Likewise, when a racemic mixture is used in the imprinting protocol, no enantioseparation can be recorded, indicating that the enantioselective property of the polymer is introduced by the imprinting process and not by the polymerization *per se*. The use of chiral monomers in the imprinting protocol, although not yet fully assessed, may also further enhance the chiral discrimination effect.

Configurations

Molecularly imprinted polymers can be prepared in a variety of configurations. The "classical" one is the synthesis of macroporous block polymers, which are subsequently ground to particles with a diameter of approximately 25 µm. These particles can be used directly as separation media in chromatography (*15*) or in batch binding assays (*9*). For chromatographic applications, however, the use of uniformly sized spherical particles is preferable for optimal column performance. Spherical polymer beads can be obtained in several different ways. One such way is through suspension polymerization, for which a rather elegant process has recently been developed in our laboratory (*29*). It relies on perfluorocarbon liquids as the suspension media, instead of the classical organic liquid-in-water suspension. This eliminates the problem of water destabilizing the pre-arranged complex during non-covalent imprinting. It has also been demonstrated that an imprinted polymer layer can be grafted onto the surface of spherical silica or TRIM particles (*30-32*).

It can sometimes be advantageous to synthesize the imprinted polymer *in situ* within an HPLC column (*33*). This approach reduces the work-up time of the polymer considerably, and the column can often be used within 2-3 hours of its preparation. Polymers prepared *in situ* have been shown to be as specific as ground bulk polymers, whereas the separation factors obtained have been somewhat poorer. A similar approach has been used to prepare imprinted polymers in capillary columns for use in capillary electrophoresis (*34*).

In applications such as equilibrium binding assays, particles smaller than those used in chromatography may be preferable. It has been reported that particles with a diameter of 1 µm and smaller, which are normally discarded during work-up of the bulk polymer, have the same binding properties as larger particles and can be used very successfully in radioligand binding assays (Haupt, this volume; (*35*)). Not only were the incubation times reduced due to shorter diffusion distances, but the small particles (fines) were found to be more practical for binding assays as they stayed in suspension longer and were easier to pipette. Moreover, if the fines are small enough, they can be fractionated according to the presence and quality of accessible binding sites using classical affinity techniques (*35*). Last but not least, if the template is expensive, it is undesirable to lose a considerable amount of polymer by discarding the fines.

Another format for molecularly imprinted polymers are thin films or membranes. Particularly if the polymer is intended as the recognition element in a sensor, it may be deposited in the form of a thin film directly onto the transducer surface (*36*). The preparation of imprinted polymer membranes has been reported by several groups (*37-39*) and the selective transport or optical resolution of target molecules demonstrated.

Imprints of larger molecules. Despite an increasing interest in extending molecular imprinting technology to larger biomolecules such as proteins, there have been only a few reports describing successful attempts to imprint proteins. Due to their size and fragility, the "classical" approach using highly cross-linked bulk polymers cannot be employed, although this protocol is still suitable for the imprinting of small polypeptides, *e.g.* enkephalin (*40*). It has therefore been suggested that imprints of proteins can be prepared in a surface-grafted thin polymer layer. One step in this direction has been made by the imprinting of the enzyme ribonuclease A on the surface of silica particles (*25*). In this case a polymerizable metal chelate was used as the functional monomer. Since ribonuclease A has two surface histidines, during pre-arrangement two metal chelate molecules were coordinated by the enzyme in the ideal case and upon polymerization, immobilized onto the activated silica surface thus creating specific binding sites (Figure 2). The affinity support thus obtained was able to specifically adsorb the original template. Metal chelate ligands have also been applied to the creation of binding sites for metal binding proteins in lipid monolayers (41). Proteins have also been imprinted in an aqueous environment via the formation of boronate esters (12).

Since proteins are large, fragile molecules, which are insoluble in apolar organic solvents and carry a large number of surface functionalities, it had initially been assumed that special functional monomers, e.g. metal chelates, which can bind to the template even in an aqueous environment, are needed in order to obtain specific imprints of proteins. However, more recent work seems to indicate that shape-specific binding sites for proteins can be obtained by imprinting them in low cross-link density acrylamide gels (11). There has even been a report describing the creation of imprints of whole bacteria cells in the surface of polymer beads using a lithographic technique (42). Although all these achievements are encouraging, imprints of larger structures are further away from real practically relevant applications than the imprints of small molecules.

Properties

Molecularly imprinted polymers offer considerable advantages over biomolecules as recognition systems in terms of their chemical and physical stability. Owing to their highly cross-linked polymeric nature, these materials are resistant not only to mechanical stresses, high pressures and elevated temperatures, but also to acids, bases, organic solvents and metal ions. They can be stored in the dry state at ambient temperatures for several years and, if necessary, regenerated and re-used more than 100 times, without loss of their molecular memory (43).

A potential shortcoming in molecular imprinting arises from the fact that a reasonable quantity of template molecule is usually required in order to prepare the imprinted polymer in the first place, and this may, in some cases, be expensive or difficult to obtain. This is of less concern if the polymer is intended for use in binding assays as outlined below, or in sensors, but more so if the applications are in preparative chromatography. However, in most cases the possibility of repeatedly using the same imprinted polymer over a long time period will more than compensate for this.

Applications

Molecularly imprinted polymers have been used in an increasing number of applications where molecular binding events are of interest (c.f. Table II). These include: i) the use of molecularly imprinted polymers as tailor-made separation materials; ii) antibody and receptor binding site mimics in recognition and assay systems; iii) enzyme mimics for catalytic applications; and iv) recognition elements in biosensors.

Separation and Isolation. To date, the application that has been the most extensively explored is the use of molecularly imprinted materials in separation and isolation. A wide range of print molecules, from small molecules, such as drugs (9,40), amino acids and carbohydrates (27,59), to larger entities, such as proteins (12,25), have been used in various imprinting protocols and separations have been pursued which employ several different techniques, such as HPLC (60), TLC (61) and CE (34).

As previously mentioned, chiral separations have been a major area of investigation, and molecularly imprinted materials have been extensively employed as chiral stationary phases in HPLC (for reviews, see: (7,44)). A characteristic of these stationary phases is the *pre-determined* elution order of the enantiomers, which

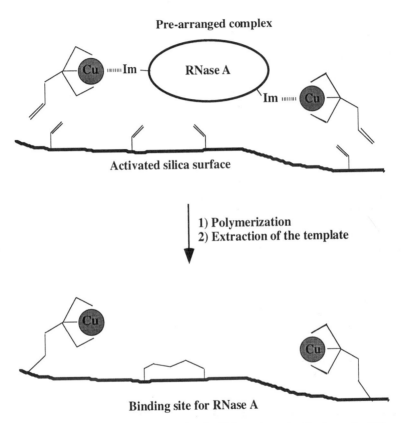

Figure 2. Creation of a Binding Site for RNase A on the Surface of a Silica Particle by Molecular Imprinting (Adapted from ref. 25.).

depend only on which enantiomeric form was used as the print molecule. For instance, when the R-enantiomer is used as print molecule, the S-form will be eluted first and *vice versa* (*21*). Since most chiral drugs on the market are still administered as racemic mixtures, racemic resolution of drugs is a major potential application (*62,63*). In the light of recent guidelines from official authorities concerning drug preparation and administration (*64*) which are forcing moves towards enantiomerically pure compounds, new and efficient techniques for enantioseparations are needed. This provides an opening for the use of molecularly imprinted materials in this area.

Table II. Applications of molecularly imprinted materials (self-assembly systems)

Application area		Reference example
Separations/isolations	Chiral separations	(*7,44*)
	Substrate-selective separations	(*24,27,45,46*)
Antibody/receptor binding mimics	Competitive ligand binding assays	(*40,45,47,48*)
	Diagnostic applications	(*9*)
Enzyme mimics/catalysis		(*19,49-51*)
Synthetic applications	Assisted synthesis	(*52,53*)
	Equilibrium shifting	(Ramström *et al.*, this volume)
Biosensor-like devices		(*36,54-56*)
Recognition studies		(*20,57,58*)

The discrimination of enantiomers is often very efficient with molecularly imprinted materials (*c.f.* Figure 3). Using self-assembly imprinting protocols, highly selective chirally discriminating sites have been prepared, and large separation factors between the enantiomers have been recorded when the molecularly imprinted materials have been used as chiral stationary phases in chromatography. As pointed out, many of the early investigations employed amino acid derivatives as model substances. In recent years, however, a great deal of emphasis has also been put on the chiral discrimination of drug compounds. Several studies involving the separation of physiologically active compounds, *e.g.* naproxen (a non-steroidal anti-inflammatory drug) (*28*), ephedrine (an adrenergic agent) (*26*), and timolol (a ß-adrenergic antagonist) (*43*) have been described.

For print molecules containing two chiral centers, all four stereoisomers may be selectively recognized by the imprinted materials. Thus, for a polymer imprinted against the dipeptide Ac-L-Phe-L-Trp-OMe, using hydrogen bonding interactions primarily, the LL-form can selectively be distinguished from the DD-, the DL,- and the LD-isomers (separation factors: α = 17.8, 14.2 and 5.21, respectively) (*65*). In systems where more than two chiral centers are involved, such as carbohydrates, these properties of molecularly imprinted materials become even more significant. For example, in a study where polymers were imprinted against a glucose derivative, very high selectivities between the various stereoisomers and anomers were recorded (*27*).

Antibody mimics. It has been shown on several occasions that molecularly imprinted polymer particles can be used as artificial recognition elements in competitive immunoassay-type binding assays. The first report in this field arose from a study within our own laboratory (9). The radioligand binding assay developed for the bronchodilator theophylline based on a molecularly imprinted polymer showed not only a very good correlation with an antibody-based EMIT assay currently used in analytical laboratories in hospitals (Figure 4), but, surprisingly, even yielded a cross-reactivity profile very similar to that of the natural antibodies. Among a range of closely related substances, only 3-methylxanthine, which lacks one methyl group as compared to theophylline, was bound to the polymer to some extent, whereas caffeine, which has one additional methyl group showed virtually no cross-reactivity (Table III). Another polymer-based assay for the tranquilizer diazepam was also very specific, with cross-reactivities comparable to those of natural antibodies (Table III). In the following years, binding assays have been developed for several other compounds, *e.g.* herbicides (45,48; Haupt, this volume), drugs (40,47,66), the neuropeptide enkephalin (40), and corticosteroids (67). The latter provides yet another demonstration of the specificity that can be obtained with molecular imprinting. A polymer imprinted with cortisol showed 36% cross-reactivity of the analogue prednisolone, which differs only in an additional double bond in the A-ring, and less than 1% cross-reactivity of cortisone, where a hydroxy group is exchanged for a keto functionality, as compared to the original template.

When compared to natural antibodies, molecularly imprinted polymers have particular advantages and drawbacks. So does the only assay format used to date, the competitive radioligand binding assay. This involves the handling of radioactive materials, which is undesirable. Alternative formats for competitive imprinted polymer-based assays could use fluorescence or an enzyme reaction for detection.

Table III Cross-reactivity of structurally related compounds for binding of the radiolabels ^{3}H-theophylline and ^{3}H-diazepam to imprinted polymers (MIP) and antibodies (Ab). (Adapted from ref. 9.)

Template	Competitive ligand	Cross-reaction, %	
		MIP	Ab
Theophylline	Theophylline (1,3-dimethylxanthine	100	100
	3-Methylxanthine	7	2
	Theobromine (3,7-dimethylxanthine)	<1	<1
	Caffeine (1,3,7-trimethylxanthine)	<1	<1
	7-ß-Hydroxythyl-1,3-dimethylxanthine	<1	<1
	Xanthine	<1	<1
	Hypoxanthine	<1	<1
	Uric acid	<1	<1
	1-Methyluric acid	<1	<1
	1,3-Dimethyluric acid	<1	<1
Diazepam	Diazepam	100	100
	Alprazolam	40	44
	Desmethyldiazepam	27	32
	Clonazepam	9	5
	Lorazepam	4	1
	Chlordiazepoxide	2	<1

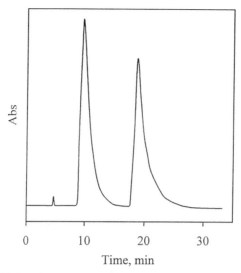

Figure 3. Molecular Imprinting Chromatography (MIC). Racemic resolution of Boc-tryptophan enantiomers using a molecularly imprinted chiral stationary phase prepared with the L-isomer. Gradient elution, solvent A: acetonitrile, solvent B: acetic acid; flow rate: 1.0 mL/min; 40 μg Boc-D,L-Trp-OH was injected in 20 μL acetonitrile; $k'_D=0.94$, $k'_L=2.53$, $\alpha=2.69$, $R_s=2.51$. (Adapted from ref. 23.)

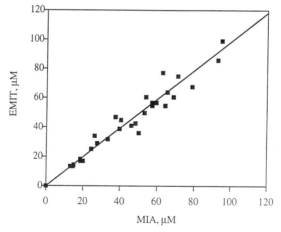

Figure 4. Comparison of the Imprinted Polymer-based Assay (MIA) for Theophylline with a Commercial Antibody-based EMIT. (Adapted from ref. 9.)

Another problem with molecularly imprinted polymers is that many of them can only be used in apolar organic solvents, which adds an extraction step to the assay method. It has recently been shown however, that some polymers which have been imprinted in the presence of an organic solvent can later be used in aqueous buffer (40,45,47). This is particularly noteworthy, and can be regarded as a major breakthrough in non-covalent molecular imprinting in general. It has been suggested that the balance of the molecular forces involved during binding of the analyte is different in organic solvents than in aqueous buffer, which accounts for the sometimes altered, but never seriously compromised, selectivity. In other situations, however, it can actually be an advantage that the assay can be performed directly in the organic solvent, for example in environmental analyses that require an extraction step during sample work-up.

It has to be added that non-covalently imprinted polymers contain a heterogeneous population of binding sites with a range of different affinities and are therefore comparable to polyclonal antibodies. In some cases, dissociation constants have been estimated using two or three-site binding models. For high affinity sites, which are the most specific ones and therefore utilized in equilibrium binding assays, K_D values in the nanomolar or micromolar range (40,47,67), have been obtained. These affinities are comparable to those of antibodies. The fact that the percentage of high-affinity sites in a non-covalently imprinted polymer is rather low (0.1-1% of the sites) is of less concern for equilibrium assays, as the concentration of radioligand applied to the polymer is also very low. It has actually been shown that through careful optimization of the assay protocol, the amount of polymer needed can be reduced to as little as 20-50 μg/ml. That means that a 5 gram batch of polymer would yield enough material for up to 100,000 assays which renders the system considerably less costly than antibody-based assays (47).

A further benefit of molecularly imprinted polymers over antibodies is that polymers can be imprinted with substances against which natural antibodies are difficult to raise, e.g. immunosuppressive drugs (66), and non-immunogenic or small compounds. The latter have to be coupled to a carrier molecule in order to raise natural antibodies, which often changes their antigenic properties considerably. Therefore artificial antibodies and receptors prepared by molecular imprinting can provide an attractive alternative or complement to natural antibodies in many applications.

Catalysis. Creation of enzyme mimics or artificial enzymes as novel catalysts has been a dream of chemists for a long time. This section primarily emphasizes the work done towards this goal in our laboratory.

One of our earliest attempts in this direction dates back to making imprints of a substrate analogue to prepare a catalytic MIP for ester hydrolysis (22):

p-nitrophenyl ester of Boc-methionine → Boc-methionine + p-nitrophenol

The metal ion Co(II) was used as a "coordinator" to bring the functional monomer 4-vinylimidazole and the template (N-protected amino acid) together in the proper alignment for anticipated catalytic activity, via coordination with both the template and monomers prior to the subsequent copolymerization with crosslinker divinylbenzene. After extraction of both the template and Co(II), the substrate (p-nitrophenyl esters of methionine or leucine) was introduced, and the resulting MIP showed a clear rate enhancement in hydrolysis, together with both substrate specificity and some degree of turnover. This example demonstrated rather well that metal coordination can be used effectively for molecular imprinting. Similar reactions have also been recently studied in imprinted polymers by several other groups (68-70).

Another early example was the preparation of a MIP which catalyzed the hydrolysis of p-nitrophenyl acetate (49):

p-nitrophenyl acetate → p-nitrophenol + acetic acid

A polymer was imprinted with the transition state analogue, *p*-nitrophenol methylphosphonate, using 4-vinylimidazole as the functional monomer and 1,4-dibromobutane as the crosslinker. Again Co(II) was used to assist monomer-template assembly. The resulting MIP showed better catalysis of the hydrolysis of *p*-nitrophenyl acetate over a reference polymer. Inhibition studies with the transition state analogue indicated that the catalysis occurred inside the imprinted cavities. This was probably the first example of making imprints of a transition state analogue in synthetic polymers and was stimulated by the work on catalytic antibodies. The same reaction has also been studied more recently by another research group using imprinted polymers (*71*). In addition, transition state analogue imprinted silica has also been studied for similar catalysis, and was termed "footprint catalysis" (*72*).

Pyridoxal phosphate is a common coenzyme for many enzymes. A MIP for a coenzyme-substrate complex analogue, *N*-pyridoxal-L-phenylalanine anilide, was prepared using MAA as the functional monomer and EDMA as the crosslinker (*73*). This imprinting process can be considered as imprinting with a product analogue. The MIP obtained did catalyze the adduct formation of pyridoxal and L-phenylalanine anilide in an enantioselective fashion, and furthermore with a three-fold rate enhancement relative to a control polymer.

pyridoxal + L-phenylalanine anilide → *N*-pyridoxal-L-phenylalanine anilide

This is a step towards the preparation of artificial enzymes utilizing coenzymes. While proteins which do not have side chains with electrophilic groups have to rely on coenzymes to catalyze group-transfer reactions, one could envisage using electrophilic monomers together with other functional monomers to make MIPs capable of catalyzing such types of reactions.

Catalytic antibodies capable of catalyzing the dehydrofluorination of 4-fluoro-4-(*p*-nitrophenyl)-butan-2-one have been generated using a substrate analogue (*74*), and subsequently, a similar strategy was followed to prepare a MIP capable of catalyzing the β-elimination reaction (*50*):

4-fluoro-4-(*p*-nitrophenyl)-butan-2-one → 4-(*p*-nitrophenyl)-2-oxo-3-butene + HF

The catalytic MIP was synthesized with a substrate analogue, *N*-benzylisopropylamine, using MAA as the functional monomer and EDMA as the crosslinker. The secondary amine of the print molecule supposedly aligns the carboxylate groups of MAA juxtaposed to the real substrate's amine group, to assist the base catalysis. The result was a 600-fold rate enhancement (*1*). A MIP catalyzing the same reaction was also prepared by Beach and Shea, where the system was "reversed", *i.e.* with the carboxylic acid as the template and the amine as the functional monomer (*19*).

A more recent example in the area involves attempts to prepare class II aldolase-mimicking MIPs, capable of catalyzing the following reaction (*51*):

acetophenone + benzaldehyde → chalcone

In this case, a reactive intermediate analogue, dibenzoylmethane, was used as a template with 4-vinylpyridine and styrene as functional monomers. The metal ion Co(II) assisted self-assembly of the functional monomers and print molecules into the correct geometry. After copolymerization with the crosslinker divinylbenzene, the template and metal ion were extracted. Then, after reloading of Co(II) for the catalysis, the MIP showed an eight-fold rate acceleration, with substrate selectivity and 138 turnovers per theoretical active site. This was the first example of catalytic carbon-carbon bond formation using MIPs.

The most recent example of catalytic activity is for the Diels-Alder reaction between tetrachlorothiophene and maleic anhydride (*75*) (Figure 5). The Diels-Alder

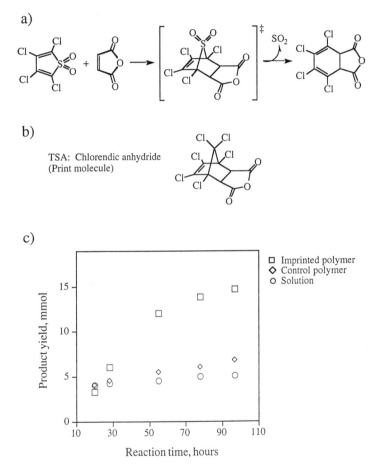

Figure 5. a) The Diels-Alder Reaction between Tetrachlorothiophene and Maleic Anhydride; b) the Print Molecule Chlorendic Anhydride; and c) the Reaction Rate curves.

reaction is a concerted cycloaddition reaction, and has a large entropic barrier. To catalyze the reaction, creation of a substrate selective cavity which functions as an "entropy trap" seems critical, and the technique of molecular imprinting has the potential to accomplish this. There is no known natural enzyme that catalyzes the Diels-Alder reaction, although catalytic antibodies have in the past been generated successfully against transition state analogues for the catalysis of these reactions (76,77). Our strategy was to use a transition state analogue, similar to the one employed for generating catalytic antibodies (76), as a template for molecular imprinting, and to synthesize an active imprinted polymer catalyst for this reaction. The inherent advantage of this reaction is that the structure of the final product changes substantially from that of the transition state due to simultaneous exclusion of the byproduct sulfur dioxide. Therefore product inhibition, which could be a serious problem for rigid MIPs, is minimized. In our system, the MIP showed a 270-fold rate enhancement and exhibited Michaelis-Menten kinetics for tetrachlorothiophene, with an apparent K_m of 42.5 mM in acetonitrile. By dividing the apparent K_{cat} value by the second order rate constant, an apparent effective molarity of 128 M per binding site was obtained. Inhibition studies showed that when chlorendic anhydride (the print molecule) was added to the reaction system as an inhibitor, the rate of the MIP catalyzed reaction decreased to only 41% of the rate of the uninhibited reaction; whereas for the control polymer, the rate was 96% of the uninhibited reaction. These results indicate that in the MIP case, the reaction did indeed occur in the cavities generated by imprinting with the transition state analogue. These experiments provide the first example of a MIP catalyzed Diels-Alder reaction, and demonstrate that MIPs may be used as an effective tool to create "entropy traps" for catalyzing bimolecular reactions, such as the Diels-Alder reaction.

To summarize, one can make catalytic MIPs using three approaches based on the template used: i) substrate analogues; ii) transition state analogues (mostly used); and iii) product analogues. If one chooses imprints with a substrate, more energy would probably be required to overcome the high affinity between the MIP and the substrate to reach the subsequent transition state, whereas if one makes the imprint with a product, product inhibition is likely to occur. Ultimately, the choice of approach taken will probably in part depend on the reaction to be studied. One point of caution which we would like to draw attention to here, involves the interpretation of kinetic data: a control polymer should ideally be prepared using a template which is highly similar in structure to the compound involved in the reaction, *e.g.* an enantiomer of the print molecule. Also, if no print molecule is present during polymerization, often the morphology of the polymer can change substantially. Furthermore, it should be kept in mind that the imprints are still heterogeneous in nature, which creates difficulties by itself in the interpretation of kinetic date. Also, since most MIPs are highly crosslinked and have little flexibility, studies towards preparing conformationally flexible catalytic MIPs capable of "induced fit" similar to enzymes are of importance.

Nevertheless, due to the superior properties of MIPs, such as their high stability and their ability to operate in organic solvents and under high temperatures and pressures, and not least that new reactions can be studied for which there is no biological counterpart available, makes the preparation of enzyme-mimicking polymers a worthwhile challenge and a potentially fruitful pursuit.

Biomimetic sensors. One of the areas where specific recognition phenomena play a key role is in sensor technology. A chemical sensor or biosensor consists of a recognition element in close contact with a transducer, which translates the signal produced upon analyte binding (and eventually conversion) into a quantifiable output signal. Many sensors for environmental monitoring, biomedical and food analysis, *etc.* rely on biomolecules, such as antibodies or enzymes, as the specific recognition elements. Given the poor chemical and physical stability of biomolecules, artificial receptors are therefore gaining increasing interest. There are several kinds of synthetic recognition systems, including rationally designed host-guest systems, receptors or ligands obtained by selection from libraries, and molecular imprinting. Molecularly imprinted systems have the advantage that the recognition sites are tailor-made directly in a solid polymeric support. Taking into account the very high specificity that can be obtained, as well as the chemical and physical stability of imprinted polymers, it is not surprising that there have been a number of attempts to construct chemical sensors based on these materials as the recognition elements.

After earlier attempts to introduce imprinted polymers into sensor technology (78-80) in which the recognition elements were not in close contact with the transducer, the first "real" biosensor based on an imprinted polymer was described by our group in 1991 (81). Subsequently, the development of a biomimetic sensor based on a field-effect transistor coated with a phenylalanine anilide imprinted polymer was reported, which allowed detection of the analyte in a qualitative manner (36). An amperometric morphine sensor described later could already measure the concentration of morphine in the range of 0.1-10 µg/ml (55). The development of molecularly imprinted polymers which also exhibited electrical conductivity (82) was a step in the same direction. A further sensor, reported recently, was based on conductometric measurements (83), where an increased local concentration of a charged analyte (due to binding to the molecularly imprinted polymer placed on a conductometric transducer) was translated into an electrical signal. However, this sensor could only be used in relatively well defined solutions with a known background conductivity. A conductometric sensor for the herbicide atrazine has also been described by another group (84).

To date, one of the most promising sensor formats involving an imprinted polymer, which is already somewhat closer to the requirements for the measurement of real samples, is, in our opinion, a recently described optical sensor (54). The fluorescence signal generated upon binding of a fluorescent model analyte (dansyl-L-phenylalanine) to an imprinted polymer was measured using fiber-optics, and the signal was found to be a function of the analyte concentration (Figure 6). Moreover, the sensor showed a certain degree of chiral selectivity for the L-form of the analyte, which was the original template molecule.

Conclusions and Future Directions

Much of what has been written on this topic in our most recent review from last year (1) is still valid and does not require re-iteration. A goal one would especially like to achieve is to be able to generate a single 1:1 complex between the print molecule and the molecularly imprinted polymer similar to the situation for a monoclonal antibody:antigen complex. This would allow better characterization with respect to the binding constant, a better understanding of recognition mechanisms, and allow the preparation of better imprinted polymers of greater practical value. Finally, it is noticeable in general that there has been a growing interest over the last 2-3 years from industry and agencies in the molecular imprinting area which bodes well for the future.

Acknowledgments

Over the years, many graduate students, post-docs and guest professors have been participating in much of the work quoted here, in addition to scientists who worked with one of us (K.M.), while professor at the ETH (Swiss Federal Institute of Technology). It gives great satisfaction to listen at this symposium to contributions by scientists who have found their own independent groups but who have either "emanated" from our laboratory, like Professor Sellergren (Mainz, Germany), or who have been starting in the area in our laboratory, like Professor Akashi (Kagoshima) and Dr. Matsui (Hiroshima-Tokyo, together with Professors Takeuchi and Karube).

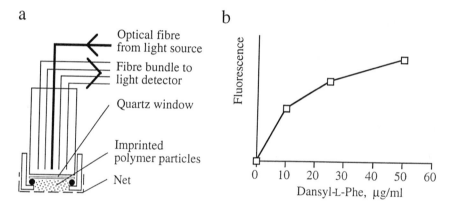

Figure 6. a) Configuration of an Optical Sensor Based on a Molecularly Imprinted Polymer as the Recognition Element. b) Binding of the Print Molecule Dansyl-L-phenylalanine Induces an Increased Fluorescence Signal. (Adapted from ref. 54.)

References

1. Mosbach, K.; Ramström, O. *Bio/Technology* **1996**, *14*, 163.
2. Wulff, G. *Angew. Chem., Int. Ed. Engl.* **1995**, *34*, 1812.
3. Shea, K. J. *Trends. Polym. Sci.* (Cambridge, UK) **1994**, *2*, 166.
4. Vidyasankar, S.; Arnold, F. H. *Curr. Opin. Biotechnol.* **1995**, *6*, 218.
5. Mosbach, K. *Trends Biochem. Sci.* **1994**, *19*, 9.
6. Ansell, R. J.; Kriz, D.; Mosbach, K. *Curr. Opin. Biotechnol.* **1996**, *7*, 89.
7. Nicholls, I.A.; Andersson, L.I.; Mosbach, K.; Ekberg, B. *Trends Biotechnol.* **1995**, *13*, 47.
8. Mosbach, K.; US Patent application; 5,110,833; 1992.
9. Vlatakis, G.; Andersson, L. I.; Müller, R.; Mosbach, K. *Nature* **1993**, *361*, 645.
10. Mosbach, K.; Mosbach, R. *Acta. Chem. Scand.* **1966**, *20*, 2807.
11. Hjertén, S.; Liao, J.-L.; Nakazoto, K.; Wang, Y.; Zamaratskaia, G.; Zhang, H.-X. *Chromatographia* **1997**, *44*, 227.
12. Glad, M.; Norrlöw, O.; Sellergren, B.; Siegbahn, N.; Mosbach, K. *J. Chromatogr.* **1985**, *347*, 11.
13. Andersson, L.; Sellergren, B.; Mosbach, K. *Tetrahedron Lett.* **1984**, *25*, 5211.
14. Sellergren, B.; Lepistö, M.; Mosbach, K. *J. Am. Chem. Soc.* **1988**, *110*, 5853.
15. Andersson, L. I.; Mosbach, K. *J. Chromatogr.* **1990**, *516*, 313.
16. Matsui, J.; Miyoshi, Y.; Takeuchi, T. *Chem. Lett.* **1995**, 1007.
17. Dunkin, I. R.; Lenfeld, J.; Sherrington, D. C. *Polymer* **1993**, *34*, 77.
18. Norrlöw, O. Doctoral Thesis, University of Lund, 1986.
19. Beach, J. V.; Shea, K. J. *J. Am. Chem. Soc.* **1994**, *116*, 379.
20. Ramström, O.; Andersson, L. I.; Mosbach, K. *J. Org. Chem.* **1993**, *58*, 7562.
21. Kempe, M.; Fischer, L.; Mosbach, K. *J. Mol. Recognit.* **1993**, *6*, 25.
22. Leonhardt, A.; Mosbach, K. *React. Polym.* **1987**, *6*, 285.
23. Yu, C.; Mosbach, K. *J. Org. Chem.* **1997**, *62*, 4057.
24. Dhal, P. K.; Arnold, F. H. *J. Am. Chem. Soc.* **1991**, *113*, 7417.
25. Kempe, M.; Glad, M.; Mosbach, K. *J. Mol. Recognit.* **1995**, *8*, 35.
26. Ramström, O.; Yu, C.; Mosbach, K. *J. Mol. Recognit.* **1996**, *9*, 691.
27. Mayes, A. G.; Andersson, L. I.; Mosbach, K. *Anal. Biochem.* **1994**, *222*, 483.
28. Kempe, M.; Mosbach, K. *J. Chromatogr.* **1994**, *664*, 276.
29. Mayes, A. G.; Mosbach, K. *Anal. Chem.* **1996**, *68*, 3769.
30. Norrlöw, O.; Glad, M.; Mosbach, K. *J. Chromatogr.* **1984**, *299*, 29.
31. Glad, M.; Reinholdsson, P.; Mosbach, K. *React. Polym.* **1995**, *25*, 47.
32. Dhal, P. K.; Vidyasankar, S.; Arnold, F. H. *Chem. Mater.* **1995**, *7*, 154.
33. Matsui, J.; Kato, T.; Takeuchi, T.; Suzuki, M.; Yokoyama, K.; Tamiya, E.; Karube, I. *Anal. Chem.* **1993**, *65*, 2223.
34. Schweitz, L.; Andersson, L. I.; Nilsson, S. *Anal. Chem.* **1997**, *69*, 1179.
35. Haupt, K.; Swedish patent application;1997.
36. Hedborg, E.; Winquist, F.; Lundström, I.; Andersson, L.I.; Mosbach, K. *Sens. Actuators, A* **1993**, 796.
37. Wang, H. Y.; Kobayashi, T.; Fujii, N. *Langmuir* **1996**, *12*, 4850.
38. Yoshikawa, M.; Izumi, J.-I.; Kitao, T.; Koya, S.; Sakamoto, S. *J. Membr. Sci.* **1995**, *108*, 171.
39. Mathew-Krotz, J.; Shea, K. J. *J. Am. Chem. Soc.* **1996**, *118*, 8154.
40. Andersson, L. I.; Müller, R.; Vlatakis, G.; Mosbach, K. *Proc. Natl. Acad. Sci. USA* **1995**, *92*, 4788.
41. Shnek, D. R.; Pack, D. W.; Sasaki, D. Y.; Arnold, F. H. *Langmuir* **1994**, *10*, 2382.
42. Aherne, A.; Alexander, C.; Payne, M. J.; Perez, N.; Vulfson, E. N. *J. Am. Chem. Soc.* **1996**, *118*, 8771.
43. Fischer, L.; Müller, R.; Ekberg, B.; Mosbach, K. *J. Am. Chem. Soc.* **1991**, *113*, 9358.
44. Kempe, M.; Mosbach, K. *J. Chromatogr. A* **1995**, *694*, 3.
45. Siemann, M.; Andersson, L.I.; Mosbach, K. *J. Agric. Food Chem.* **1996**, *44*, 141.
46. Matsui, J.; Doblhoff-Dier, O.; Takeuchi, T. *Chem. Lett.* **1995**, *6*, 489.

48

47. Andersson, L. I. *Anal. Chem.* **1996**, *68*, 111.
48. Muldoon, M.; Stanker, L. *J. Agric. Food Chem.* **1995**, *43*, 1424.
49. Robinson, D. K.; Mosbach, K. *J. Chem. Soc., Chem. Commun.* **1989**, 969.
50. Müller, R.; Andersson, L. I.; Mosbach, K. *Makromol. Chem., Rapid Commun.* **1993**, *14*, 637.
51. Matsui, J.; Nicholls, I. A.; Mosbach, K. *J. Org. Chem.* **1996**, *61*, 5414.
52. Byström, S. E.; Börje, A.; Åkermark, B. *J. Am. Chem. Soc.* **1993**, *115*, 2081.
53. Mosbach, K.; Nicholls, I. A.; Ramström, O. PCT patent application; WO 9414835; 1993.
54. Kriz, D.; Ramström, O.; Svensson, A.; Mosbach, K. *Anal. Chem.* **1995**, *67*, 2142.
55. Kriz, D.; Mosbach, K. *Anal. Chim. Acta* **1995**, *300*, 71.
56. Kriz, D.; Ramström, O.; Mosbach, K. *Anal. Chem.* **1997**, *69*, A 345.
57. Andersson, L. I.; O'Shannessy, D. J.; Mosbach, K. *J. Chromatogr.* **1990**, *516*, 167.
58. Dhal, P. K.; Arnold, F. H. *Macromolecules* **1992**, *25*, 7051.
59. Sellergren, B.; Ekberg, B.; Mosbach, K. *J. Chromatogr.* **1985**, *347*, 1.
60. O'Shannessy, D. J.; Ekberg, B.; Andersson, L. I.; Mosbach, K. *J. Chromatogr.* **1989**, *470*, 391.
61. Kriz, D.; Berggren Kriz, C.; Andersson, L.I.; Mosbach, K. *Anal. Chem.* **1994**, *66*, 2636.
62. Caldwell, J. *Chem. Ind.* **1995**, *6 March*, 176.
63. Caldwell, J. *J. Chromatogr. A* **1996**, *719*, 3.
64. Food and Drug Administration ; Fed. Reg., 1992; Vol. 57; pp 102.
65. Ramström, O.; Nicholls, I. A.; Mosbach, K. *Tetrahedron: Asymmetry* **1994**, *5*, 649.
66. Senholdt, M.; Siemann, M.; Mosbach, K.; Andersson, L. I. *Anal. Lett.* **1997**, *30*, 1809.
67. Ramström, O.; Ye, L.; Mosbach, K. *Chem. Biol.* **1996**, *3*, 471.
68. Sellergren, B.; Shea, K. *Tetrahedron Asymmetry* **1994**, *5*, 1403.
69. Ohkubo, K.; Funakoshi, Y.; Urata, Y.; Hirota, S.; Usui, S.; Sagawa, T. *J. Chem. Soc., Chem. Commun.* **1995**, *20*, 2143.
70. Karmalkar, R. N.; Kulkarni, M. G.; Mashelkar, R. A. *Macromolecules* **1996**, *29*, 1366.
71. Ohkubo, K.; Urata, Y.; Hirota, S.; Honda, Y.; Fujishita, Y.; Sagawa, T. *J. Mol. Catal.* **1994**, *93*, 189.
72. Shimada, T.; Hirose, R.; Morihara, K. *Bull. Chem. Soc. Jpn.* **1994**, *67*, 227.
73. Andersson, L. I.; Mosbach, K. *Makromol. Chem., Rapid Commun.* **1989**, *10*, 491.
74. Shokat, K. M.; Leumann, C. J.; Sugasawara, R.; Schultz, P. G. *Nature* **1989**, *338*, 269.
75. Liu, X.-C.; Mosbach, K. *Makromol. Chem., Rapid Commun.* **1997**, *18*, 609.
76. Hilvert, D.; Hill, K. W.; Nared, K. D.; Auditor, M. M. *J. Am. Chem. Soc.* **1989**, *111*, 9261.
77. Braisted, A. C.; Schultz, P. G. *J. Am. Chem. Soc.* **1990**, *112*, 7430.
78. Andersson, L.I.; Mandenius, C.; Mosbach, K. *Tetrahedron Lett.* **1988**, *29*, 5437.
79. Andersson, L. I.; Miyabayashi, A.; O'Shannessy, D. J.; Mosbach, K. *J. Chromatogr.* **1990**, *516*, 323.
80. Piletskii, S.; Parhometz, Y.; Lavryk, N.; Panasyuk, T.; El'skaya, A. *Sens. Actuators, B* **1994**, *18-19*, 629.
81. Mosbach, K.; Andersson, L. I., Swedish patent application; SE 9102843-1; 1991.
82. Kriz, D.; Andersson, L. I.; Khayyami, M.; Danielsson, B.; Larsson, P.-O.; Mosbach, K. *Biomimetics* **1995**, *3*, 81.
83. Kriz, D.; Kempe, M.; Mosbach, K. *Sensors and Actuators B* **1996**, *33*, 178.
84. Piletsky, S. A.; Piletskaya, E. V.; Elgersma, A. V.; Yano, K.; Karube, I.; Parhometz, Y. P.; El'skaya, A. V. *Biosens. Bioelectron.* **1995**, *10*, 959.

Chapter 4

Important Considerations in the Design of Receptor Sites Using Noncovalent Imprinting

Börje Sellergren

Department of Inorganic and Analytical Chemistry, Johannes Gutenberg University Mainz, J. J. Becherweg 24, 55099 Mainz, Germany

Receptor sites capable of distinguishing between molecules having minor structural differences can be prepared by noncovalent imprinting of templates in network polymers. Initially, these polymers are prepared considering functional group complementarity between a functional monomer and the template. The performance of these „first generation" materials is often unacceptable and a careful optimization of the variables in the imprinting process is therefore needed in order to reach the desired level of affinity and selectivity for the target compound. Depending on the separation requirements, the imprinted materials are further associated with problems, *i.e.* non-linear adsorption isotherms, strong medium dependence, slow mass transfer kinetics and a low sample load capacity, that may need to be solved. In this review some guidelines for achieving the desired level of recognition is given based on the existing knowledge of noncovalent imprinting.

A number of technologically important areas in chemistry rely on molecular recognition where one or a few compounds are bound to a binding site in preference to a number of structurally related compounds (*1-3*). This binding event can be followed by: a) a release event such as in chemical separations (*1*); b) a measurable signal such as in chemical sensing (*2*); or c) a chemical transformation, such as in catalysis (*3*). Depending on the function, the required efficiency of recognition varies. Usually a strong, but reversible, and highly selective binding of the target compound is desirable. The ability of recognizing molecules has long been associated with biological receptors, *i.e.* antibodies, receptors and enzymes, that also have found use in the aforementioned processes. Due to their limitations, *i.e.* limited stability in harsh environments, lack of reproducibility in their preparation, limited reusability and lack of design elements, synthetic molecular recognition elements have been the focus of several research groups (*4-6*). By molecular design, low molecular binding sites can be constructed which are capable of selective binding and catalysis although rarely with the same efficiency as those observed in the biological systems. With a considerably smaller effort, mimicking of biological molecular recognition can now be achieved by molecular imprinting of polymeric network materials (*7-12*). In one of the simpler imprinting routes (Figure 1), a template (usually the target molecule to be recognized) and a functional monomer are allowed to self assemble in solution followed by copolymerization with a crosslinking monomer (*9-12*). After crushing and sieving of the resultant polymer, the template is removed through a simple washing procedure. The generated binding sites are capable of rebinding the template with a high affinity, as shown from chromatography experiments in which the polymers are used as chromatographic packings (*13-15*). The rebinding in this noncovalent imprinting approach resembles to a large extent the biological recognition processes and has been successfully employed to generate antibody-like binding properties for a number of target compounds (*16-19*). The objective of this review is to give the reader an idea of the present possibilities and limitations of the technique, which compounds can be imprinted using the present imprinting protocols, what is the optimum medium for rebinding, how to systematically optimize the selectivity, affinity and performance, as well as some further practical considerations depending on the separation requirements.

Noncovalent Molecular Imprinting.

The first example of noncovalent imprinting was shown by Dickey who imprinted silica gel with a homologues series of alkyl orange dyes (Figure 2) (*20,21*). In spite of the lack of well-defined interactions between the template and the monomers or the binding sites, the imprinted silicas were able to efficiently discriminate the imprinted dye from their analogues. Thus, this shape recognition showed us already 50 years ago that the simple addition of a template molecule during the formation of an adsorbent may be enough in order to generate binding sites for the same template. Some limitations of Dickey's approach were apparent. Only compounds that adsorbed to nonimprinted silica gel gave rise to the imprinting effect, the stability was limited, and the reproducibility questioned.

The concept introduced by Wulff and coworkers in the early 1970s considerably extended the versatility of molecular imprinting (Figure 1) (*22-24*). In this method a monomer, properly functionalized to form covalent or noncovalent bonds to complementary binding sites on the template molecule, is copolymerized with a crosslinking monomer. After removal of the template, binding sites were generated containing functional groups at defined positions. In addition to simple shape complementarity, specific functional group complementarity is now responsible for the molecular recognition. Also this approach allows two or more different functional groups to be introduced into the binding sites, a feature of particular importance in the construction of catalytic sites. The model system developed by Wulff and his group was based on reversible covalent bond formation between the monomer and the template.

The noncovalent route was developed by the group of Mosbach (*9-12*). After demonstration of the concept in the imprinting of acrylamides or coatings of acrylamides or organosilanes on wide pore silica with dyes (*25-27*) a more versatile imprinting system was developed using carboxylic acid containing

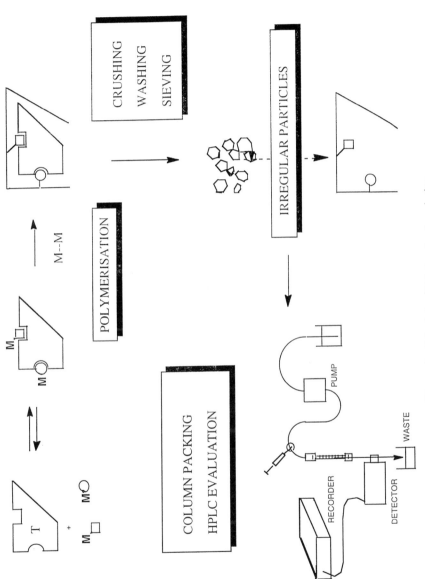

CRUSHING
WASHING
SIEVING

POLYMERISATION

IRREGULAR PARTICLES

COLUMN PACKING
HPLC EVALUATION

Figure 1. Principle of Molecular Imprinting.

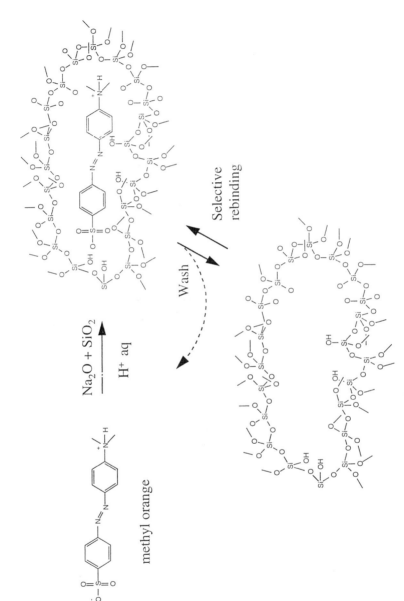

Figure 2. Imprinting of Alkylorange Dyes in Waterglass according to Dickey (20,21)

olefins as functional monomer and di- or tri- methacrylates as the crosslinking monomer. In the initial work, derivatives of amino acid enantiomers were used as templates for the preparation of stationary phases for chiral separations (CSPs). The procedure applied to the imprinting with L-phenylalanine anilide (L-PheNHPh) is outlined in Figure 3 (28). In the first step, the template (L-PheNHPh), methacrylic acid (MAA) and ethyleneglycol dimethacrylate (EDMA) are dissolved in a poorly hydrogen bonding solvent (porogen) of low to medium polarity. The free radical polymerization is then initiated with azobisisobutyronitrile (AIBN) either by photochemical homolysis below room temperature or thermochemically at 60°C or higher. Lower thermochemical initiation temperatures down to 40°C may be obtained using azobisdivaleronitrile (ABDV) instead of AIBN as initiator (29). In the final step, the resultant polymer is crushed, Soxhlet-extracted in methanol, and sieved to a suitable particle size for either chromatographic (25-38 µm) or batch (150-250 µm) applications (28). Using L-PheNHPh as template, the Soxhlet-extraction alone only result in about 70 % recovery of the template. Additional extraction in the chromatographic mode, using an acidic mobile phase, results in additional recovery with a final recovery of around 90 %. Other templates are recovered in up to 99 % yield.

Chromatographic columns are packed and the rebinding properties evaluated chromatographically by comparing the retention time or capacity factor (k') for the template with that of structurally related analogues (Figure 4). An advantage of using enantiomers as templates is that nonspecific binding, which affects both enantiomers equally, will cancel out. Therefore the separation factor (α) uniquely reflects the contribution to binding from the enantioselectively imprinted sites. Except for the retention and selectivity, the efficiency of the separations are routinely characterized by estimating a number of theoretical plates (N), a resolution factor (R_s) and a peak asymmetry factor (A_s). These quantities are affected by the quality of the packing and mass transfer limitations as well as of the amount and distribution of the binding sites.

The potential of the noncovalent imprinting technique is obvious. A robust material exhibiting antibody-like molecular recognition properties can be prepared by simply mixing commercially available components followed by heating or UV-irradiation. The preparation, work up and chromatographic evaluation of the imprinted polymer may be carried out in less than two days.

Structure-Selectivity Relationships.

Imprinted CSPs. The number of racemates which have been successfully resolved on imprinted CSPs is continuously increasing. A few representative examples are shown in (Table I). These include amino acid derivatives (13-15,30-33), peptides (18,19,32), carboxylic acids (30,33,34), amines (35), amino alcohols (36,37) and monosaccharides (38) and include a number of compounds with therapeutic importance. Generally speaking good recognition is obtained for templates containing functional groups with Brönsted-basic or hydrogen bonding properties close to the stereogenic center, provided the template is soluble in the monomer mixture. Some points should be emphasized. The separation factors are high and higher than those observed for many of the widely used commercial CSPs (39). Since these CSPs are tailor-made, the data given in Table I refers to each racemate resolved on its complementary column, i.e. each column is only able to resolve a limited number of racemates. Although the separation factors

54

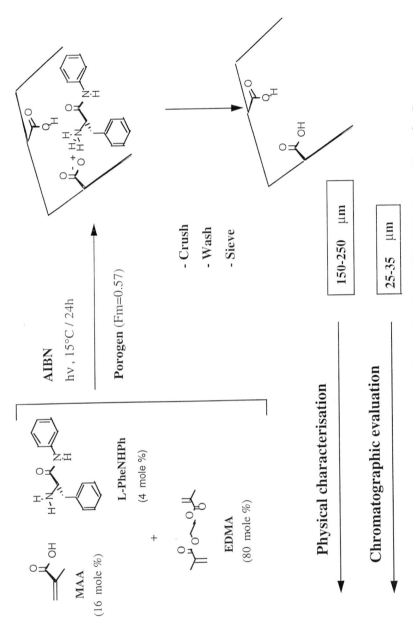

Figure 3. Scheme for Preparing an Imprinted Chiral Stationary Phase for L-Phenylalanine anilide (L-PheNHPh) using Methacrylic Acid as the Functional Monomer.

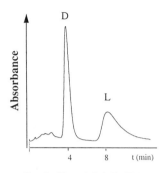

Retention and selectivity:
Capacity factor: $k'_L = (t_L - t_0)/t_0$
Separation factor: $\alpha = k'_L / k'_D$

Efficieny:
Plate number: N (as for Gaussian peaks)
Resolution factor: R_s
Asymmetry factor: A_s

Sample: 10 nmole D,L-PheNHPh
Eluent: MeCN / 0.05M potassium phosphate, pH 4 : 7/3 (v/v)
Temperature: 60°C, $h_D = 12$, $h_L = 35$

Figure 4. Chromatographic Characterisation of an L-PheNHPh-imprinted Polymer.

Table I. Tailor-made Chiral Stationary Phases

Chiral target:	α	R_s	Note	Ref.
Aminoacid derivatives				
	16	1.5	Load: 6 nmole	49
Peptides:				
	3.6	4.2	Crosslinker: Trim	32
Drugs: (Naproxen)	1.7	0.8	Comonomer: 4-VPY	34
Amino acids:				
(Phenylalanine)	1.6	1.5	CE, L-PheNHPh imprinted	43

are high, the resolution factors are low, *i.e.* the column efficiency of the imprinted CSPs is usually poor. However due to the high selectivities, baseline resolutions are obtained in many cases. Within a low sample load regime, the retention on these phases is extremely sensitive to the amount of sample injected indicating overloading of a small amount of high energy binding sites (*40*).

High Selectivity. In Tables II and III are presented examples of imprinted polymers that are able to recognize subtle structural differences in the template molecule. We compared a polymer imprinted with L-phenylalanine anilide (**1**) and one with L-phenylalanine-N-methylanilide (**2**) (comparing a secondary and tertiary amide as template) for their abilities to resolve the template racemate or a template analogue racemate (*14*). Both polymers efficiently separated the enantiomers of the racemic template while the analogue racemate was less retained and only poorly resolved. In other words, the strongest retention was observed in both cases for the original template. Similar results were obtained when comparing a polymer imprinted with L-phenylalanine ethyl ester (**1**) and one with its phosphonate analogue (**2**) (Table III). In summary, the imprinted polymers are able to efficiently discriminate enantiomers of a secondary and tertiary amide that are different in a single amide methyl group and a carboxylate and a phosphonate ethyl ester. Similar effects are observed comparing a primary and a tertiary amine different in two amino methyl groups and two diacids which differ only in one methylene group in the alkyl chain (*41*).

Since the recognition effect also was apparent in polymers imprinted with bulkier substrates, selective binding must be due to either a close shape complementarity between the site and the substrate, which provides a complementary Van der Waals surface, or large conformational differences between the derivatives. In the secondary-tertiary amide system (Table II), we concluded on the basis of ^{1}H-NMR NOE experiments and molecular mechanics calculations that these templates have large conformational differences. This was seen in the different torsional angles between the anilide ring plane and the amide plane, as well as in the E-Z preference over the amide bond. In the carboxylate-phosphonate system (Table III) the α-ammonium-phosphonate unit is known to form a stable intramolecular hydrogen bond between the ammonium proton and the phosphonyl oxygen. This may stabilize a conformation, different from the most stable carboxylate conformer thereby increasing the difference in shape between the templates. The discrimination between the different substrates may also result from a different disposition of functional groups at the site. This may be the explanation for the discrimination between the two and three-carbon alkyl chain diacid aspartic and glutamic acid (bound by carboxylic acid-carboxylic acid hydrogen bonding) (*41*).

Low Selectivity. Some examples have also been reported where some structural variations are tolerated without seriously compromising the efficiency of the separation. For instance, a polymer imprinted with L-phenylalanine anilide resolved amino acid derivatives with different side chains or amide substituents (*42*). Anilides of all aromatic amino acids were here resolved as well as β-naphtylamides and p-nitroanilides of leucine and alanine (Table IV). Apparently substitution of groups that are not involved in potential binding interactions only leads to a small loss in enantioselectivity. Also it was noted that the dipeptide, D,L-phenylalanylglycine anilide was resolved, while glycyl-D,L-phenylalanine

Table II. Discrimination between secondary and tertiary amides (14)

| Polymer selective for | Injected racemate | | | |
| | D,L-1 | | D,L-2 | |
	k'(L)	α	k'(L)	α
1	6.6	4.2	1.05	1.07
2	1.7	1.4	2.1	2.0

Mobile phase: acetonitrile/acetic acid: 90/10 (v/v)
Sample: 0.2 µmol racemate/g

Table III. Discrimination between carboxylic acid and phosphonic acid esters (12)

| Polymer selective for | Injected racemate | | | |
| | D,L-1 | | D,L-2 | |
	k'(L)	α	k'(L)	α
1	2.4	2.0	0.9	1.3
2	0.4	1.1	0.8	2.3

Mobile phase: acetonitrile/water/acetic acid: 96.3/1.2/2.5 (v/v)

58

anilide was not. This observation emphasizes the importance of the spatial relationship between the functional groups at the sites and indicates that substitutions made at some distance from the center of chirality are allowed. Recently a base line resolution of underivatized aminoacids (phenylalanine) on polymers imprinted with L-PheNHPh was reported (43). This emphasizes the importance of a complete investigation of the medium dependence of these polymers before excluding their use for resolving a particular racemate.

Binding Site Model. ^{1}H-NMR spectroscopy and chromatography were used to study the association between MAA and the template L-PheNHPh in solution as a mimick for the pre-polymerisation mixture (13). The ^{1}H-NMR chemical shifts of either the template or the monomer versus the amount of added MAA as well as the chromatographic retention of PheNHPh versus the amount of acid in the mobile phase, varied in accordance with the formation of multimolecular complexes between the template and the monomer. A 1:2 template-monomer complex was proposed to exist prior to polymerisation. Based on these results, hydrogen bond theory, and the assumption that the solution structure was essentially frozen by the polymerisation a structure of the template bound to the site was proposed (Figure 5). Since these initial studies, a number of other examples support this model *i.e.* the recognition is due to functional group complementarity and a correct positioning of the functional groups in the sites as well as steric fit in the complementary cavity. Rebinding to sites formed of residual nonextracted template may also contribute to the observed recognition. In most imprinted systems however, rebinding selectivity or catalytic efficiency increase with increasing recovery of the template.

Antibodylike Recognition. Particularly impressive recognition has been obtained for N-heterocycles and other rigid structures (Figure 6) with functional groups which engage in hydrogen bonding with MAA, in many cases in the form of stable cyclic hydrogen bonds (16-19,44-46). For instance, the templates theophyllin, diazepam, morphine and enkephalin gave rise to imprinted polymers which exhibit antibody-like molecular recognition properties in competitive assays with reference to the corresponding immunoassay (17-19). In another convincing example, a derivative of the DNA-base adenine was used as template (16). A strong binding and a pronounced selectivity was observed only for compounds containing the adenine function. In the competitive assays the crossreactivities were often similar to those observed using the immunoassay reference method. This analogy suggests that the polymers may be used in drug development for screening of chemical libraries for binding to a particular receptor (37,47).

Binding Isotherm and Site Distribution.

A feature of the molecular imprinting technique is the lack of uniform binding sites. In noncovalent imprinting, primarily two effects contribute to the binding site heterogeneity. Due to the amorphous nature of the polymer, the binding sites are not identical, a situation that for an immunologist is expressed as a polyclonal preparation of antibodies. Secondly, this effect is reinforced by the incompleteness of the monomer-template association. In reality most of the functional monomer exists in a free or dimerized form, not associated with the

Table IV. **Resolution of Amino acid anilides on an L-PheNHPh imprinted stationary phase (*42*)**

Racemate	k'(L)	α
PheNHPh	*3.5*	*2.3*
TyrNHPh	2.9	2.2
TrpNHPh	2.4	2.0
PheNHPNP [a]	3.1	2.1
LeuNHPNP [a]	2.1	1.6
AlaNHPNP [a]	2.0	1.6

a) PNP=p-nitrophenyl

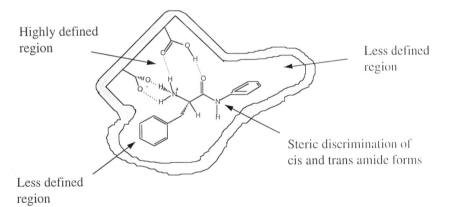

Figure 5. Current Model of the L-PheNHPh Binding Site (*13*).

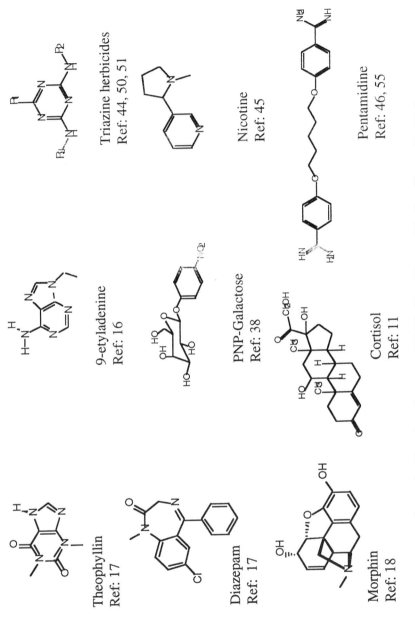

Figure 6. Template Structures Which Give Rise to Binding Sites with High Affinity and Selectivity for the Template.

Triazine herbicides
Ref: 44, 50, 51

Nicotine
Ref: 45

Pentamidine
Ref: 46, 55

9-etyladenine
Ref: 16

PNP-Galactose
Ref: 38

Cortisol
Ref: 11

Theophyllin
Ref: 17

Diazepam
Ref: 17

Morphin
Ref: 18

template. As a consequence, only a part of the template added to the monomer mixture gives rise to selective binding sites. This contrasts with covalent imprinting where theoretically all of the template split from the polymer should be associated with a specific binding site (*48*). The poor yield results in a strong dependence of selectivity and binding on sample load at least within the low sample load regime (Figure 7). In spite of these limitations, the sample load capacity for imprinted CSPs is comparable to some commercial CSPs used for preparative purposes (*39*).

For estimating the binding site density and the binding affinity, Langmuir adsorption is assumed and the concentrations of bound and free solute as determined from a batch equilibrium rebinding experiment or frontal analysis are entered into the Eady-Scatchard expression to yield the binding constant and the site density from the slope and the intercept, respectively. This form of representation clearly reveals the existence of multiple site classes. The binding sites generated by noncovalent imprinting commonly are described by 2- or 3-site models where a majority of the sites are responsible for nonspecific binding, *i.e.* they are not templated. The number of specific high affinity sites range from < 1% (Figure 8) to about 35% (Figure 9) of the amount of template added to the monomers. This figure as well as the measured binding constant depends on the degree of template-functional monomer complementarity, thermal treatments, the matrix monomer, the polymerization solvent, the polymerization temperature and the rebinding medium

The Monomer-Template Assemblies.

It is of obvious importance that the functional monomers strongly interact with the template prior to polymerization, since the solution structure of the resulting assemblies presumably defines the subsequently formed binding sites. By stabilizing the monomer-template assemblies, it is possible to achieve a large number of imprinted sites. At the same time the number of nonspecific binding sites will be minimized, since free functional monomer not associated with the template is likely to be accessible for binding (*28*). Considering one particular binding site we identify four factors that affect the recognition properties of the site (Figure 10).

Number of Interactions. Increasing the number of interaction sites on the template will lead to sites of higher specificity. This was observed in our study of the molecular imprinting of enantiomers of phenylalanine derivatives (Figure 11) (*14*). Starting with L-phenylalanine ethyl ester (**1**) as the template, interactions with carboxylic acids in acetonitrile should consist of the ammonium carboxylate ion pair, as well as a weak ester-carboxylic acid hydrogen bond (indicated by arrows). By replacing either the ester group with the stronger hydrogen bonding amide group in (**2**) or by introducing an aromatic amino group as in (**3**), which allows an additional hydrogen bond interaction with another carboxylic acid group, the enantiomeric selectivity increased. Replacing the ethyl amide substituent with the more rigid anilide group as in L-PheNHPh (**4**) produced an additional increase in selectivity. In this case a contribution from different hydrogen bonding properties of the two amide groups cannot be excluded. As expected the derivative containing all three interaction sites, p-amino-phenylalanine-anilide (**5**), gave the highest separation factor ($\alpha=5.7$) (*49*). Similar

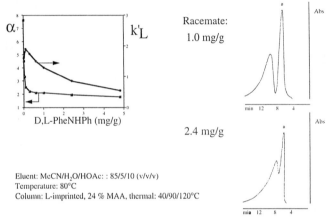

Figure 7. Sample Load Capacity of an L-PheNHPh-imprinted Polymer (*15*).

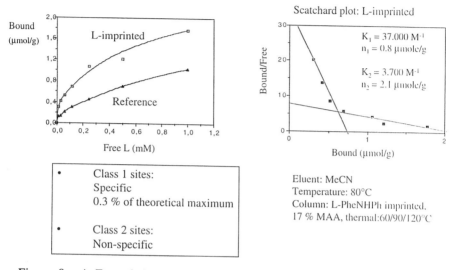

Figure 8. A Frontal Analysis Binding Isotherm for L-PheNHPh on an L-imprinted and a Non-imprinted Polymer (29).

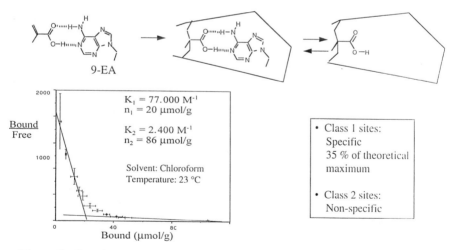

Figure 9. Scatchard Plot of the Batch Rebinding of 9-Ethyladenine (EA) to a 9-EA-imprinted Polymer (16).

• Number of interactions - "Handles"

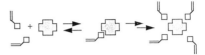

• Nature and position of the interactions

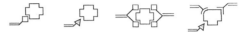

• Shape of the template: van der Waal surface

• Rigidity of the template and functional monomer

Figure 10. The Monomer-template Assemblies.

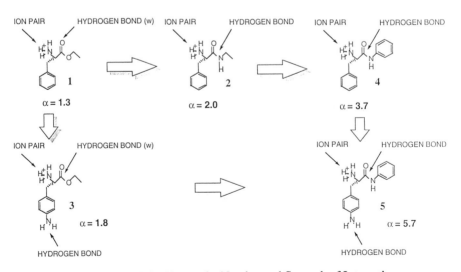

Figure 11. Selectivity Versus the Number and Strength of Interactions.

observations have been made in imprinting for a number of different classes of compounds. Thermodynamic evidence for the existance of multiple additive interactions in the sites came recently from a study, comparing the binding constants of a number of pyridine templates to their imprinted polymers (45).

Strength of Interactions. The strength and positioning of each of the template-monomer interactions is also of importance for a high affinity rebinding. Rather simple considerations can help to dramatically improve the recognition. For instance, in the case of proton donor-acceptor interactions, such as the interaction between carboxylic acids and nitrogen bases in noncovalent imprinting, the electrostatic contribution to binding increases with the acidity of the donor and the basicity of the acceptor. Thus by increasing the basicity of the template (50) or the acidity of the functional monomer (51) (Figure 12), stronger monomer-template assemblies may form which may in turn promote a larger number of selective binding sites. Even stronger interactions are found in the cyclic hydrogen bonded ion pairs formed between amidines or guanidines and carboxylic acids (52). For a comparison of some relevant solution assemblies see Figure 13 (52-54). These are strong enough to allow successful imprinting of amidines in aqueous medium (Figure 14) (46,55).

Other Functional Monomers. Other functional monomers have also been used successfully in noncovalent imprinting (Figure 15). The 2- or 4- vinylpyridine (VPY) are particularly well-suited for the imprinting of carboxylic acid templates and provide selectivities of the same order as those obtained using MAA for basic templates. In the imprinting of carboxylic acids and amides even higher selectivities can be obtained using acrylamide (AAM) as functional monomer (31) whereas diethylaminoethylmethacrylate (DEAEMA) has given high selectivity for alcohols (56). The use of designed multidentate monomers chosen from the area of host-guest chemistry is an alternative approach when the interactions between the template and monomers are too weak (57). Furthermore a combination of two or more functional monomers, giving terpolymers or higher, has in a few cases given polymers with better recognition ability than the recognition observed from the corresponding copolymers. This system is particularly complex when the monomers constitute a donor-acceptor pair, since monomer-monomer association will strongly compete with template-monomer association if neither of the monomers have a particular preference for the template. The encouraging results however shows that this route may be beneficial (33,51,58) and suggests a combinatorial approach to optimization of the binding sites.

Template Shape. As previously suggested, the template shape itself may be sufficient to create the necessary steric complementarity for efficient discrimination between two molecules. One example, where the recognition is controlled by only one electrostatic interaction is seen in Table V (59). Benzylamine is retained more than twice as much on its complementary polymer as on a reference polymer imprinted with L-PheNHPh. Interactions between the carboxylic acid group and the π-system or the benzylic protons can not be excluded as additional stabilizing interactions in this system. Pronounced shape complementarity and size exclusion effects was recently observed in the imprinting of amino acids with different N-protecting groups (31).

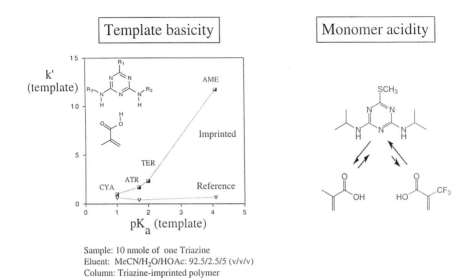

Sample: 10 nmole of one Triazine
Eluent: MeCN/H$_2$O/HOAc: 92.5/2.5/5 (v/v/v)
Column: Triazine-imprinted polymer

Figure 12. Effect of ΔpKa (Monomer-Template) in the Imprinting of Triazines (*50, 51*).

			Ref:
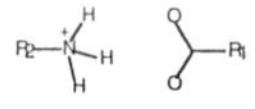	K ≈ 30 M^{-1}	MeCN	*13*
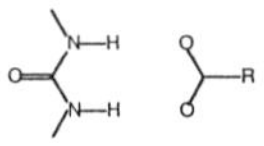	K = 45 M^{-1}	DMSO	*52*
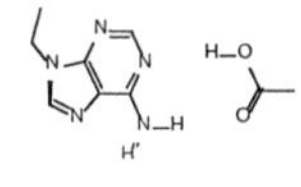	K = 160 M^{-1}	CHCl$_3$	*53*
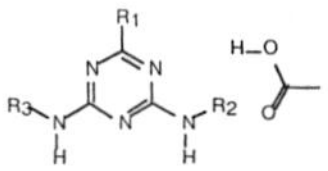	K = 212 M^{-1}	CCl$_4$	*54*
	K = 12000 M^{-1}	DMSO	*52*
	ΔG ≈ 6 kcal/mol		

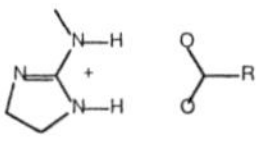

Figure 13. Carboxylic acid - N-Base Donor-acceptor Interactions.

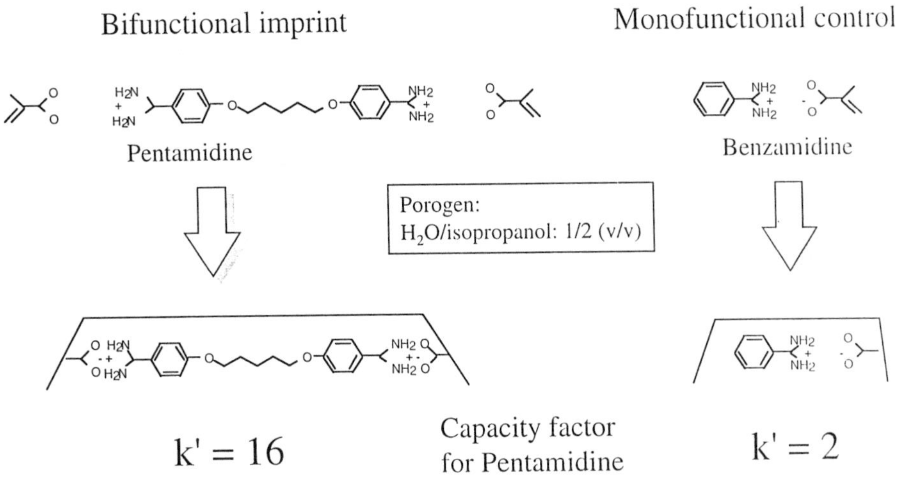

Figure 14. Imprinting of Amidines in Aqueous Media (*46*).

Mobile phase: MeCN/potassium phosphate buffer 0.05 M, pH 5 (7:3, v/v)
Column: Pentamidine or benzamidine *in situ* imprinted stationary phase

Table V. Recognition of template shape (*59*)

Template	Retention (k')	
	Benzylamine[a]	L-PheNHPh[b]
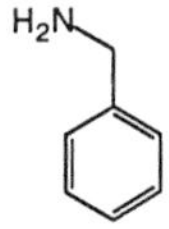 Benzylamine	28	5
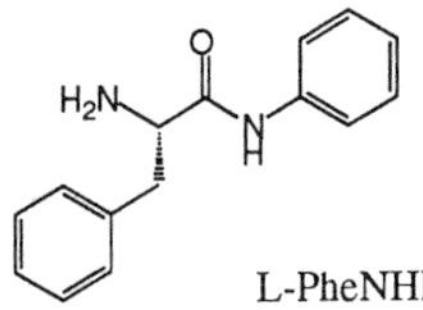 L-PheNHPh	14	25

Conditions: Sample, 100 nmole substrate; eluent,
MeCN/0.1 M potassium phosphate buffer (7:3 v/v) in a) pH 8.5 and in b) pH 6.5

68

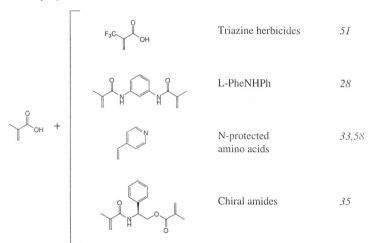

Figure 15. Functional Monomers for Molecular Imprinting.

The Matrix.

For the formation of defined recognition sites, the structural integrity of the monomer-template assemblies has to be preserved during polymerization to allow the functional groups to be fixed in space in a stable arrangement complementary to the template. This is achieved by use of a high level of crosslinking. In Figure 16 it is seen that at least 50% of the total monomer in the coMAA-EDMA system has to be EDMA for recognition to be seen (29). The role of the polymer matrix, however, is not only to contain the binding sites in a stable form but to provide porosity allowing easy access for the template to all sites (60). Most of the crosslinked network polymers used for molecular imprinting have a wide distribution of pore sizes associated with various degrees of diffusional mass transfer limitations. Based on the above criteria, *i.e.* site accessibility, integrity, and stability, the sites can be classified in different types (Figure 17). The sites associated with meso- and macro-pores (> 20Å, site A) are expected to be easily accessible compared to sites located in the smaller micropores (< 20Å, site B) where the diffusion is slow. The number of the latter may be higher since the surface area, for a given pore volume, of micropores are higher than that of macropores. One undesirable effect of adding an excess of template is the loss of site integrity due to coalescence of the binding sites (site D). The optimum amount of template is usually about 5% of total amount of monomer. For most applications in liquid media, accessible meso- and macro- pores are preferred over micropores.

Polymer Morphology. The porosity is determined by the morphology of the material which can be influenced by controlling the degree of crosslinking, the type of inert solvent (porogen) used during polymerization, and various post treatments, such as heat treatment and hydrolysis (60). A complication in noncovalent imprinting arises from the fact that the first two variables also affect the stability of the monomer-template assemblies and thereby the amount of nonselective binding sites (site F) relative to selective binding sites (sites A-E). This is clearly seen in a comparison of polymers imprinted with L-PheNHPh with different porogens under otherwise identical conditions (Figure 18) (28). It is seen that only the materials prepared using poor hydrogen bonding porogens, with less ability to compete with MAA for the binding sites on the template, give high enantiomeric separation factors. At the same time, these materials show different swellings and pore volumes reflecting structural differences. The differences in swelling between the materials may have several causes. First it appears to be unrelated to the number of unreacted double bonds since the infrared spectra showed the intensities of the C=C stretch at 1638 cm^{-1} to be similar for the materials. Furthermore CP-MAS-NMR results showed that the number of unreacted double bonds was less than 10 % in polymers prepared using THF as porogen. Rather it is related to the heterogeneity of the crosslink density which in turn will affect the stiffness of the chains linking the agglomerates or microspheres together during phase separation (61,62). This is influenced by the ability of the porogen to solvate the growing polymer chains. For instance in a poor solvent, agglomeration is promoted leading to the formation of dense crosslinked microspheres where the phase separation occurs when the microspheres precipitate, giving a relatively open pore structure with strong links

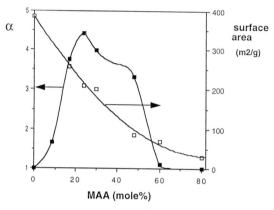

Template: L-PheNHPh
Porogen: MeCN
Selectivity determined
chromatographically
Eluent: MeCN + 5% HOAc
Temp: 80 °C

Figure 16. Effect of Crosslinking Level with Constant Porogen (29).

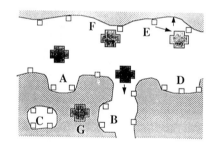

- **Site A.** Macropores. High accessibility and rapid mass transfer.

- **Site B.** Micropores. Diffusion limitations. High surface area and sample load capacity.

- **Site C.** Embedded. Low accessibility. High selectivity (?) and stability.

- **Site D.** Site coalescence. Low selectivity. Excess template.

- **Site E.** Induced binding site. Low stability.

- **Site F.** Nonselective site.

- **Site G.** Nonextracted template ?

Goal : Large pores and high surface area.
  Sites: A and B, minimize F

- Stabilize monomer template assemblies.
- Crosslinking + Porogen

Figure 17. Site Accessibility, Integrity, and Stability.

CH_2Cl_2 MeCN

Porogen	H-bond capacity	Swelling (ml/ml)	Pore volume (ml/g)	Separation factor ($\alpha = k'L/k'D$)
MeCN	P (= poor)	1.36	0.60	2.6
CH_2Cl_2	P	2.01	0.007	2.4
DMF	M (=moderate)	1.97	0.17	NR
THF	M	1.84	0.24	1.5
HOAc	S (=strong)	1.45	0.52	NR
Degree of crosslinking: ca. 83 %		Template: L-PheNHPh		NR = not resolved

Figure 18. Effect of Porogen with a Constant Level of Crosslinking (*28*).

between the microspheres. Such materials exhibit permanent porosity and low swelling. With a good solvent, intermolecular crosslinks are favored and a material is obtained which is built up of loosely linked grains of solvent swollen gel particles. These materials exhibit low pore volume and high swelling. The swollen state morphology may vary less between the materials. In view of the small variation in the separation factor, the binding sites are apparently located in the more dense, less swellable domains of the matrix.

The Crosslinking Monomer. The chemical and physical properties of the matrix are also of primary importance with respect to nonspecific binding and flexibility. One illustration of these effects is seen when comparing imprinting in styrene-based and methacrylate-based resins. Resins prepared from divinylbenzene (DVB) exhibit lower selectivity, poorer chromatographic performance, and lower thermal stability than the methacrylate-based resins (*63*). It is noted that the former are less solvated by the solvents used in the rebinding experiments and are more rigid as seen in the lower swelling when comparing otherwise identically prepared resins (in methanol the swelling was 1.27 and 1.64 for polymers prepared from DVB and EDMA respectively. Porogen: acetonitrile/benzene: 1/1). Ethyleneglycol dimethacrylate (EDMA) is the most commonly used crosslinker for the methacrylate-based systems primarily because it provides mechanical and thermal stability, good wettability in most rebinding media, and rapid mass transfer with good recognition properties. Except for the trimethacrylate trimetylolpropane trimethacrylate (TRIM) none of the other methacrylate-based crosslinkers provide similar recognition properties. TRIM has recently proven to give higher sample load capacity and performance than EDMA in the imprinting of peptides (*32*).

Medium Effects.

In analytical method development involving molecularly imprinted polymers, the medium used in the rebinding step must be carefully optimized to fully exploit the MIPs ability to recognize the target template. Based on the growing database available on the dependence of retention and selectivity in various media on the template structure, predictability will increase in the future. In the imprinting protocol using MAA as the functional monomer, the molecular recognition exhibited by the imprinted polymer can be driven by hydrogen bonding, ion exchange and/or the hydrophobic effect depending on the template and the medium.

The Solvent Memory Effect. For templates interacting with the functional monomer mainly by hydrogen bonding, optimum recognition is often seen in the exact same solvent used in the polymerisation step (*31,64*). Provided the same solvent is used in the rebinding as in the polymerisation step the strength and selectivity of the template rebinding increase with decreasing polarity and hydrogen bond capacity of the solvent. For instance a polymer imprinted with 9-ethyladenine using chloroform as porogen exhibited higher affinity for the template in a chloroform medium than that of a polymer prepared using acetonitrile as porogen evaluated in an acetonitrile medium (*64*). Interestingly, when evaluating the latter polymer in the less polar chloroform medium a much lower affinity was observed. Likewise, the polymer prepared in chloroform showed a lower rebinding affinity in acetonitrile. Apparently, the polymer remembers the solvent used in its preparation.

The Hydrophobic Effect. A good example of medium effects was seen in the evaluation of materials imprinted with triazine herbicides of different basicity and hydrophobicity (*50*). Thus in aqueous poor media the affinity and selectivity correlated with the template basicity (Figure 12) whereas in aqueous rich media a correlation with the template hydrophobicity was seen (Figure 19). Apparently, depending on the template hydrophobicity the hydrophobic effect contributes specifically or nonspecifically to the recognition. Recognition in aqueous media, presumably driven by the hydrophobic effect, has been observed in a number of other systems (*18,19*).

Ion Exchange. Templates containing protolytic functional groups may bind to the sites by an ion-exchange mechanism. This was observed in the mobile phase optimization for the resolution of D,L-PheNHPh on an L-imprinted polymer (*59*). On going from an organic mobile phase to an aqueous-organic mobile phase two effects were observed. First a dramatic improvement in the chromatographic efficiency accompanied by a lower sample load capacity was observed (Figure 20). This is probably due to shrinkage of the polymer leading to the closing of micropores with slowly exchanging binding sites. Secondly, retention maxima were seen at a pH which turned out to closely match the pKa value of the solutes in the same solvent system (determined potentiometrically). On the other hand, the selectivity was high in the low pH region but dropped when the pH exceeded the pKa value (Figure 21). The correlation between pKa and the maximum retention made us believe that a simple ion exchange process was operating. According to this theory, the retention can be expressed as being proportional to the product of the degree of ionization ($\alpha*$) of the amino groups of the solute ($\alpha*1$) and the carboxylic acid groups of the polymer ($\alpha*2$) as $k'=C(\alpha*1)(\alpha*2)$ (*65*). Based on the data obtained from potentiometric titrations, $(\alpha*1)(\alpha*2)$ was calculated and plotted against the pH of the mobile phase. A striking agreement between the experimental and simulated pH retention profile was obtained (Figure 21). Therefore the retention seems to be controlled by a simple cation-exchange process. The high selectivity observed at low pH values was in agreement with the finding of a lower average pKa of the imprinted polymer compared to the blank nonimprinted polymer. This led us to propose an electrostatic model which accounts for the changes in retention and selectivity when the mobile phase pH was varied (Figure 22).

Concluding Remarks

A number of conditions will directly influence the development of a new MIP. The availability of the template in preparative amounts will determine whether it will have to be recycled or a template analogue will have to be used. The latter alternative should also be considered in cases where the template is unstable or poorly soluble in the monomer mixture. Depending on the format of the separation, the polymer must meet certain requirements. If the material is going to be used as a HPLC stationary phase, monodisperse spherical particles are

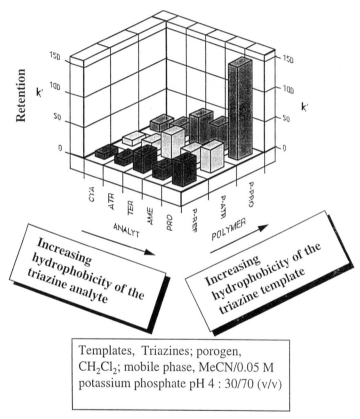

Figure 19. Chromatographic Retention of Triazines on Triazine-imprinted Polymers Using an Aqueous Rich Mobile Phase (*50*).

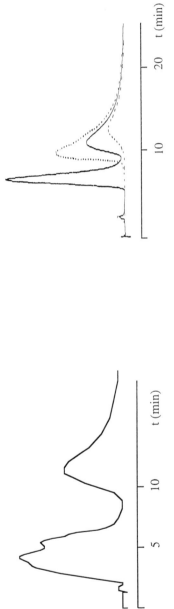

Organic
MeCN/H$_2$O/HOAc : 92.5/2.5/5 (v/v/v)

Aqueous
MeCN / 0.05M potassium phosphate pH 7 : 7/3 (v/v)

Template, L-PheNHPh; porogen, CH$_2$Cl$_2$; swelling, 2.0 ml/ml (MeCN); sample, 100 nmole D,L-PheNHPh

- Slow mass transfer
- Sample load capacity 1-2 mg/g

- Faster mass transfer
- Sample load capacity 0.1 mg/g

Figure 20. Resolution of D,L-PheNHPh on an L-imprinted Polymer in Two Different Mobile Phases (40).

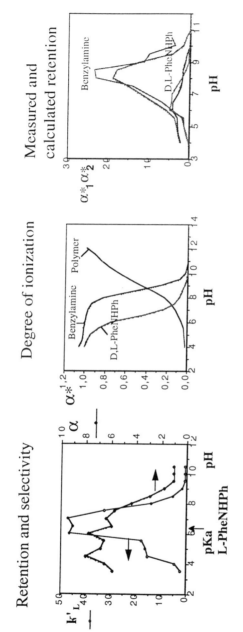

Figure 21. pH-Effects and Ion-Exchange in the Separation of D- and L-PheNHPh on on L-imprinted Polymer (59).

Polymer, L-PheNHPh-imprinted; sample, 10 nmol D,L-PheNHPh or 100 nmol benzylamine; mobile phase, acetonitrile/potassium phosphate buffer 0.05 M: 70/30 (v/v)

desirable and rapid adsorption-desorption of the template to the sites is necessary for high performance separations (*66,67*) Band broadening and assymmetry effects due to the heterogeneous distribution of binding sites is an important problem that may need to be addressed. In solid phase extraction, the important criteria are the sample load capacity, affinity, and selectivity for the compound in question as well as its recovery (*68,55*). On the other hand, in capillary chromatography or capillary electrophoresis, materials that can be fabricated in situ in capillaries offer certain advantages (*69,47*). Finally for continuous separations, imprinted membranes may need to be developed (*70-72*).

The choice of monomer should be made considering functional group complementarity and the option of terpolymers should be considered. A combinatorial approach may present advantages provided that rapid preparation and evaluation can be performed. Some predictions of the recognition properties may be done by investigating the monomer-template association using NMR. Furthermore, this technique may be used to determine the optimum composition of the prepolymerisation mixture.

An important part of the optimization process is the stabilization of the monomer-template assemblies by thermodynamic considerations (Figure 23). Knowledge of the enthalpic and entropic contributions to the association is informative of how the association will respond to changes in the polymerization temperature (*29*). Knowledge of the change in free volume of interaction is informative of how the association will respond to changes in polymerization pressure (*73*) and finally knowledge of how the solvent solvates the monomer-template assemblies relative to the free species will indicate which solvent will stabilize the monomer-template assemblies the most (*28*). Here each system has to be optimized individually. A final option is to simply increase the concentration of the monomer or the template. In the former case, a problem is that the crosslinking as well as the potentially nonselective binding will increase simultaneously. In the latter case, the site integrity will be compromised.

The final optimization concerns the medium used in the rebinding step. Some general observations have been made. For nonprotolytic templates which interact weakly with the monomer, efficient recognition is usually seen in media of low-to-medium polarity which resemble the media used in the polymerization. For Brönsted-basic or acidic templates, a higher chromatographic efficiency is usually seen in aqueous media where the retention is controlled by an ion-exchange mechanism. Finally for templates containing hydrophobic substituents, a specific hydrophobic contribution to binding may be observed at higher aqueous contents.

Due to the many variables involved in molecular imprinting, a systematic approach to the preparation and evaluation of a new material is necessary to fully exploit the potential of this technology.

Literature Cited

1. *Highly Selective Separations in Biotechnology*; Street, G. (Ed.), Blackie: New York, 1994.
2. *Biosensors and Chemical Sensors*; Edelman, P. G.; Wang, J. (Eds.) ACS Symposium Series 487, American Chemical Society: Washington, DC, 1992
3. Schultz, P. G. *Angew. Chem. Int. Ed. Engl.* 1989, 28, 1283
4. *Comprehensive Supramolecular Chemistry*; Lehn, J.-M. (Ed.); Elsevier: 1996; Vol. 1-3

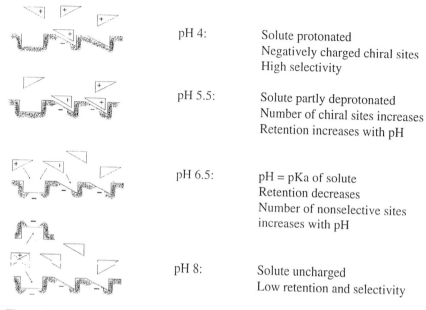

pH 4:	Solute protonated
	Negatively charged chiral sites
	High selectivity
pH 5.5:	Solute partly deprotonated
	Number of chiral sites increases
	Retention increases with pH
pH 6.5:	pH = pKa of solute
	Retention decreases
	Number of nonselective sites
	increases with pH
pH 8:	Solute uncharged
	Low retention and selectivity

Figure 22. Electrostatic Model Accounting for the Changes in Retention and Selectivity on an L-imprinted Polymer in Response to the pH in the Mobile Phase (59).

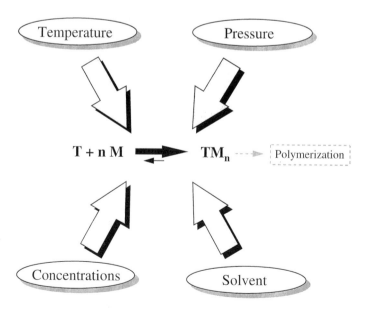

Figure 23. Stabilization of Template-monomer (TM) Assemblies.

5.   Rebek, J. Jr. *Acc. Chem. Res.* 1990, 23, 399
6.   Pirkle, W. H.; Welch, C. J.; Lamm, B. *J. Org. Chem.* 1992, 57, 3854
7.   For a comprehensive review see: Wulff, G. *Angew. Chemie. Int. Ed. Engl.* 1995, 34, 1812
8.   Shea, K. J. *Trends Polym. Sci.* 1994, 2, 166
9.   Mosbach, K. *Trends Biochem. Sci.* 1994, 19, 9
10.   Sellergren, B. *Trends Anal. Chem.* 1997, 16, 310
11.   Mayes, A. G.; Mosbach, K. *Trends Anal. Chem.* 1997, 16, 321
12.   Sellergren, B. In *A Practical Approach to Chiral Separations by Liquid Chromatography*; Subramanian, G. (Ed.); VCH, Weinheim, 1994, p. 69
13.   Sellergren, B.; Lepistö, M.; Mosbach, K. *J. Am. Chem. Soc.* 1988, 110, 5853
14.   Lepistö, M.; Sellergren, B. *J. Org. Chem.* 1989, 54, 6010
15.   Sellergren, B. *Chirality* 1989, 1, 63
16.   Shea, K. J.; Spivak, D.; Sellergren, B. *J. Am. Chem. Soc.* 1993, 115, 3368
17.   Vlatakis, G.; Andersson, L. I.; Müller, R.; Mosbach, K. *Nature* 1993, 361, 645
18.   Andersson, L. I.; Müller, R.; Vlatakis, G.; Mosbach, K. *Proc. Natl. Acad. Sci.* 1995, 92, 4788
19.   Andersson, L. I. *Anal. Chem.* 1996, 68, 111
20.   Dickey, F. H. *Proc. Natl. Acad. Sci.* 1949, 35, 227
21.   Dickey, F. H. *J. Phys. Chem.* 1955, 59, 695
22.   Wulff, G.; Sarhan, A.; Zabrocki, K. *Tetrahedron Lett.* 1973, 4329-4332
23.   Wulff, G.; Vesper, W.; Grobe-Einsler, R.; Sarhan, A. *Makromol. Chem.* 1977, 178, 2799
24.   Wulff, G.; Poll, H.-G. *Macromol. Chem.* 1987, 188, 741
25.   Arshady, R., Mosbach, K., *Makromol. Chem.* 1981, 182, 687-692
26.   Norrlöw, O., Glad, M., Mosbach, K. *J. Chromatogr.* 1984, 299, 29-41
27.   Glad, M.; Norrlöw, O.; Sellergren, B.; Siegbahn, N.; Mosbach, K. *J. Chromatogr.* 1985, 347, 11-23
28.   Sellergren, B.; Shea, K. J. *J. Chromatogr.* 1993, 635, 31
29.   Sellergren, B. *Macromol. Chem.* 1989, 190, 2703
30.   Kempe, M.; Mosbach, K. *J. Chromatogr.* 1995, 691, 317
31.   Yu, C.; Mosbah, K. *J. Org. Chem.* 1997, 62, 4057
32.   Kempe, M. *Anal. Chem.* 1996, 68,1948
33.   Ramström, O.; Andersson, L. I.; Mosbach, K. *J. Org. Chem.* 1993, 58, 7562
34.   Kempe, M.; Mosbach, K. *J. Chromatogr. A* 1994, 664, 276
35.   Hosoya, K.; Yoshizako, K.; Shirasu, Y.; Kimata, K.; Araki, T.; Tanaka, N.; Haginaka, J. *J. Chromatogr.* 1996, 728, 139.
36.   Fischer, L.; Müller, R.; Ekberg, B.; Mosbach, K. *J. Am. Chem. Soc.* 1991, 113, 9358
37.   Ramström, O.; Yu, C.; Mosbach, K. *J. Mol. Rec.* 1996, 9, 691
38.   Mayes, A. G.; Andersson, L. I.; Mosbach, K. *Anal. Biochem.* 1994, 222, 483
39.   *A Practical Approach to Chiral Separations by Liquid Chromatography*; Subramanian, G. (Ed.), VCH, Weinheim, 1994
40.   Sellergren, B.; Shea, K.J. *J. Chromatogr.* 1995, 690, 29
41.   Andersson, L. I., Mosbach, K., *J. Chromatogr.* 1990, 516, 313-322
42.   O'Shannessy, D. J., Andersson, L. I., Mosbach, K. *J. Mol. Recogn.* 1989, 2, 1-5

43. Lin, J. M.; Nakagama, T.; Uchiyama, K.; Hobo, T. *Chromatographia* 1996, 43, 585
44. Muldoon, M. T.; Stanker, L. H. *J. Agric. Food Chem.* 1995, 43, 1424
45. Andersson, H. S.; Koch-Schmidt, A.-C.; Ohlson, S., Mosbach, K. *J. Mol. Recogn.* 1996, 9, 675
46. Sellergren, B. *J. Chromatogr.* 1994, 673, 133
47. Schweitz, L.; Andersson, L. I.; Nilsson, S. *Anal. Chem.* 1997, 69, 1179
48. Whitcombe, M. J.; Rodriguez, M. E.; Villar, P.; Vulfson, E. N. *J. Am. Chem. Soc.* 1995, 117, 7105.
49. Sellergren, B., Nilsson, K., *Meth. Molec. Cell. Biol.* 1989, 10, 183
50. Dauwe, C.; Sellergren, B. *J. Chromatogr.* 1996, 753,191
51. Matsui, J.; Miyoshi, Y.; Takeuchi, T. *Chem. Lett.* 1995, 1007
52. Fan, E.; Van Arman, S. A.; Kincaid, S.; Hamilton, A. D. *J. Am. Chem. Soc.* 1993, 115, 369
53. Lancelot, G. *J. Am. Chem. Soc.* 1977, 99, 7037
54. Welhouse, G. J.; Bleam, W. F. *Environ. Sci. Technol.* 1993, 27, 500
55. Sellergren, B. *Anal. Chem.* 1994, 66, 1578
56. Levi, R.; McNiven, S.; Piletsky, S. A.; Cheong, S.-H.; Yano, K.; Karube, I. *Anal. Chem.* 1997, 69, 2017
57. Tanabe, K.; Takeuchi, T.; Matsui, J.; Ikebukuro, K.; Yano, K.; Karube, I. *J. Chem. Soc. Chem. Comm.* 1995, 2303
58. Kempe, M.; Fischer, L.; Mosbach, K. *J. Mol. Recogn.* 1993, 6, 25
59. Sellergren, B.; Shea, K. J. *J. Chromatogr.* 1993, 654,17
60. Guyot, A. in *Synthesis and Separations Using Functional Polymers*; Sherrington, D. C.; Hodge, P., (Eds.); Wiley-Interscience: New York, 1988
61. Galina, H.; Kolazz, B. N.; Wieczorek, P. P. Wojczynska, M. *Brit. Polym. J.* 1985, 17, 215.
62. Reinholdsson, P.; Hargitai, T.; Isaksson, R.; Tornell, B. *Angew. Macromol. Chem.* 1991, 192, 113-132.
63. Wulff, G., Kemmerer, J., Vietmeier, J., Poll, H.-G., *Nouv. J. Chim.* 1982, 6, 681-687
64. Spivak, D.; Gilmore, M. A.; Shea, K. J. *J. Am. Chem. Soc.* 1997, 119, 4388
65. Pietrzyk, D. J. in *Packings and Stationary Phases in Chromatographic Techniques*; Unger, K. K. (Ed) Chromatographic Science Series, New York: Marcel Dekker Inc., 1990, vol. 47, pp 585.
66. Snyder, L. R.; Kirkland, J. J., *Introduction to Modern Liquid Chromatography*. Wiley, New York, 1979
67. Mayes, A. G.; Mosbach, K. *Anal. Chem.* 1996, 68, 3769
68. Berrueta, L. A.; Gallo, B.; Vicente, F. *Chromatographia* 1995, 40, 474
69. Baba, Y.; Tsuhako, M. *Trends Anal. Chem.* 1992, 11, 280
70. Piletsky, S. A.; Parhometz, Y. P.; Lavryk, N. V.; Panasyuk, T. L.; El'skaya, A. V. *Sens. Actuators B* 1994, 18-19, 629
71. Mathew-Krotz, J.; Shea, K.J. *J. Am. Chem. Soc.* 1996, 118, 8154
72. Wang, H. Y.; Kobayashi, T.; Fuji, N. *Langmuir* 1996, 12, 4850
73. Sellergren, B.; Schneider, T.; Dauwe, C. *Macromolecules* 1997, 30, 2454

MOLICULAR RECOGNITION
WITH ORGANIC-BASED POLYMERS

Chapter 5

Recent Advances in the Use of Molecularly Imprinted Materials in Separation and Synthesis

Olof Ramström, Lei Ye, Cong Yu, and Per-Erik Gustavsson

Department of Pure and Applied Biochemistry, Center for Chemistry and Chemical Engineering, Lund University, P.O. Box 124, S-221 00 Lund, Sweden

Molecularly imprinted materials, prepared using self-assembly imprinting protocols using only non-covalent interactions, can be used as chromatography media in aqueous phase. The recognition properties are highly dependent on the interacting species used in the imprinting protocol. Ionic interactions, together with strong hydrogen bonding, represent useful means of obtaining recognition. With increasing levels of water in the recognition media, the hydrophobic effect comes into play. Molecularly imprinted materials can furthermore be used as auxiliary agents in enzymatic syntheses in water-saturated organic phases. When materials molecularly imprinted for the reaction product were applied to the thermolysin-catalyzed aspartame synthesis, an increase in yield was observed. Introduction of a thermodynamic trap provided by the molecularly imprinted matrices allowed a non-favorable equilibrium for the thermolysin reaction to be pushed in the forward direction.

Molecularly imprinted materials possess a high potential for use in a variety of applications, such as chromatographic stationary phases, immunoassay-type analyses, and sensor technologies (1-5). The stability and endurance of these materials are attractive features entailed by these materials which are focusing attention on their use as recognition elements besides naturally occurring matrices, such as antibodies and receptors.

One of the current challenges for molecularly imprinted materials is their preparation and/or usage in aqueous environments (6). When self-assembly imprinting protocols using only non-covalent interactions are employed, water reduces the binding strength of such interactions. Since self-assembly protocols have been shown to be of great importance, which produce faster binding kinetics and a wider range of appropriate systems, this challenge needs to be overcome.

One approach to meeting the challenge is to prepare the materials in an organic phase and subsequently use them in an aqueous phase (*7-9*).

Although the aqueous phase challenge is an important one, these materials may also be employed in various applications in water-poor phases. A multitude of applications has been envisioned for such materials in organic phases, the foremost being in the separation area. In some biotechnological applications, water-poor phases are prerequisite to the success of the process. For example, some enzymatic syntheses proceed well in water-poor phases, which control or even reverse the reaction equilibrium (*10*). In these cases, molecularly imprinted polymers (MIPs) may serve as auxiliary recognition elements for an enhanced control of the process.

In this work we describe some recent advances in the use of imprinted materials in recognition and separation applications. Molecularly imprinted polymers are used as stationary phases in aqueous-phase chromatographic systems and as equilibrium traps in enzymatic synthesis.

Recognition Properties of Molecularly Imprinted Polymers in Aqueous Phases

In biological systems, water plays a dominant role as the surrounding medium. Therefore, mimics of natural binding events are most effectively demonstrated in aqueous systems. However, the influence of water as a porogen in the imprinting process or recognition media greatly diminishes the energy of non-covalent interactions (*11*). For techniques currently in use, where small, often monofunctional, monomers are employed for recognition of the imprint species in solution prior to polymerisation, the disrupting effect of water on ionic, hydrogen bonded, and van der Waals' interactions intrinsically leads to very weak recognition effects. Also, the solubility of the necessary ingredients, such as crosslinkers and monomers, may be too low in an aqueous or partially aqueous medium to allow for the formation of solid polymers.

To overcome the bond-breaking effects of water, molecularly imprinted polymers have first been prepared in organic media, where the interactions are strong, and subsequently been used in aqueous environments (*12-13*). With this protocol, the designed sites formed in the imprinting process may lead to a concerted action of participating functional groups to selectively recognize the ligands.

The recognition properties of molecularly imprinted polymers in aqueous phases are dependent on the nature and quality of the interacting species. The choice of functional monomers together with the arrangement of functional groups of the imprint species are two of the main factors responsible for recognition.

This was clearly demonstrated when molecularly imprinted polymers were prepared with the protected amino acid benzyloxycarbonyl-L-tyrosine (Cbz-L-Tyr) and either 2-vinylpyridine (2VPy) or methacrylic acid (MAA), or a combination of both, as functionally interacting monomers in a self-assembly imprinting protocol which involved only non-covalent interactions. These systems have previously been shown to possess excellent recognition properties in organic media (*14*). In particular, a terpolymer system, which incorporated both functional monomers, exhibited a very high discrimination effect between the enantiomers of Cbz-Tyr. The separation factor (α) was 4.32 in acetonitrile containing 1% acetic acid (v/v). A combination of ionic interactions between the pyridinyl moieties of the polymer and the carboxyl functionalities of the imprinted species, together with hydrogen bonding capabilities of the carboxy functionalities of the matrix are responsible for recognition. In the present study, these polymer systems were analyzed in the HPLC-mode for their performance to discriminate in binding between the enantiomers of the imprint species in aqueous media. The influences of the pH of the mobile phase and the level of added organic solvent on the separation behavior were investigated. It was found that the polymers containing 2-vinylpyridine performed well in aqueous media at pH≤5 (Figures 1 and 2). The amount of acetonitrile in the eluent was found to have a strong influence. When methacrylic acid was used as a monomer in the absence of 2-vinylpyridine, no chiral selectivity was detected. These results are indicative of the importance of strong ionic bonds in conjunction with hydrophobic interactions in the formation and maintenance of complexes between the analyte and the polymer. At a pH below the pK_a of the pyridinyl groups of the matrix ($pK_{a(app)}$ ~ 4.7) most of the nitrogen atoms are protonated, thus contributing to ion pair formation. In addition, the acidity of the carboxyl moiety of the imprint species is in the same pH-range (pK_a ~ 3.5) and is largely deprotonated. When only methacrylic acid was used as functional monomer, the pK_a of the polymer matrix varies. Due to the different microenvironments of the carboxyl moieties in the matrix, the corresponding ionization values may cover a wide range of pK_a-values. However, when the pH of the mobile phase is increased, there is a counteracting effect from the enhanced deprotonations of both the imprint species and the carboxyl moieties of the matrix. This eliminates any recognition. With an enhanced proportion of aqueous in the eluent, binding to the polymer matrices increased. However for these systems, the non-specific binding and the specific binding increased simultaneously so there was no increase in specificity of the matrices.

A further example of separation in aqueous media was provided by a study in which only non-ionic interactions were employed in the imprinting protocol. Here, acrylamide was chosen as functional monomer. This hydrogen bonding monomer has proven very useful when utilized in organic phases (15). When the prepared molecularly imprinted materials were applied in water-rich phases, a remarkable dependence on the water/organic solvent ratio was recorded. With low proportions of water, the recognition properties were substantially reduced. On the other hand, a higher proportion of water led to an ameliorated recognition quality of the matrices and an excellent separation was obtained (Figure 3). When the percentage of aqueous phase in the eluent was increased from 50% to 70% of 10 mM glycine (pH 3.0) in acetonitrile for a matrix imprinted with Boc-L-Trp, the separation factor increased from 1.55 to 1.74 and the resolution factor increased from 1.30 to 2.15. In addition, the effect was found to follow the hydrophobicity of the ligand. The more non-polar the ligand - the better the separation effect. These results suggest a build-up of a hydrophobic effect with the increased content of the aqueous phase which augments the effect of the polar groups.

In this context, it needs to be stated that the effect of water is double-edged. At low concentration, water acts as an organic phase modifier that weakens the overall binding by reducing the strength of polar interactions. At high concentration, the binding increases again due to the hydrophobic effect (Figure 4).

MIPs as Auxiliary Agents in Enzymatic Synthesis

The possibility of shifting a thermodynamically unfavorable enzymatic equilibrium towards product formation in the presence of a highly specific molecularly imprinted adsorbent has been investigated. The commercially interesting enzymatic condensation of benzyloxycarbonyl-L-aspartic acid (Cbz-L-Asp) with L-

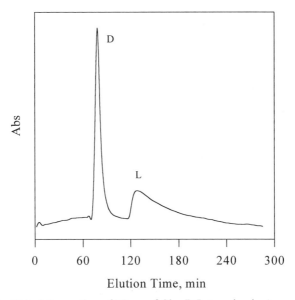

Figure 1. Chiral Separation of 10 μg of Cbz-D,L-tyrosine in Aqueous Media by a MIP Based on 2-Vinylpyridine as the Functional Monomer. Conditions: mobile phase: 10 mM phosphate buffer, pH 4.0/acetonitrile (3:1, v/v); flow rate = 0.5 mL/min; α=1.70; R_S=2.47

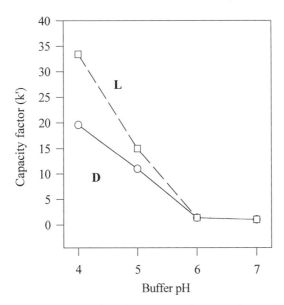

Figure 2. Influence of Buffer Phase pH on the Capacity Factors (k') for the Enantiomers of Cbz-tyrosine for a MIP Based on 2-Vinylpyridine as the Functional Monomer.

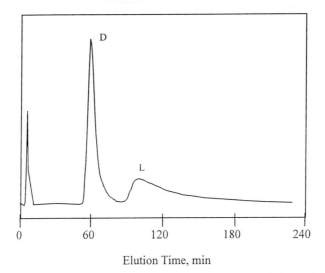

Elution Time, min

Figure 3. Separation of Boc-D,L-Trp Enantiomers on a Polymer Imprinted with Boc-L-Trp. Conditions: mobile phase = 10 mM glycine, pH 3.0, in acetonitrile (7:3); 10 μg of sample was injected; flow rate =1.0 ml/min. k'_D=10.12, k'_L=17.76, α=1.75, R_s=2.14.

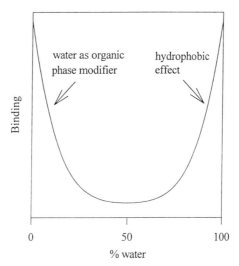

Figure 4. Phase Diagram for the Role of Water in Binding of Ligands by MIPs Prepared by Self-Assembly. A switch in the binding regime occurs on going from systems containing a low percentages of water, where water diminishes the strength of polar interactions, to systems containing mostly water, where the hydrophobic effect comes into play.

phenylalanine methyl ester (L-Phe-OMe) to form the sweetener aspartame was chosen as model system (*16-17*).

As shown in Figure 5, the route to aspartame using thermolysin is a straightforward process. It starts from Cbz-L-aspartic acid and L-phenylalanine methyl ester and leads to Cbz-α-aspartame in a single condensation step (Path 1). The actual sweetener, α-aspartame, is subsequently produced by removal of the *N*-protecting group. This enzymatic reaction is widely used in industry. It is fully reversible and the process proceeds until the reactants reach an equilibrium state with the product. The use of thermolysin makes the reaction very selective and only Cbz-α-aspartame is formed. In contrast to this 'clean' preparation, impurities of Cbz-ß-aspartame can also be formed to some extent in chemical peptide synthetic schemes (Path 3). On the other hand, the chemical approach does not suffer from the equilibrium problem to the same degree as the enzymatic approach, since the reaction is easier to push in the forward direction by removal of water.

In the selected model system, molecularly imprinted polymers were utilized to *shift the reaction equilibrium* of the enzymatic synthesis towards product formation (Path 2). The Cbz-α-aspartame product was continuously trapped from the enzymatic reaction, thereby pushing the reaction to increased product formation.

Figure 6 shows the yields for the enzymatic synthesis in the absence and presence of different MIPs used as thermodynamic traps. The polymers were incubated together with the reactants, Cbz-L-Asp and L-Phe-OMe, and the enzyme thermolysin, using ethyl acetate saturated with aqueous buffer solution as the reaction solvent. The concentration of L-Phe-OMe was kept twice that of Cbz-L-Asp in all experiments. The free enzyme reaction was evaluated using the same protocol in the absence of any polymers. After reaction for 48 hours, the products were extracted from collected solids and monitored by reverse phase HPLC.

The message of this investigation is that the use of such specific, sterilizable adsorbents should be considered for enzymatic or fermentation processes to increase the yield. They also have potential for use in the continuous removal of toxic compounds that may be formed. Direct isolation of a product formed by the retrieval of the adsorbents (especially magnetic) carrying the product can be envisaged.

Conclusions and Future Prospects

In this study we have pointed to the possibility of using molecularly imprinted polymers as recognition matrices in aqueous phases and as thermodynamic traps in reversible enzymatic syntheses. To overcome the effects of water with respect to reduced strength of non-covalent interactions, molecularly imprinted materials may be produced in organic phases for subsequent utilization in aqueous media. In this way, the strength is maintained by the non-polar surrounding in the imprinting protocol, leading to sufficiently high recognition to withstand the aqueous phase effect during rebinding. We believe that this approach is potentially very useful for the preparation of bio-recognition mimics, such as antibody or receptor binding mimics, for real-sample use in aqueous environments.

By use of a specific reaction component, either a product or a reactant, as the imprint species, we have also demonstrated the possibility of using molecularly imprinted polymers to shift a reaction equilibrium in a desired direction. In view of the attractive physical features entailed by molecularly imprinted polymers, such as high pressure and temperature stability which allow sterilization, we believe that this new methodology may find use in a variety of synthetic applications. With the

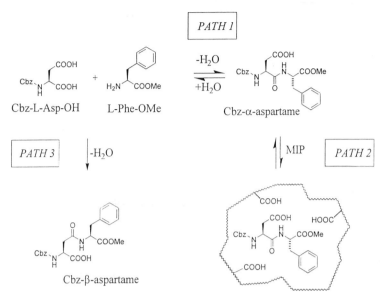

Figure 5. Synthetic Pathways to *N*-benzyloxycarbonyl-aspartame (Cbz-aspartame). In the enzymatic synthesis, catalyzed by thermolysin, the reactants Cbz-L-aspartic acid and L-phenylalanine methyl ester are condensed in a single step leading to the formation of Cbz-α-aspartame (Path 1). By use of MIPs, templated with the product, a new equilibrium is introduced and formation of Cbz-α-aspartame is increased by continuous entrapment of the product (Path 2). In chemical synthetic protocols, the β-form (Cbz-β-aspartame) may also be formed to some extent (Path 3).

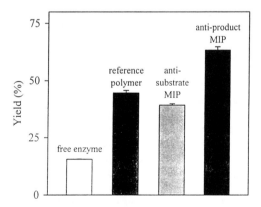

Figure 6. Yields from the Thermolysin-Catalyzed Cbz-Aspartame Reaction in 48 Hours. MIPs used: anti-product MIP, prepared with Cbz-aspartame; anti-substrate MIP, prepared with Cbz-L-Asp and L-Phe-OMe; reference polymer, prepared without any print species.

development of novel imprinted protocols resulting in polymers with higher loading capacity and further improved polymerization methodology, molecularly imprinted polymers may find additional applications in reaction systems as auxiliary reagents and possibly also for product isolation.

Acknowledgements

The authors wish to express their sincere gratitude to Dr. Karsten Haupt and Dr. Richard Ansell for valuable discussions.

Literature Cited

1. Mosbach, K.; Ramström, O. *Bio/Technology* **1996**, *14*, 163.
2. Ansell, R. J.; Ramström, O.; Mosbach, K. *Clin. Chem.* **1996**, *42*, 1506.
3. Wulff, G. *Angew. Chem., Int. Ed. Engl.* **1995**, *34*, 1812.
4. Shea, K. J. *Trends Polym. Sci. (Cambridge UK)* **1994**, *2*, 166.
5. Vidyasankar, S.; Arnold, F. H. *Curr. Opin. Biotechnol.* **1995**, *6*, 218.
6. Ramström, O.; Ansell, R. J. *Chirality* **1997**, *in press*.
7. Andersson, L. I.; Müller, R.; Vlatakis, G.; Mosbach, K. *Proc. Natl. Acad. Sci. U.S.A.* **1995**, *92*, 4788.
8. Andersson, L. I. *Anal. Chem.* **1996**, *68*, 111.
9. Ramström, O.; Gustavsson, P.-E. *submitted* **1996**.
10. Zaks, A.; Klibanov, A. M. *J. Biol. Chem.* **1988**, *263*, 3194.
11. Schneider, H.-J. *Angew. Chem., Int. Ed. Engl.* **1991**, *30*, 1417.
12. Sellergren, B., Shea, K. J. *J. Chromatogr. A*, **1993**, *654*, 17.
13. Andersson, L. I. *Anal. Chem.*, **1996**, *68*, 111.
14. Ramström, O.; Andersson, L. I.; Mosbach, K. *J. Org. Chem.* **1993**, *58*, 7562.
15. Yu, C.; Mosbach, K. *J. Org. Chem.* **1997**, *62*, 4057.
16. Nakanishi, K.; Kamikubo, T.; Matsuno, R. *Bio/Technology*, **1985**, *3*, 459.
17. Sheldon, R. A. in Enzymatic Reactions in Organic Media; Koskinen, A. M. P.; Klibanov, A. M., Eds.; Blackie Academic & Professional: Glasgow, UK, 1996.

Chapter 6

Applications of Molecular Imprinting to the Recognition and Detection of Bioactive Molecules

Scott J. McNiven, Soo-Hwan Cheong, Raphael Levi, Yohei Yokobayashi, Takeshi Nakagiri, Kazuyoshi Yano, and Isao Karube

Research Center for Advanced Science and Technology, The University of Tokyo, 4-6-1 Komaba, Meguro, Tokyo 153, Japan

Research in our laboratory has been focused on the development of molecularly imprinted polymers (MIPs) which exhibit selectivites for various biologically active molecules.   In an effort to create steroid-selective sensors we have thoroughly investigated the effect of various polymerization conditions on the selectivity of MIPs for testosterone. Separation factors (α) ranging from 3 to 10 were observed for a suite of similar steroids.   Furthermore, we report a novel post-polymerization technique by which the selectivity of this type of MIP may be further enhanced.   Diastereoselective recognition of peptide derivatives has also been achieved using a novel polymerizable valine derivative as a functional monomer. We have also developed an FIA-based sensor for the antibiotic chloramphenicol which relies upon the displacement of a chloramphenicol-dye conjugate from an MIP by the drug.   The sensor is effective over a wide range of concentrations encompassing the therapeutic dosage.

Molecular imprinting (*1-4*) is now recognized as a technique for the creation of materials with predetermined affinity for particular substrates. While there are several variations on the theme, *non-covalently imprinted* polymers (with which we are mainly concerned) are usually prepared by the radical polymerization of a solution of the template, functional monomer, and excess cross-linker in an organic solvent.   After crushing and extraction, the highly crosslinked, macroreticular particles possess microcavities complementary in size, shape, and chemical functionality to the original template molecule. The persistence of these rigid, well-defined cavities containing accessible functionalities allows the MIPs to be used as chiral stationary phases for enantioseparation (*5,6*), catalysts (*7,8*), microreactors for regio- or stereoselective reactions (*9*), artificial receptors for use in drug assays (*10*), and as recognition elements in sensors (*11*).

The analogy between interactions which occur in MIPs and those between enzyme-substrate, antibody-antigen, and receptor-hormone complexes is often noted.   This is particularly apt when the functional monomers incorporate binding groups resembling those of the naturally occurring matrices. While MIPs may not always achieve the

---

[1]Corresponding author.

exquisite selectivity of the proteinaceous matrices, they possess advantages in terms of cost, ease of production, tolerance to extreme conditions, and a greater number of active sites per unit weight. Moreover, their stability in organic solvents makes the use of MIPs a complementary approach to that involving artificial enzymes and antibodies which function in aqueous media.

In the broadest definition, biosensors are devices which combine a biomolecule and a transducer (*12*). The biological component (enzyme, microbe, antibody, *etc.*) functions as both the recognition element and the source of a signal which can be detected by the transducer. Our work in this field has led us to seek novel methods for the recognition and detection of biologically active molecules in an attempt to surmount the difficulties arising from the use of the often delicate biomolecules. We report herein some of our recent results.

A Testosterone Receptor Binding Mimic

In addition to their established role in hormone replacement therapy and in the treatment of breast cancer in women, the recent successes of hormone therapy in alleviating the deleterious physical and mental effects of aging has raised awareness of the importance of androgenic steroids (*13*). The continued illegal use of anabolic steroids by athletes also demands the development of rapid, sensitive, and inexpensive methods of detecting the drugs and their metabolites. Furthermore, the steroids are a family of large, rigid molecules with a variety of substituents at the extremities, making them ideal candidates with which to study those factors which influence their affinity for the MIP. The differing behavior of various probes allows us to determine those substituents which most strongly affect the binding affinity and this information can in turn be used for the design of receptors for different steroids. The structures of the steroids used in this work are shown in Figure 1. While there have been studies regarding the optimization of formulations and polymerization conditions for other substrates (*14,15*), very few reports have appeared dealing with the molecular imprinting of steroids (*16,17*). Thus we created a series of polymers under different polymerization conditions in order to optimize their affinity for testosterone.

Experimental Conditions. All polymers were synthesized by slight variations (noted in the tables) from the procedure used to synthesize **P2** (*18*). Briefly, the template (1.0 mmol) was dissolved in chloroform (7.50 mL) in a 50-mL glass flask and 2,2'-azobis(2,4-dimethyl)valeronitrile (V-65, 50 mg), ethylene glycol dimethacrylate (EGDM, 5.00 mL), and the functional monomer methacrylic acid (MA, 8.0 mmol) were added. The samples were then degassed and polymerized at 40 °C. Non-imprinted polymers were synthesized simultaneously under exactly the same conditions with the omission of the template. The covalently imprinted polymer was prepared using testosterone methacrylate (TM) in place of the template and functional monomer. Hydrolyses were carried out in refluxing aqueous methanolic 1 M NaOH (MeOH-H_2O, 1:1) and the polymers were reprotonated with either acetic acid or MeOH-HCl (10:1).

Particles were wet sieved in ethanol and the 25-45 μm fraction was packed into 100 x 4.6 (i.d.) mm HPLC columns. Analyses were performed using MeCN as eluent at a rate of 2.0 mL/min, using acetone as a void marker. In addition to the well known capacity factor k', separation factor α, and resolution R_S, we define an *imprinting factor* $I = k'_{imp}/k'_{non}$, where k'_{imp} and k'_{non} are the capacity factors for the same compound on the imprinted and non-imprinted polymers. Thus I is a measure of the effect of the imprinting process. Consequently, comparison of I values for the various steroids gives a measure of how well the inherent selectivity of the polymer improves upon imprinting and we define a *selectivity factor* $S = I_x/I_{template}$.

92

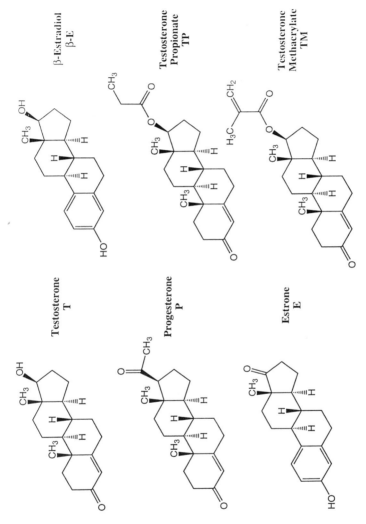

Figure 1. Structures of the steroids used in this work. (Adapted from Reference 18.)

Effect of Template/Monomer Ratio. Perhaps the most important factor in determining the selectivity of an MIP is the template/functional monomer ratio. The results of our investigations are shown in Table I. As the proportion of monomer increases, so do the values of k'_{imp} and α_{imp} for all of the steroids. However, k'_{non} and α_{non} increase also, but to a lesser extent. This is reflected in the increase in I values for all of the steroids as the proportion of template is increased. Thus, the affinity of both the imprinted and non-imprinted MIPs for the steroids gradually increases as the amount of MA in the formulation is enhanced. The pertinent point is, however, that the *relative* affinity of the matrices for testosterone does reach a maximum at a template/functional monomer ratio of 1:8, as reflected in the low selectivity factors for the other steroids.

Table I. Parameters Used to Choose the Template/Monomer Ratio

Polymer	Template/ Monomer Ratio	Steroid	k'_{imp}	α_{imp}	k'_{non}	α_{non}	I	S
P1	1:4	T	1.39	1[a]	0.795	1	1.75	1
		β-E	1.13	1.24	0.928	0.857	1.21	0.691
		TP	0.398	3.50	0.331	2.40	1.20	0.686
		P	0.530	2.63	0.464	1.71	1.14	0.651
		E	0.464	3.00	0.398	2.00	1.17	0.669
P2	1:8	T	5.10	1	1.26	1	4.05	1
		β-E	1.52	3.35	1.19	1.06	1.28	0.316
		TP	0.597	8.55	0.464	2.72	1.28	0.316
		P	0.795	6.42	0.597	2.11	1.33	0.328
		E	0.530	9.63	0.464	2.72	1.14	0.281
P3	1:12	T	6.56	1	1.59	1	4.12	1
		β-E	1.92	3.42	1.26	1.26	1.52	0.369
		TP	0.729	9.00	0.464	3.43	1.57	0.381
		P	1.06	6.19	0.663	2.40	1.60	0.388
		E	0.597	11.0	0.398	4.00	1.50	0.364

[a]By definition $\alpha = 1$, $S = 1$ for testosterone. SOURCE: Adapted from Reference 18.

Effect of Altering the Polymerization Procedure. Having determined that polymer **P2** exhibited optimal selectivity for testosterone we then proceeded to investigate the effects of slight variations in the polymerization conditions on the selectivity of the resulting polymers. The results are shown in Table II and exemplify some of the intuitive notions and "rules of thumb" developed for the synthesis of MIPs.

The porogen should be the least polar solvent able to dissolve the components of the mixture so as to maximize the non-covalent interactions between the template and functional monomer. Polar solvents are known to disrupt hydrogen bonding (*17*). The polymerization should be carried out at low temperatures rather than high to shift the equilibrium in solution toward the formation of template-functional monomer complexes (*14,15*). The poor results obtained using UV polymerization suggest incomplete polymerization, perhaps due to the power of the source or the opacity of the polymerizing mixture. The success of post-polymerization heat treatment indicates that the polymeric products formed under these conditions contain residual unreacted monomers.

Table II. Polymerization Conditions and Selected Data for Imprinted Polymers

Polymer	Feature[a]	k_{imp}' Steroid					α_{imp} Steroid[b]			
		T	β–E	TP	P	E	β–E	TP	P	E
P4	DAM as monomer	0.464	2.78	0.266	0.331	1.92	0.167	1.74	1.40	0.241
P5	THF as porogen	1.99	1.06	0.530	0.729	0.331	1.87	3.75	2.73	6.01
P6	Polymerized at 60°C	1.52	0.862	0.331	0.464	0.266	1.77	4.60	3.28	5.73
P7	Polymerized at room temp using UV	2.39	1.19	0.464	0.597	0.398	2.00	5.14	4.00	6.00
P8	PGDM as cross-linker	0.729	1.06	0.199	0.266	0.398	0.687	3.66	2.74	1.83
P9	Covalently imprinted, hydrolyzed 6h	1.26	1.26	0.398	0.597	0.530	1.00	3.17	2.11	2.38

[a]Standard polymerization conditions are given in the text. Only deviations from this procedure are listed. DAM = 2-(diethylamino)ethyl methacrylate, PGDM = propylene glycol dimethacrylate. [b]By definition α = 1 for testosterone. SOURCE: Adapted from Reference 18.

The strong binding of the phenolic steroids β-estradiol and estrone to polymers made using 2-(diethylamino)ethyl methacrylate (DAM) (and also to deprotonated MA polymers) reinforces the fact that the functional monomer must be chosen with great care. Estrone was the compound most rapidly eluted from MA-functionalized MIPs and all attempts to produce an estrone-imprinted MA polymer failed. We are currently exploiting the affinity of the phenolic steroids for DAM-based polymers to develop sensors for these steroids. Since it constitutes the bulk of the polymer, the choice of crosslinker is of great importance. In our case, the insertion of a single methylene link between the polymerizable groups of EGDM resulted in great reductions in the separation factors of the steroids. We attribute this to the greater flexibility of propylene glycol dimethacrylate compared with EGDM. Alternative rigid crosslinkers, such as trimethylolpropane trimethacrylate (*19,20*) and aromatic bis (meth)acrlylamides (*21*) may offer greater benefits than those of EGDM.

Attempts to produce a testosterone-selective MIP by covalent imprinting of testosterone methacrylate were unsuccessful. Such a polymer was expected to more closely resemble a natural receptor as covalently imprinted polymers exhibit a narrow range of binding affinities (*17*) (*vide infra*). However, the convenience and success of the non-covalent imprinting technique make it the method of choice.

Origin of the Selectivity of the MIPs. Generally, testosterone and β-estradiol are retained to similar extents on non-imprinted polymers while the former has a greater affinity for the matrix in imprinted polymers. The other steroids, especially estrone, are eluted relatively quickly, indicating that the binding of both testosterone and β-estradiol is predominantly due to the C-17 hydroxyl group. Since the C and D rings of these two compounds are identical, the difference in binding strengths is due to the ability of the MIP to recognize and discriminate between the different structures of the A and B rings of the steroids. Participation of the A and B rings of testosterone in binding is illustrated by comparison of the relative affinities of testosterone propionate and progesterone with that of estrone. The first two compounds possess the same A and B ring structures as testosterone and are retained more strongly than the latter, which has no functionality in common with the template. These comparisons provide an excellent demonstration of the fidelity with which the MIP records structural and functional information about the template.

Conclusions. We have established polymerization conditions for the synthesis of matrices with an appreciable level of selectivity for testosterone. Basic studies of this kind provide much useful information regarding those factors which determine the affinity and degree of selectivity of MIPs for their template molecules. While some further optimization of the formulation and polymerization conditions is usually advisable, the basic tenets of molecular imprinting illustrated here provide a sound starting point for the synthesis of materials capable of acceptable levels of molecular recognition. We are currently engaged in the extension of this work to the construction of steroid-selective sensors.

Enhancing the Selectivity of MIPs

One of the few drawbacks to the non-covalent approach to molecular imprinting is that a large excess of functional monomer is used in the polymer formulation. In the "prearrangement phase" prior to polymerization, molecules of the functional monomer are in dynamic equilibrium with the template, interacting in a number of different modes. Interactions between the species are weak, being predominantly hydrogen bonds, so a large excess of the functional monomer is required to promote the formation of template-monomer complexes in the mixture. This results in the formation of a myriad of binding sites which have a wide range of affinity for the template, depending on the extent of their physical and chemical complementarity (22). By virtue of the similarity of MIPs and antibodies, this has been termed "polyclonality" (1) and is responsible for the exaggerated tailing of peaks when MIPs are used as stationary phases in chromatography. The vast majority of binding sites have little affinity for the template, but there are a diminishing number of sites with increasing binding strengths (23). In contrast with covalently-imprinted matrices, Scatchard plots obtained using non-covalently imprinted polymers are non-linear, indicating the presence of a continuum of sites with smoothly varying association constants. We arbitrarily term the stronger binding sites "specific" since, presumably, they arose as a result of the formation of template-functional monomer complexes, while the weaker binding sites are termed "non-specific". Adjunct to our attempts to optimize the selectivity of MIPs for their templates, we have developed a novel post-polymerization technique whereby, upon judicious methylation of MA-based polymers in the presence of the template, we not only increase the *average* binding strength of the sites but also enhance their selectivity.

Experimental Procedure and Results. Polymers were synthesized (18) by the method utilized for the preparation of **P2** (*vide supra*) but on a larger scale, resulting in a small decrease in performance. After allowing the polymers to equilibrate with one theoretical equivalent of the template (testosterone) overnight at 8 °C, they were esterified by treatment with increasing amounts of methyl iodide (MeI) and 1,8-diazabicyclo[5.4.0]undec-7-ene (DBU) for 30 minutes (24). They were then filtered, washed thoroughly with acetonitrile and water, reprotonated with 1 % acetic acid in acetonitrile, and packed into HPLC columns.

The results of the chromatographic analyses are summarized in Table III. As the amount of MeI/DBU reagent added was increased, the retention times of the three steroids reach a maximum with **P10** and decreases for **P100**, indicating that the methylated polymers have increasingly higher *average* binding strengths. Addition of one theoretical equivalent of the reagent (based on the amount of MA in the formulation) drastically reduces the affinity of the MIPs for the steroids. Examination

Table III. Selected Data for the Polymers Studied in This Work.

Polymer	Methylation conditions[a]	k'_{TP}	k'_P	k'_T	α_{TP}	α_P
P_0	Untreated polymer	0.30	0.41	1.58	5.26[b]	3.87
$P_{0.1}$	0.1% MeI + tetosterone	0.30	0.40	1.65	5.43	4.07
P_1	1 % MeI + testosterone	0.33	0.44	1.91	5.80	4.38
P_{10}	10% MeI + tesosterone	0.39	0.54	2.25	5.73	4.17
P_{100}	100 % MeI + testosterone	0.35	0.47	1.71	4.90	3.66
P_{1A}	1% MeI, no testosterone	0.33	0.43	1.68	5.16	3.87

[a]All methylations were carried out for 30 min in the presence of the indicated theoretical percentage of methyl iodide. [b]α refers to the separation of the indicated steroid from testosterone. 10 µL of a 0.2 mM solution were measured at 240 nm, 1.0 mL/min in acetonitrile.

of the separation factors shows a similar trend; the α values for both testosterone propionate and progesterone increase to a maximum with P_1 and then diminish. As shown above, the MIPs which give the greatest capacity factors do not necessarily afford the highest selectivity. Moreover, the separation factors are lower for P_{100} than for the untreated polymer and P_{100} was unable to separate the template from either of the other two steroids (see Figure 2).

Origin of the Enhancement Effect. Our interpretation of these results is as follows: Equilibration of the MIP with the template over a long period of time at low temperature ensures that the system reaches thermodynamic equilibrium, wherein the strongest accessible binding sites (*i.e.* those exhibiting a high level of physical and functional complementarity with the template) are occupied. Addition of the reagent then results in esterification of the exposed carboxylic acid residues in the unoccupied- or partially occupied "non-specific" binding sites, which are subsequently no longer available to the template. This is shown schematically in Figure 3. Since the esterification is a kinetically controlled process, only the most readily accessible carboxylate groups are available to the reagent and the groups in the "specific" binding sites, which are bound strongly to the template, are protected. As the weaker binding sites are being blocked upon methylation, the capacity factors increase because the *average* binding strength of the sites increases. Similarly, the separation factors increase because the most highly complementary binding sites are preserved.

Further support for this interpretation was obtained by methylating the polymer in the absence of the template. The capacity and separation factors of the resulting polymer P_{1A} are much less than those of P_1 and the chromatograms in Figure 2 illustrate this graphically. The separation factors for this polymer are approximately equal to those of the untreated polymer although the capacity factors are slightly higher, indicating that some non-preferential methylation has occurred.

Thus, we contend that this method preferentially methylates the "non-specific" binding sites within the polymer but, as the concentration of MeI/DBU is increased, the "specific" sites begin to be methylated. This is a cooperative process because the esterification of one of the carboxyl groups in a binding site weakens its affinity for the template and makes the other functionality(ies) more prone to attack. Simply put, discreet methylation of the MIPs in the presence of the protecting template masks the weakest binding sites and improves the performance of the polymer, but too much is deleterious. The widespread use of MA-based MIPs and the generality of this technique should allow significant advances in the molecular imprinting field.

Stereoselective Peptide Recognition Using a Novel Chiral Monomer. Naturally occurring receptors, such as antibodies and enzymes, rely upon multiple interactions of different types (hydrogen bonding, electrostatic interactions, *etc.*) between the substrate and host molecule. The substrates are held in clefts or pockets constructed by the arrangement of stable secondary structural units, such as α-helices, β-sheets, *etc.* and are accessible to catalytic functional groups positioned within these cavities. Moreover, the individual amino acids of which these structures are comprised are themselves chiral and this is yet another reason for the admirable selectivity exhibited by these natural receptors. While there are several reports dealing with the stereoselective properties of MIPs toward amino acid and peptide derivatives, most of these use MA as the functional monomer (*19,25-29*). The addition of *N*-acetyl-L-valine-*tert*-butylamide to the mobile phase during liquid chromatography was shown to facilitate the optical resolution of amino acid derivatives (*30,31*) so in an attempt to create MIPs with binding characteristics which more closely mimic those interactions

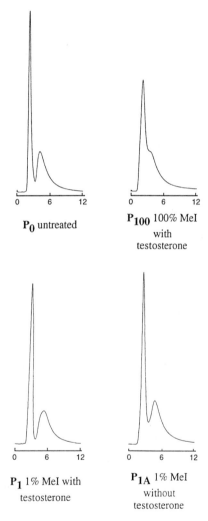

Figure 2. Selected chromatograms for solutions of progesterone and testosterone (0.2 mM each). (Conditions: 10 μL sample; 1 mL/min of acetonitrile; detection at 240 nm; and retention time in minutes.)

exploited in natural systems, we synthesized *N*-methacryloyl-L-valine-*tert*-butylamide (**1**), a novel functional monomer capable of multiple hydrogen bonding interactions.

Experimental Conditions. Peptide derivatives and the functional monomer were synthesized by standard methods. Unless otherwise noted, the polymer formulations consisted of a solution of the template (0.25 mmol), functional monomer (0.5 mmol), EGDM (12.5 mmol), and V-65 (20 mg) in chloroform (5 mL). The mixtures were polymerized for 24 h at 45 °C and were then crushed, extracted, and wet sieved and particles from 26-63 μm were packed into 10 cm columns for HPLC analyses.

Effect of Template/Monomer Ratio. The first step in preparing MIPs is to determine the optimal ratio of template to functional monomer and our results for Boc-D-Ala-L-Ala-pNA (**2**) and its L,L isomer **3** are shown in Table IV. While MA-based polymers usually require higher (*ca.* 5 equivalents) ratios of monomer to template, we found that good selectivity was observed with lower ratios. This, we infer, is a result of both the inherent chirality of the functional monomer and its ability to engage the template with multiple hydrogen bonds. A schematic representation of this is shown in Figure 4.

Table IV. Selected Data for Polymers with *N*-Methacryloyl-L-valine-*tert*-butylamide as the Functional Monomer

Template	Template: Monomer Ratio	Porogen CHCl₃-hexane	Capacity Factor				Separation Factor[a]		
			k'_2	k'_3	k'_4	k'_5	$\alpha_{2/3}$	$\alpha_{4/5}$	$\alpha_{6/7}$
Boc-D-Ala-L-Ala-pNA (**2**)	0:1	100/0	1.06	0.81			1.31	1.26	
	1:1	100/0	3.66	2.48			1.48		
	1:2		2.55	1.50			1.70	1.25	1.21
	1:3		4.06	2.49			1.63		
	1:4		1.41	1.10			1.28		
	1:2	60/40	4.95	2.54			1.95		
	1:2	40/60	6.39	2.49			2.57		
Boc-D-Phe-L-Ala-pNA (**4**)	0:2	100/0			0.54	0.43		1.26	
	1:2				2.95	1.04	1.28	2.84	1.55
Boc-D-Trp-L-Ala-pNA (**6**)	1:2						1.42	2.05	1.70

[a]The separation factor given is that between the D,L-template and its L,L-isomer.

Effect of the Porogen. With a template-monomer ratio of 1:2 and Boc-D-Ala-L-Ala-pNA (**2**) as the template, we then examined the effect of the polarity of the porogen on the performance of the MIPs. These results (Table IV) reinforce the general observation that less polar solvents enhance the hydrogen bonding between the template and functional monomer, resulting in more selective polymers.

Complementarity of the Template. To investigate the effect of the side chains of the amino acids on the selectivity of MIPs, we synthesized a similar polymer using

Equilibration of Imprinted Polymer with Template

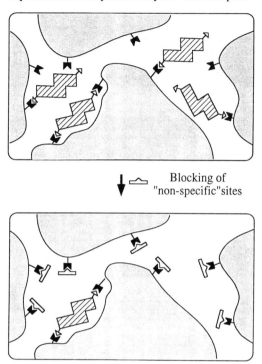

Figure 3. Preferential methylation of the "non-specific"binding sites within a MIP.

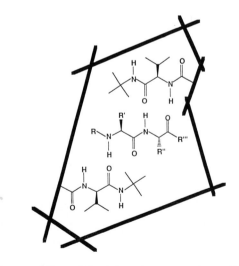

Figure 4. Representation of the interactions occurring within the binding site.

Boc-D-Phe-L-Ala-pNA (**4**) as the template. This MIP exhibited greater selectivity for the template than did the Boc-D-Ala-L-Ala-pNA-imprinted polymer, as evinced by the larger separation factors (Table IV). Since the configurations of the two chiral centers in each peptide are the same, this must be due to the extra bulk of the phenylalanine side chain. The peptide Boc-L-Ala-D-Phe-pNA, which contains a D-Phe residue at the carboxyl terminal, was barely retained which further exemplies the selectivity of this polymer. We then synthesized another polymer using Boc-D-Trp-L-Ala-pNA (**6**), which has an even bulkier side chain, as a template. In this case, however, the separation factor with respect to the corresponding L,L diastereomer **7** decreased, but this polymer still provided the best separation of these diastereomers. Surprisingly, unlike the other polymers, this MIP was more able to separate the corresponding diastereomers of Boc-D-Phe-L-Ala-pNA than those of its template. We expected that this MIP could separate these diastereomers due to the similarity between the side chains of Phe and Trp, but not to such an extent.

Conclusions. The use of functional monomers derived from amino acids offers the possibility of creating MIPs which more closely resemble naturally occurring receptors than do polymers based on MA. The polymers described here were able to distinguish between quite similar diastereomers and their selectivity is comparable with previously described MIPs. By judicious selection of the functional monomer it may be possible to construct MIPs specific for certain peptides. In addition, studies such as this should afford valuable insight into the nature of biological interactions.

An FIA-Based Chloramphenicol Sensor

While the majority of the reports in the literature are concerned with the ability of MIPs to act as molecular recognition elements, very few describe the use of these polymers in sensing devices. Among these are sensors are examples utilizing MIPs in conjunction with field effect devices (*32*), conductometry (33), amperometry (*11*), and fluorimetry (*34*). Theoretically, any MIP with sufficient selectivity could be used in radioimmunoassay-type techniques (35) or other methods based upon the displacement of a suitably labeled analog of the template molecule.

Chloramphenicol (CAP) is a broad spectrum antibiotic used in both human and veterinary medicine. It is a potent drug for the treatment of childhood meningitis and typhoid fever. However, since it can produce toxic effects, such as bone marrow suppression, aplastic anemia, and "Grey baby" syndrome, its use has been banned in food-producing animals (*36*). Thus the monitoring of drug levels in patients' blood and in foodstuffs is essential. Currently used methods include chromatographic (*37*), microbiological (*38*), enzymatic (*39*), and immunoassays (40), but these can be tedious and expensive so simpler and more rapid analytical methods are required. We have developed an FIA-based system for the spectrometric determination of CAP which relies upon the displacement of a chloramphenicol-dye conjugate from a CAP-imprinted polymer, as shown schematically in Figure 5. The MIP is first allowed to equilibrate with a mobile phase containing the chloroamphenicol-methyl red dye conjugate (CAP-MR). Since the MIP has a higher affinity for the template molecule (CAP) than for the dye conjugate (CAP-MR), the latter is displaced from the polymer upon injection of samples of CAP. The resulting increase in absorbance at 460 nm (due to the displaced dye; the other species have negligible absorbance at this wavelength) is thus related to the concentration of CAP in the sample.

Experimental Conditions. The MIP was prepared by the heat-initiated polymerization of a solution of CAP (1.0 mmol), DAM (2.0 mmol), EGDM (5 mL) and 2,2'-azobis(isobutyronitrile) (150 mg) in tetrahydrofuran (5 mL). Particles from

25 to 63 μm were packed into 10 cm HPLC columns for the initial analyses (Table 5) and 5 cm columns for subsequent experiments. The CAP-MR was synthesized as previously described (41)

Table V. Influence of the Flow Rate on k, I, and α Values for the Various Analytes as Determied by HPLC Analysis Using a Non-imprinted or CAP-imprinted Polymer as the Stationary Phase and Acetonitrile as the Mobile Phase.

Analyte	k'_{non}	k'_{imp}	I	α
		2 mL/min		
CAP	1.75	17.4	9.92	1.00
CAP-DA	0.00	0.14	-	124
TAM	1.14	9.09	7.97	1.91
CAP-MR	0.83	5.32	6.40	3.26
		3 mL/min		
CAP	1.54	15.4	10.0	1.00
CAP-DA	0.00	0.15	-	102
TAM	0.99	8.40	8.48	1.83
CAP-MR	0.77	4.75	6.16	3.24
		4 mL/min		
CAP	1.40	14.4	10.3	1.00
CAP-DA	0.00	0.13	-	110
TAM	0.84	7.33	8.72	1.96
CAP-MR	0.70	4.51	6.44	3.19
		5 mL/min		
CAP	1.15	12.0	10.4	1.00
CAP-DA	0.00	0.16	-	74.9
TAM	0.65	7.06	10.9	1.70
CAP-MR	0.49	3.94	8.04	3.04

Effect of Flow Rate on Chromatographic Parameters. The various CAP analogs used in this study (Figure 6) were chromatographed using a range of flow rates and the results are shown in Table V. As expected, CAP was the most strongly retained compound, followed by thiamphenicol (TAM) and CAP-MR. Chloramphenicol diacetate (CAP-DA) was eluted very rapidly, indicating that the 1,3-diol moiety of these molecules was responsible for the majority of the binding. Although the imprinting factor increased with the flow rate, the separation factor did not change appreciably.

Optimization of the Sensor Response. Having established the behavior of the various compounds on the MIP, the polymer was packed into a 5 cm column and allowed to equilibrate with a mobile phase containing CAP-MR. The shorter column was used to reduce the time taken for each assay, but the system still took approximately 5 min to return to equilibrium after injection of the analytes. Solutions of 20 μg/mL CAP (approximately the toxic level) were injected and the sensor response was recorded with various concentrations of CAP-MR in the mobile phase. Figure 7a shows that high responses were obtained using 0.4-0.6 μg/mL of CAP-MR so the effect of the flow rate was then examined using a concentration of 0.6 μg/mL in the mobile phase. As shown in Figure 7b, the response decreased as the flow rate was increased, indicating that a slower flow rate should enhance the sensitivity of the system. However, slower flow rates result in broader peaks and make the analyses

CAP imprinted polymer crushed, extracted, and
packed into HPLC columns

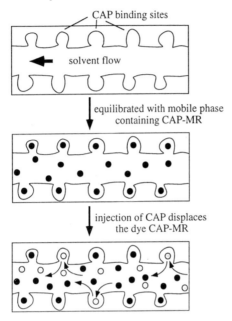

Figure 5.   Displacement of the dye CAP-MR from a CAP-imprinted polymer.

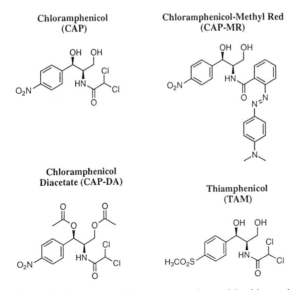

Figure 6. Structures of the compounds used in this work.

less accurate. Thus a compromise was made between speed of analysis, sensitivity and accuracy, and the optimal flow rate was deemed to be 2 mL/min.

Cross Reactivity and Sensitivity. The response of the sensor to the various compounds was determined under the optimized conditions and the resulting calibration curves are shown in Figure 8. CAP-DA gave no response even at concentrations of 1000 µg/mL, while the response to CAP was about 40% higher than that to TAM, an almost identical analog. Thus, while the sensor cannot distinguish between the latter two compounds, it does show a linear response to both over an extremely wide concentration range (3 -1000 µg/mL). The detection limit for CAP was approximately 3µg/mL, which is well below the therapeutic level of 10-20 µg/mL of serum. In addition, the response of the sensor to 20 µg/mL of CAP was not affected by the presence of CAP-DA at concentrations up to 1000 µg/mL (Figure 9).

The signals produced upon injection of CAP are only approximately 10 % of those obtained upon injection of an equivalent concentration of CAP-MR. This is the main factor determining the sensitivity of the system, but is of little consequence as the detection limit is already very low. Considering the enormous excess of CAP-MR in the mobile phase, this figure is not surprising.

Analysis of Bovine Serum Samples. To establish the practical utility of our sensor, we analyzed bovine serum samples which were spiked with varying concentrations of CAP. The samples were divided in two, extracted as previously described (*42*) and each was analyzed at least three times. The responses (Figure 10) were linear (correlation coefficient 0.990) up to 1000 µg/mL CAP, with a detection limit of *ca.* 5 µg/mL and standard errors of less than 10 %.

Conclusions. The simplicity and success of this method heralds a new method of constructing tailor-made MIP-based sensors. The fundamental requirement is that the analyte may be derivatized with an appropriate label without interfering with its ability to bind to the MIP.

Outlook

Molecular imprinting provides a simple, inexpensive method with which to create matrices with a high affinity for an enormous range of compounds. Having established the guidelines for the successful synthesis of the polymers and determined the capabilities of the imprinting technique, it is imperative that we take the next step and exploit the remarkable properties of these materials. The utility of these materials in separation technology, as tailor-made catalysts and as recognition elements in sensor technology has been demonstrated and further progress in this expanding field is nothing but inevitable.

Acknowledgments

This project was financially supported by the Ministry of Education, Science and Culture of Japan: Large-Scale Research Projects under the New Program in Grants-in-Aid for Scientific Research. S.M., S.-H.C. and R.L. are grateful to the Japan Society for the Promotion of Science for postdoctoral fellowships. We would like to thank Maiko Kato for her superlative assistance.

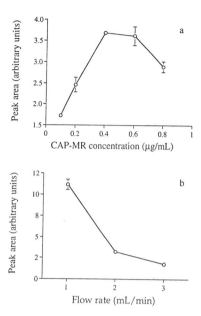

Figure 7. Influence of the a) CAP-MR concentration in the mobile phase and b) flow rate on the response of the sensor. (Conditions: 20μL of 20 μg/mL CAP solutions injected on a 50 mm x 4.6 mm (i.d.) column packed with CAP-imprinted polymer. Each point represents the mean of three measurements and the error bars indicate the standard errors.) (Adapted from Reference 41.)

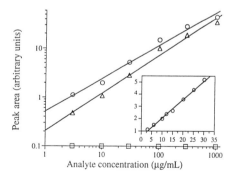

Figure 8. Calibration curves for CAP (E), TAM (C) and CAP-DA (G). The inset shows detail over the clinically significant concentration range. (Conditions: mobile phase, 0.6 μg/mL of CAP-MR in acetonitrile at 2 mL/min; 20 μL samples. Each point represents the mean of three measurements and the error bars indicate the standard errors.) (Adapted from Reference 41.)

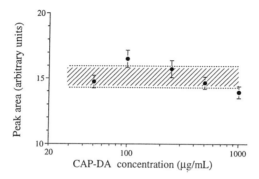

Figure 9. Response of the sensor to 20 μL injections of 20 μg/mL CAP solutions containing CAP-DA. The shaded region corresponds to the response due to CAP alone. (Conditions: mobile phase, 0.6 μg/mL of CAP-MR in acetonitrile; 2 mL/min. Each point represents the mean of three measurements and the error bars indicate the standard errors.) (Adapted from Reference 41.)

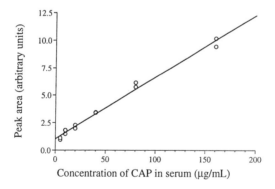

Figure 10. Determination of CAP in bovine serum. Each sample was halved and analyzed separately. (Conditions: mobile phase, 0.6 μg/mL of CAP-MR in acetonitrile: 2 mL/min, 20 μL samples.) (Adapted from Reference 41.)

Literature Cited

1. Mosbach, K.; Ramström, O. *Bio/Technology* **1996**, *14*, 163.
2. Shea, K. J. *Trends Polym. Sci.* **1994**, *2*, 166.
3. Steinke, J.; Sherrington, D.; Dunkin, I. *Adv. Polym. Sci.* **1995**, *123*, 80.
4. Wulff, G. *Angew. Chem., Int. Ed. Engl.* **1995**, *34*, 1812.
5. Andersson, L.; Ekberg, B.; Mosbach, K., In *Molecular Interactions in Bioseparations;* Ngo, T., Ed.; Plenum Press: New York, 1993, pp. 383-394.
6. Nicholls, I. A.; Andersson, L. I.; Mosbach, K.; Ekberg, B. *Trends Biotechnol.* **1995**, *13*, 47.
7. Sellergren, B.; Shea, K. J. *Tetrahedron Asymm.* **1994**, *5*, 1403.
8. Matsui, J.; Nicholls, I. A.; Takeuchi, T. *J. Org. Chem.* **1996**, *61*, 5414.
9. Byström, S.; Börje, A.; Akermark, B. *J. Am. Chem. Soc.* **1993**, *115*, 2081.
10. Vlatakis, G.; Andersson, L. I.; Müller, R.; Mosbach, K. *Nature* **1993**, *361*, 645.
11. Kriz, D.; Mosbach, K. *Anal. Chim. Acta* **1994**, *300*, 71.
12. Karube, I., In *Handbook of Measurement Science;* Sydenham, P. H., Thorn, R., Eds.; Wiley: New York, 1992, Vol. 3, pp. 1721-1755.
13. Hoberman, J. M.; Yesalis, C. E., in *Scientific American*. 1995, pp. 76-81.
14. Sellergren, B. *Makromol. Chem.* **1989**, *190*, 2703.
15. Sellergren, B.; Shea, K. J. *J. Chromatogr.* **1993**, *635*, 31.
16. Ramström, O.; Andersson, L. I.; Mosbach, K. *J. Org. Chem.* **1993**, *58*, 7562.
17. Whitcombe, M.; Rodriguez, M.; Villar, P.; Vulfson, E. *J. Am. Chem. Soc.* **1995**, *117*, 7105.
18. Cheong, S. H.; McNiven, S.; Rachkov, A.; Levi, R.; Yano, K.; Karube, I. *Macromolecules* **1996**, *30*, 1317.
19. Kempe, M.; Mosbach, K. *Tetrahedron Lett.***1995**, *36*, 3563.
20. Kempe, M. *Anal. Chem.* **1996**, *68*, 1948.
21. Shea, K.; Stoddard, G.; Shavelle, D.; Wakui, F.; Choate, R. *Macromolecules* **1990**, *23*, 4497.
22. Nicholls, I. A. *Chem. Lett.* **1995**, 1035.
23. Matsui, J.; Miyoshi, Y.; Doblhoff-Dier, O.; Takeuchi, T. *Anal. Chem.* **1995**, *67*, 4404.
24. Rao, C. G. *Org. Prep. Proc. Int.* **1980**, *12*, 225.
25. Alexander, C.; Smith, C.; Vulfson, E., In *Separations for Biotechnology;* 1994, Vol. 3, pp. 22-28.
26. Andersson, L.; Müller, R.; Vlatakis, G.; Mosbach, K. *Proc. Natl. Acad. Sci. USA* **1995**, *92*, 4788.
27. Andersson, L. I.; Müller, R.; Mosbach, K. *Macromol. Rapid Commun.* **1996**, *17*, 65.
28. Nicholls, I.; Ramström, O.; Mosbach, K. *J. Chromatogr. A* **1995**, *691*, 349.
29. Ramström, O.; Nicholls, I.; Mosbach, K. *Tetrahedron Asymm.***1994**, *5*, 649.
30. Dobashi, A.; Hara, S. *Tetrhedron Lett.***1983**, *24*, 1509.
31. Dobashi, A.; Hara, S. *Anal. Chem.* **1983**, *55*, 1805.
32. Hedborg, E.; Winquist, F.; Lundström, I.; Andersson, L. I.; Mosbach, K. *Sensors and Actuators A* **1993**, *37*, 796.
33. Piletsky, S. A.; Parhometz, Y. P.; Lavryk, N. V.; Panasyuk, T. L.; El'skaya, A. V. *Sensors and Actuators B* **1994**, *18-19*, 629.
34. Kriz, D.; Ramström, O.; Svensson, A.; Mosbach, K. *Anal. Chem.* **1995**, *67*, 2142.
35. Andersson, L. *Anal. Chem.* **1996**, *68*, 111.
36. Shalit, I.; Marks, M. I. *Drugs* **1984**, *28*, 281.
37. Berry, D. J. *J. Chromatogr.* **1987**, *385*, 337.
38. de Louvois, J. *J. Antimicrob. Chemother.* **1982**, *9*, 253.

39. Lietman, P. S.; White, T. J.; Shaw, W. V. *Antimicro. Agents Chemother.* **1976**, *10*, 347.
40. Dalbey, M.; Gano, C.; Izutsu, A.; Collins, C.; Jaklitsch, A.; Hu, M.; Fischer, M. *Clin. Chem.* **1985**, *31*, 933.
41. Levi, R.; McNiven, S.; Piletsky, S.; Cheong, S.-H.; Yano, K.; Karube, I. *Anal. Chem.* **1997**, *69*, 2017.
42. Yamamoto, S.; Sugihara, H.; Shimada, K. *Chem. Pharm. Bull. (Tokyo)* **1985**, *38*, 2290.

Chapter 7

Chiral Ligand Exchange Adsorbents for Amines and Underivatized Amino Acids: 'Bait-and-Switch' Molecular Imprinting

Frances H. Arnold, Susanne Striegler, and Vidyasankar Sundaresan

Division of Chemistry and Chemical Engineering 210-41, California Institute of Technology, Pasadena, CA 91125

Ligand-exchange adsorbents for the chiral separation of underivatized α-amino acids and amine analogs have been prepared by molecular imprinting with a polymerizable copper complex and the amino acid as template. Because the complexation equilibria in these mixed-ligand systems involve numerous species, it is extremely important to characterize the template:monomer assemblies prior to polymerization. Isothermal titration calorimetry and classical species distribution diagrams have been used to determine the conditions under which the amino acids (phenylalanine) form strong 1:1 mixed-ligand complexes with the copper-iminodiacetate (CuIDA) functional monomer. In contrast, a diamine template ethylenediamine does not form 1:1 complexes with CuIDA; the diamine instead competes with IDA for the metal ion, and a stronger chelating group is required. Adsorbents prepared using enantiomers of phenylalanine as template show enantioselectivity for phenylalanine and tyrosine as well as for a chiral amine analog of phenylalanine, α-methylphenethylamine. The use of the amino acid as the template in this 'bait-and-switch' imprinting approach allows creation of chirally selective binding sites suitable for separations of chiral amines.

The numerous applications of ligand exchange chromatography (LEC) (*1–3*) attest to the versatility of metal ion coordination/chelation interactions for the separation, purification and analysis of a wide range of compounds. This versatility can be exploited to great advantage for the preparation of ligand exchange adsorbents by molecular imprinting, which requires strong interactions during imprinting (to optimize the fidelity of imprinting) and much weaker interactions and/or fast kinetics during chromatographic separation and other applications. Metal ions are particularly well-suited for applications in aqueous media, where electrostatic and hydrogen-bonding interactions are compromised. Selective ligand exchange adsorbents have been prepared by molecular imprinting with metal ion complexes (*4–7*). Recent applications of this approach include supports for the chiral resolution of underivatized amino acids and amines (*8*) and a robust, glucose-sensing polymer based on glucose complexation by copper-triazacyclononane (Cu^{2+}-TACN) and the consequent release of protons under alkaline conditions (*9*). The Borovik group has recently shown that immobilization of Co(II) complexes by molecular imprinting effectively isolates the metal complexes and leads to novel binding properties and chemistry (Krebs, J. F.; Borovik, A. S. *J. Am. Chem. Soc.*, submitted).

A key step in molecular imprinting is the 'preorganization' of the desired functional monomer:template assembly in solution, which is incorporated into the polymer matrix by crosslinking polymerization. Depending on the polymerization reaction conditions, the monomer:template assembly can exist in a number of states, some fraction of which could lead to the nonspecific incorporation of functional monomer. Incorporation of functional monomer not associated with template will of course result in nonspecific adsorption sites in the imprinted material and reduced selectivity. A stoichiometric excess of functional monomer in the imprinting reaction mixture, often used to compensate for weak interactions, necessarily leads to nonspecific sites (*10*). Ideally, the interaction between functional monomer and template will be sufficiently strong that all monomer is properly associated with a template molecule. Although covalent monomer:template assemblies (*11*) clearly fulfill this criterion, subsequent use of the material may require additional chemical manipulation to generate active binding sites, which can introduce heterogeneity, or the kinetics of bond formation and breakage may be too slow for practical application.

Metal ions exhibit a wide range of binding affinities under different solvent conditions. Furthermore, they can be replaced with relative ease in an imprinted polymer, often simply by treatment with a chelating agent and soaking in a solution of the new metal ion species. Thus a strongly-interacting metal ion (*e.g.* Cu^{2+}) used for imprinting can be replaced with a weaker-interacting metal ion (Zn^{2+}) for separation (*6*). We have termed this approach 'bait-and-switch' imprinting (*10*). An alternative 'bait-and-switch' approach is to use a strongly interacting template analog during imprinting. We have recently demonstrated this approach with polymers prepared using chiral metal complexes of underivatized amino acids; it was shown that these polymers are able to separate the related chiral amines, which, lacking the additional carboxylate ligand, bind the copper complexes more weakly (*8*).

Identification of a strongly interacting metal ion is not itself sufficient for successful molecular imprinting. Metal complexation equilibria involve multiple species (metal ion, immobilizing ligand(s), target molecule, as well as their protonated and deprotonated forms) and can yield highly complex species distribution diagrams (Figure 1) (*12*). Here we will address the key issue of the nature of the functional monomer:template assembly during the preorganization step of molecular imprinting. We have found isothermal titration calorimetry (ITC) to be a particularly useful tool for determining association constants and therefore the optimal conditions for formation of the desired chelating monomer:metal ion:template assembly. In the absence of the binding information required to calculate the species distributions, molecular imprinting is likely to yield materials exhibiting a high degree of heterogeneity and nonspecific binding. This conclusion is not limited to imprinting with metal complexes--weak noncovalent interactions necessarily result in nonspecific functional monomer incorporation. Optimal design of an imprinted polymer will consider the distribution of all species in the reaction mixture.

Chiral Separations of Underivatized Amino Acids: Preorganization of Cu-VBIDA-Amino Acid Complex

We recently described the preparation of molecularly-imprinted adsorbents based on the achiral Cu(II)-[*N*-(4-vinylbenzyl)imino]diacetic acid (Cu(VBIDA) (**1**) monomer that are able to separate *underivatized* amino acids (*8*). The amino acid-imprinted materials are also able to separate racemic mixtures of corresponding amines.

a

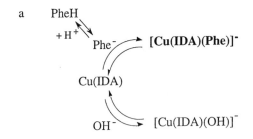

b

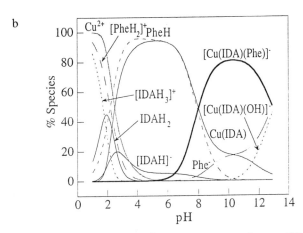

Figure 1. (a) Major species in Cu^{2+}-IDA-phenylalanine equilibrium above pH 7. (b) Distribution diagram for 1:1:1 Cu^{2+}:VBIDA:phenylalanine at 25°C.

1

For molecular imprinting the goal is to form 1:1:1 (ternary) complexes of the chelating monomer, metal ion, and template during preorganization and maintain that during crosslinking polymerization. The tridentate IDA ligand binds Cu^{2+} tightly ($K \sim 10^{11}$ M^{-1}), leaving two (or three) coordinations sites for binding to the template amino acid. The complexation products, however, depend strongly on the solution pH, as illustrated in the species distribution diagram calculated for the system of (1:1:1) IDA, Cu^{2+} and phenylalanine (Figure 1). The major species present above pH 7 are shown in Figure 1a. At 1:1:1 stoichiometry and pH 11, the maximum amount of ternary complex [Cu(VBIDA)(Phe)] corresponds to 82% of the total metal ion-monomer concentration. The remaining copper is present as Cu(VBIDA)(H_2O)$_2$, which would be incorporated nonspecifically in the imprinted polymer.

Addition of excess template improves the percentage of metal-complexing monomer associated with template. At 1:1:2 stoichiometry (100% molar excess of template) more than 95% of the monomer is present as the ternary mixed-ligand species (Figure 2). The pH range over which the desired ternary complex forms also increases to pH 9-12. Further addition of template brings the ternary complex to virtually 100% of the monomer (not shown).

To calculate the species distribution diagram one must know the association constant for formation of the ternary complex, i.e. template binding to the metal-complexing monomer. These data are often not available. We have found isothermal titration calorimetry (ITC) to be a convenient method for obtaining these association constants. The ITC results for l-phenylalanine binding to CuVBIDA at pH 9.5 shown in Figure 3 give an excellent fit to a one-site binding model (lower panel), yielding a binding constant of 2.7 ($\pm$ 0.3) x 10^4 M^{-1} (8). Identical results were obtained for the binding of d-phenylalanine to Cu(VBIDA), as expected.

The importance of fully characterizing the monomer:template assembly is well illustrated with a second example involving a model diamine template (ethylenediamine) and Cu(IDA). As shown in Figure 4, the ITC data for ethylenediamine binding to Cu(IDA) do not fit the model for a two-state system, which immediately indicates that there is a problem with this system. While UV-vis spectroscopic binding experiments indicate that ethylenediamine binds to Cu(IDA), they are inconclusive as to whether it is a two-state system. In this case, the diamine template binds copper with sufficient affinity to compete with IDA. The species involved in this equilibrium and the species distribution diagram for 1:1:1 stoichiometry are shown in Figure 5.

At 1:1:1 stoichiometry in the preorganization mixture, the ternary complex corresponds to only 80% of the total metal ion. Both free IDA and Cu(IDA) are present and would be incorporated nonspecifically into an imprinted polymer. Unfortunately, addition of excess diamine does not change the situation (not shown).

In order to avoid nonspecific incorporation of monomer, a chelating ligand that binds the metal ion more tightly than IDA is needed. A good choice might be monomer 2, based on triazacyclononane, which we have used in the preparation of polymers for recognition of carbohydrates and other diols at alkaline pH (9). The species diagram for the TACN:Cu^{2+}:ethylenediamine system (Figure 6) shows that 100% of the monomer is associated with ethylenediamine at 1:1:3 stoichiometry.

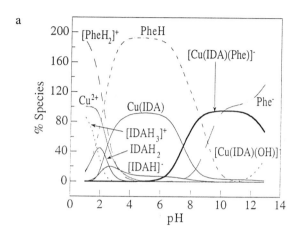

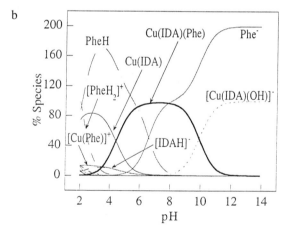

Figure 2. Distribution diagram for 1:1:2 Cu^{2+}:VBIDA:phenylalanine at (a) 25°C, (b) 40°C.

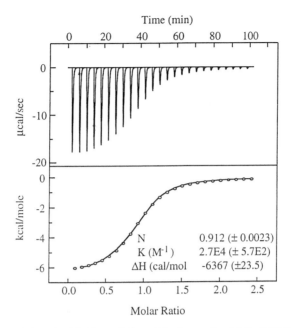

Figure 3. ITC data for binding of phenylalanine to 1 mM Cu(VBIDA) at 25°C, pH 9.5.

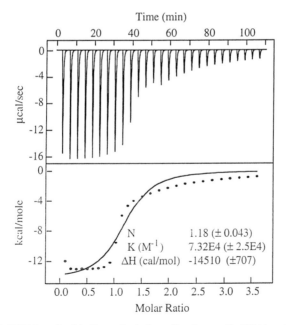

Figure 4. ITC data for binding of ethylenediamine to Cu(IDA) at 25°C, pH 9.5.

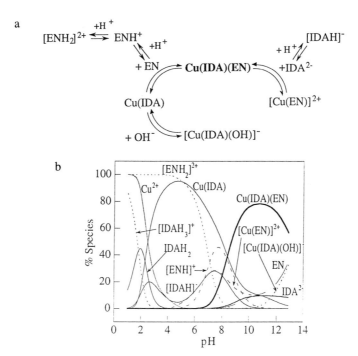

Figure 5. (a) Major species in copper-IDA-ethylenediamine equilibria above pH 7. (b) Distribution diagram for 1:1:1 Cu^{2+}:IDA:ethylenediamine at 25°C. Binding constants for this calculation were obtained from (*13*).

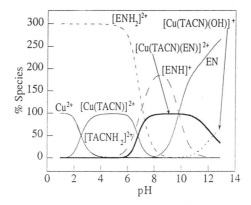

Figure 6. Species distribution diagram for 1:1:3 Cu^{2+}:TACN:ethylenediamine.

2

Effects of Concentration and Temperature on the Equilibrium Species Distributions

The equilibrium species distributions shown here have all been calculated for aqueous solutions at 25°C and at the concentrations at which the polymerizations are carried out. For the Cu(IDA)-amino acid systems, the binding affinities are such that the relative percentages of the metal-containing species are not very sensitive to changes in concentration over a range from 1 mM to 20 mM of total metal concentration. The only species that are affected significantly over this range are the various protonated forms of the IDA and amino acid ligands.

Temperature, however, has a much larger effect on the distribution of equilibrium species. Since the polymerizations are carried out at elevated temperatures, the distribution of species in the actual reaction mixture will differ from those calculated at room temperature. The distribution of equilibrium species at the temperature of polymerization can be calculated if the binding heats of all the equilibria involved are known. Figure 2b shows the distribution for the Cu^{2+}:VBIDA:phenylalanine system at 1:1:2 stoichiometry and 40°C. The pH range over which formation of the template assembly is favored shifts to lower pH than at 25°C (Figure 2a). More than 90% of the required ternary complex is obtained in the pH range 6-8.5. (Because raising the temperature to 40°C also decreases pH, there is no need to adjust the pH of solutions prepared at room temperature.)

Preparation and Characterization of Molecularly-Imprinted Polymers: Separation of Underivatized Amino Acids

Using a greater than twofold excess of various template amino acids to CuVBIDA to form the monomer:template assemblies, a series of bulk polymers was prepared by crosslinking polymerization with EGDMA (95:5 EGDMA to functional monomer) and 4,4'-azobis(4-cyanovaleric acid) in MeOH-H_2O at 40°C, as described (8). For chromatography, polymers imprinted with phenylalanine were prepared as surface coatings on 10 μm porous silica (LiChrosphere) particles under similar conditions. The imprinted adsorbents could distinguish the enantiomers of underivatized α-amino acids with varying degrees of selectivity, depending on the template used and on the amino acid targeted for separation. Adsorbents prepared from amino acids with larger, aromatic side chains exhibited the highest selectivities (α = 1.65 for the chromatographic separation of d,l-phenylalanine). Cross-selectivity for similar amino acids also depends on side chain size: materials templated with l- or d-phenylalanine exhibit good enantioselectivity for d,l-tyrosine (α ~ 1.54) and much reduced enantioselectivities towards d,l-tryptophan or aliphatic amino acids, as measured in equilibrium binding studies and chromatographically. Materials imprinted with alanine showed no selectivity.

Because the imprinted polymers contained no chiral components (other than the template, which was almost quantitatively removed), the selectivity is believed to reflect the formation of chirally-selective binding sites during polymerization. As illustrated in Figure 7, a binding site formed in the presence of the template (l-)enantiomer can preferentially destabilize binding of the d-enantiomer.

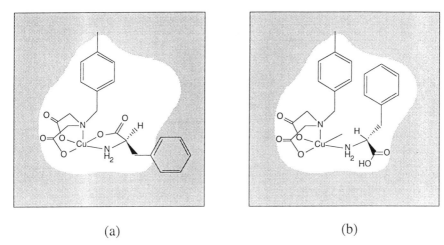

(a)   (b)

Figure 7. Enantioselectivity in imprinted ligand-exchange materials. Molecular imprinting with *l*-phe gives a cavity that is selective for *l*-phe (not shown). The *l*-isomer can simultaneously chelate to metal ion and fit into the shape-selective cavity. Rebinding of the *d*-isomer is hindered because (a) chelation of the metal ion by the *d*-isomer is sterically unfavorable. (b) Alternately, if the molecule fits into the cavity, it cannot chelate Cu(II). This highly idealized picture is probably true only for a fraction of the binding sites.

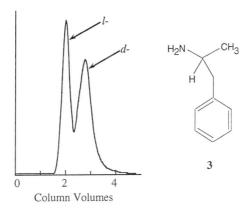

Figure 8. Chromatographic separation of *d,l* α-methylphenethylamine on column packed with phenylalanine-imprinted polymer-coated silica (4.6 mm id x 50 mm). Sample size: 100 μL of 1 mM solution. Running conditions: 1 mL/min, 50°C, 1.5 mM acetate, pH 8.

Separation of a Chiral Amine on an Amino Acid-Templated Polymer

Chiral amine **3** does not bind CuIDA with sufficient affinity to form high concentrations of the desired ternary complex for molecular imprinting. However, a racemic mixture of α-methylphenethylamine could be separated on the imprinted chromatography adsorbent prepared using *d*-phenylalanine (Figure 8).

For chiral molecules that bind to metal ions in a monodentate fashion, the metal coordination interaction can provide only one of the three 'points of interaction' necessary for enantioselectivity (*14*). Thus, in order to successfully resolve molecules such as chiral amines or carboxylic acids, molecular imprinting has to contribute the equivalent of at least two other points of interaction between the polymer matrix and the substrate. The ability of the material imprinted with a phenylalanine enantiomer to resolve the chiral amine indicates that the steric interaction between the side group and the binding cavity involves more than one point of contact. We believe this is strong evidence that molecular imprinting with a chiral template has created binding sites selective for the molecular shape of the template enantiomer.

Acknowledgments

This research is supported by the National Science Foundation (BES-9416915).

Literature Cited

1. Davankov, V. A.; Rogozhin, S. V. *J. Chem. Soc.* **1971**, 490.
2. Caruel, H.; Rigal, L.; Gaset, A. *J. Chromatogr.* **1991**, *558*, 89-104.
3. Davankov, V. A. In *Complexation Chromatography*; Caignant, D. (Ed.); Chromatographic Science Series; Marcel Dekker, New York, NY, 1992, Vol. 57; pp. 197-245.
4. Dhal, P. K.; Arnold, F. H. *J. Am. Chem. Soc.* **1991**, *113*, 7417-7418.
5. Dhal, P. K.; Arnold, F. H. *Macromolecules* **1992**, *25*, 7051-7059.
6. Plunkett S. D.; Arnold, F. H. *J. Chromatogr.* **1995**, *708*, 19-29.
7. Fujii, Y.; Matsutani, K.; Kikuchi, K. *J. Chem. Soc. Chem. Commun.* **1985**, 415-417.
8. Vidyasankar, S.; Ru, M.; Arnold, F. H. *J. Chromatogr.* **1997**, *775*, 51-63.
9. Chen, G.; Guan, Z.; Chen. C.-T.; Sundaresan, V.; Fu, L.; Arnold, F. H. *Nature Biotech.* **1997**, *15*, 354-357.
10. Vidyasankar, S.; Arnold, F. H. *Curr. Opin. Biotech.* **1995**, *6*, 218-224.
11. Wulff, G. *Angew. Chem. Int. Ed. Engl.* **1995**, *34*, 1812-1832.
12. Martell, A. E.; Smith, R. M. *Critical Stability Constants*, Plenum Press, New York, NY, 1974.
13. Rao, A. K.; Kumar, G. N.; Mohan, M. S.; Kumari, Y. L. *Ind. J. Chem A*, **1992**, *31*, 256-259.
14. Davankov V. A.; Kurganov A. A. *Chromatographia*, *17*, **1983**, 686-690.

Chapter 8

Recognition of Drugs and Herbicides: Strategy in Selection of Functional Monomers for Noncovalent Molecular Imprinting

Toshifumi Takeuchi and Jun Matsui

Laboratory of Synthetic Biochemistry, Faculty of Information Sciences, Hiroshima City University, 3-4-1 Ozuka-higashi, Asaminami-ku, Hiroshima 731-31, Japan

Molecular imprinting studies are described which focus upon the selection, design, and synthesis of functional monomers for the preparation of high affinity polymeric materials. The use of methacrylic acid and 2-(trifluoromethyl)acrylic acid as functional monomers is illustrated with biologically active molecules and agricultural and pharmaceutical compounds as model templates. Potential applications are explored with biologically active compounds as analytical targets to probe the utility of molecularly imprinted affinity polymers.

An attractive aspect of molecular imprinting [1] lies in its concept as "tailor-made" synthesis. In molecular imprinting, a target molecule is present as a template in the polymerization mixture and is subsequently extracted from the resultant polymer to leave behind a "tailor-made" cavity, *i.e.* a complementary binding site (Figure 1). In principle, this technique has the potential for preparing a receptor for any target chemical species by employing that chemical species as a template. However, when the objective is the synthesis of highly selective binding sites, successful molecular imprinting may not be a simple matter . The most critical factor is the selection of the functional monomer, that is a polymerizable molecule with functional groups which form reversible, covalent or non-covalent bonds with the template to be imprinted in the polymeric material (Figure 1). The structure of the adduct of the template and the functional monomer(s) is reflected in the resultant binding site structure, *i.e.* the arrangement of functional residues derived from the functional monomers employed. In other words, the adducts function as a medium for the introduction of structural and chemical information about the template molecule into the resulting polymer network. Therefore, it is of prime importance to select a functional monomer based on the design of the template-functional monomer adduct.

The strength of the binding force between a template and functional monomer(s) is an important factor with considerable influence on the performance of the resulting polymer. Use of an appropriate functional monomer which is capable of strong binding with the template generates the stable template-functional monomer adduct that is essential to provide "precise" molecular imprinting. In this sense, functional monomers which form covalent bonds are preferable, although such monomers have

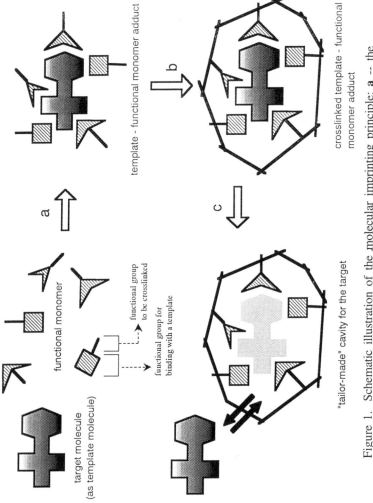

target molecule
(as template molecule)

functional monomer

functional group
to be crosslinked

functional group for
binding with a template

a

template - functional monomer adduct

b

c

crosslinked template - functional
monomer adduct

"tailor-made" cavity for the target

Figure 1. Schematic illustration of the molecular imprinting principle: **a** -- the template molecule and functional monomer(s) are covalently or non-covalently bound; **b** -- after addition of the crosslinking agent, the template-functional monomer adduct structure is "frozen" in the polymer network; and **c** -- the template is removed to yield a template-fitting cavity as a complementary binding site.

found only a few applicable target species, *e.g.* sugars, ketones [*1*], and plant hormones [2]. On the other hand, weaker interactions, such as hydrogen bonding and electrostatic interaction, are more often utilized for a wide range of molecules even though the template-functional monomer adducts based on such non-covalent interactions are inherently less stable than those with covalent bonding. Methacrylic acid (MAA) (Figure 2) has been extensively utilized as the functional monomer. Use of MAA has led to a number of satisfactory results, notably because the monomer can function as both a proton donor and an acceptor in forming double hydrogen bonds to provide a stable adduct. However, efforts continue to seek improved functional monomers with high affinities.

When a functional monomer is anticipated to act as only a proton donor, MAA may not be the best choice. Since a more acidic monomer could be a more powerful functional monomer for imprinting chemical species with proton accepting functionalities, we introduced the monomer 2-(trifluoromethyl)acrylic acid (TFMAA) (Figure 2). This monomer, with a trifluoromethyl group in place of the methyl group of MAA, is expected to be more acidic than MAA and thereby interact strongly even with weak proton acceptors during the imprinting process. Furthermore, such a functional monomer is expected to provide inherently stronger interactions as components of the binding site even after the molecular imprinting.

In this chapter, we describe studies of molecular imprinting which focus upon the selection, design, and synthesis of functional monomers for achieving high affinity polymer materials. Use of MAA and 2-(trifluoromethyl)acrylic acid are compared. To evaluate the properties of our affinity polmers, potential practical applications are studied with biologically active molecules and agricultural and pharmaceutical compounds as analytical targets.

A MAA-Based Atrazine-Imprinted Polymer

The herbicide atrazine [*3-5*] was selected as a target molecule because it is a compound of environmentally analytical importance and is known to form hydrogen bonds with acetic acid [6]. This suggests that a functional monomer with a carboxylic acid group could form complexes with atrazine by hydrogen bonding. Therefore, MAA was used as the functional monomer for imprinting atrazine (Figure 3). In the same manner as acetic acid, the carboxyl group of MAA can function as both a hydrogen bond donor and acceptor to form simultaneous double hydrogen bonds with atrazine. Such hydrogen bonding is expected to be favorable in terms of entropy for formation of the binding site in the imprinting process.

The atrazine-imprinted polymer gave a capacity factor of 15.5 in chromatographic tests, while a blank polymer prepared without atrazine exhibited a capacity factor of 0.55. This demonstrates that the affinity were enhanced in the imprinted polymer. The selectivities for various agrochemicals was examined and the results are summarized in Figure 4. Triazine herbicides were retained to the highest degree, while other structurally unrelated chemicals were eluted quickly. This suggests that the atrazine-imprinted polymer recognizes the core structure of triazine herbicides, *i.e.* 1,3,5-triazine with two alkylamino side chains, and consequently exhibits a group selectivity for triazine herbicides. Furthermore, atrazine was retained longer than other triazine herbicides. This atrazine-selective binding property is proposed to result from atrazine-imprinting, since some other triazine herbicides were found to be retained longer than atrazine in the non-imprinted blank polymer.

Further chromatographic studies were performed with different ratios of acetonitrile-chloroform as the eluent. Longer retention was observed when the less polar eluent was used (Figure 5). Also retention was drastically decreased by addition of water to an acetonitrile eluent. These results suggest that retention of atrazine arises from hydrogen bonding with carboxyl residues. Polar solvents interfere with such hydrogen bond formation. With a high water content in the eluent, strong retention

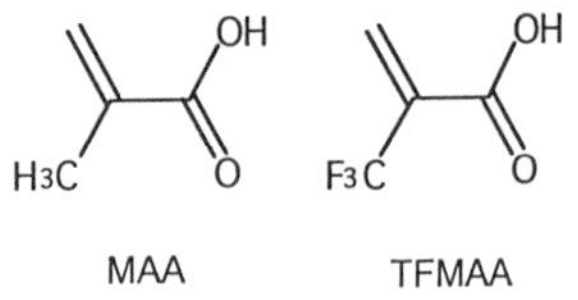

MAA   TFMAA

Figure 2. Structures of the functional monomers methacrylic acid (MAA) and 2-(trifluoromethyl)acrylic acid (TFMAA).

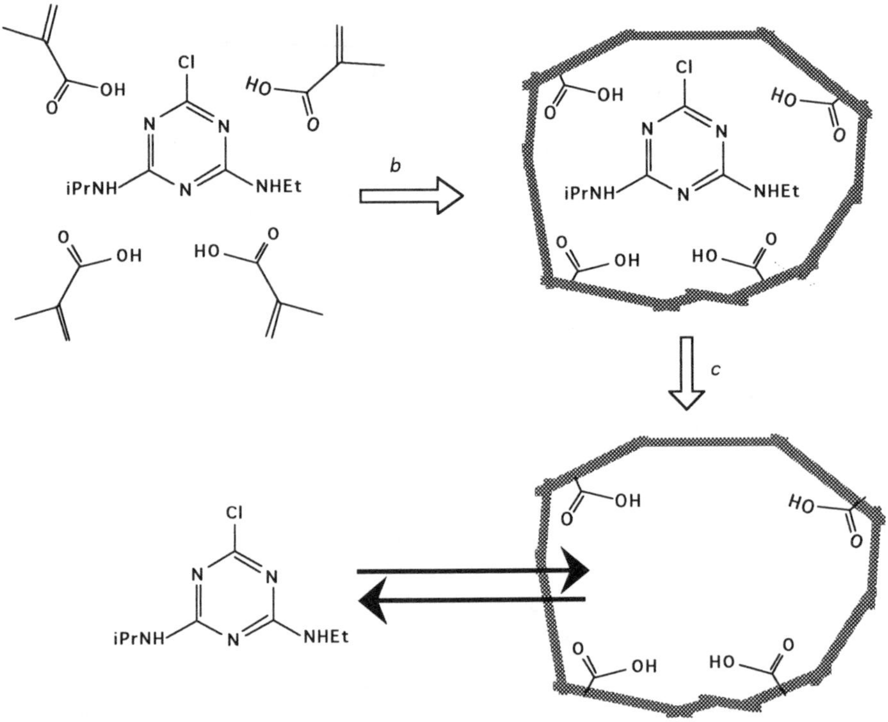

Figure 3. Schematic representation of atrazine imprinting: a) Into a chloroform (25 mL) solution of atrazine (360 mg) and MAA (575 mg), the ethylene glycol dimethacrylate crosslinker (9.35 g), and the 2,2'azobis(isobutyronitrile) initiator were added. The mixture was irradiated with UV light at 3 °C to obtain the bulk rigid polymer. b) The polymer was crushed, ground, and sieved, then washed with methanol-acetic acid to remove the atrazine. The resultant cavities in the polymer are expected to function as a complementary binding site for atrazine.

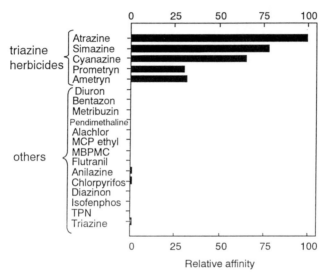

Figure 4. Selectivity of the atrazine-imprinted polymer. Relative affinity is the relative capacity factor for a sample when the capacity factor for atrazine is taken to be 100. The liquid chromatography was performed with acetonitrile-chlrorform (1:1, v/v) as the eluent.

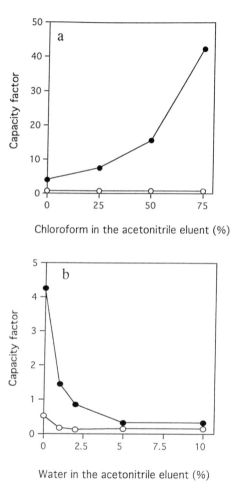

Figure 5. Effect of eluent polarity on the retention ability of the atrazine-imprinted polymer with filled circles for the atrazine-imprinted polymer and open circles for the non-imprinted blank polymer.

was observed. Presumably this is due to hydrophobic forces within the polymer network, although the reason for the selectivity is unknown.

To estimate the dissociation constant, saturation tests were performed by a batch procedure with chloroform as the solvent. In chloroform, strong hydrogen bonding of atrazine with the binding sites in the imprinted ploymer is expected. In the concentration range of 50 μM - 3.0 mM, binding reached saturation and Scatchard analysis indicates that the polymer possesses heterogeneous binding sites with different binding abilities, like polyclonal antibodies. The dissociation constant for the binding sites with the highest affinity was estimated to be 10^{-5} M for atrazine.

A More Acidic Functional Monomer TFMAA

Although MAA showed excellent performance in hydrogen bonding-based molecular imprinting for atrazine, more acidic functional monomers may be more suitable when compounds with basic functionality are molecularly imprinted. There are two major advantages when such monomers are used: i) stronger interaction is expected to form more stable template-functional monomer adducts for the polymerization, which leads to an enhancement of the imprint effects in the molecular imprinting process; and, ii) higher affinity may be produced in the resultant polymers due to stronger interactions between the polymers and the template even without an imprinting effect. We examined 2-(trifluoromethyl)acrylic acid (TFMAA) as a model functional monomer of higher acidity than MAA [7,8]. Nicotine was employed as a model template compound with basic functionality to evaluate the effectiveness of TFMAA [7].

As shown in Figure 6, TFMAA provides good performance for nicotine imprinting. When the content of TFMAA in the polymer was increased, the affinity of the resultant polymers was significantly enhanced. Use of TFMAA only gave the highest affinity compared to the combined use of TFMAA and MAA or the use of MAA alone. The imprinting effect for the TFMAA-based imprinted polymers was greater than that of the corresponding MAA-based polymers. Here, the imprinting effect is defined as the ratio of affinity of the imprinted polymer to that of the control polymer prepared in the absence of the template. This means that template-functional monomer complexes with TFMAA are more stable during the polymerization which results in more precise binding site construction and in higher selectivity in the resultant polymers.

However, we found that TFMAA is not always superior to MAA in imprinted polymers. Whether TFMAA or MAA or a combination of both functional monomers gives better results depends upon the template to be imprinted. For example, MAA gave good results in the imprinting of atrazine as described above, while the imprinting effect dimished with the use of TFMAA for atrazine imprinting as shown in Figure 7a. In contrast in the imprinting of prometryn, another triazine-herbicide and a stronger base than atrazine, the cooperative use of TFMAA and MAA provided superior affinity and selectivity compared to the use of either TFMAA or MAA alone [8] (Figure 7b). Although the factors which affect the imprinting effect are not yet completely understood, these results provide important indications on the relationship between the chemical properties of the functional monomer(s) and the structural preference of the template in the imprinting process.

As mentioned above, the carboxyl group of MAA is expected to act as both a proton donor (-OH) and an acceptor (-C=O) simultaneously when the hydrogen bonds are formed. Therefore MAA is advantageous for imprinting of templates that have double hydrogen bond-formable structures, such as atrazine in which vicinally hydrogen-bondable positions are available. TFMAA is a stronger acid than MAA and should be less favorable for double hydrogen bonding because the stronger acidity would give a lesser proton acceptor property. This means TFMAA provides a single, but strong, interaction instead of a MAA-type of weak double hydrogen bonding. Since there are no vicinal points available for double hydrogen bonding in a nicotine

126

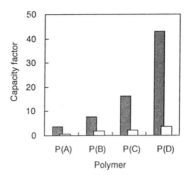

Figure 6. Retention abilities of the nicotine-imprinted (dark) and non-imprinted blank (white) polymers. P(A) was prepared with MAA (6 equivalents relative to the template), P(B) with MAA (4 equivalents) and TFMAA (2 equivalents), P(C) with MAA (2 equivalents) and TFMAA (4 equivalents), and P(D) with TFMAA (6 equivalents).

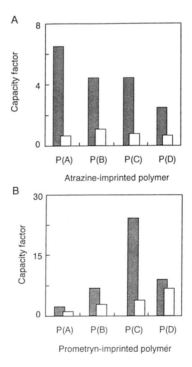

Figure 7. Capacity factors of atrazine (A) and prometryn (B) in the imprinted (dark) and non-imprinted (white) polymers. P(A) was prepared with MAA (6 equivalents relative to the template), P(B) with MAA (4 equivalents) and TFMAA (2 equivalents), P(C) with MAA (2 equivalents) and TFMAA (4 equivalents), and P(D) with TFMAA (6 equivalents).

template, TFMAA should be suitable for nicotine imprinting in which the formation of strong single hydrogen bond or electrostatic interaction would be achieved. In the case of atrazine, the double hydrogen bond could be formed with MAA; however, TFMAA may compete with MAA for complex formation with atrazine when TFMAA is added to the MAA system. The results can be explained if the double hydrogen bonding provided by MAA is more suitable for atrazine than the single hydrogen bonding provided by TFMAA. For prometryn, the situation is more complicated. For this template, both MAA and TFMAA would be accessible simultaneously without competition. This is why the cooperative use of MAA and TFMAA was superior. The addition of excessive TFMAA would lead to competition with MAA for MAA's preferred binding sites, resulting in a decrease of the binding capability for prometryn. To confirm these speculations, NMR studies on weak interaction-based complexation are necessary. Such studies are currently underway.

A Multiple Hydrogen Bond-Formable Functional Monomer 2,6-Bis-(acrylamido)pyridine

The studies with MAA and TFMAA described above suggested that employment of simultaneous multiple interactions between a template and a functional monomer would also be favorable. Recently, we reported a functional monomer capable of multiple hydrogen bond formation for barbiturate imprinting [9]. Figure 8 shows the monomer, 2,6-bis(acrylamido)pyridine, and a possible structure of binding sites constructed by molecular imprinting of cyclobarbital. This imprinted polymer strongly bound cyclobarbital. From selectivity experiments, it was determined that the polymer recognized the malonylurea structure of cyclobarbital. Since selective binding was not observed in protic solvents, hydrogen bonding was indicated as a dominant interaction between the functional monomer and the template. While detailed examination of the effect of multiple hydrogen bonding on the imprinting effect has not been performed, complementary alignment in the binding sites seems to be easily constructed by the use of this functional monomer system. Compared with functional monomers having a single hydrogen bondable site, a multiple hydrogen bonding system appears to afford higher affinity and selectivity .

Affinity-Type Solid Phase Extraction of Herbicides

In the previous sections, strategies for high affinity materials were demonstrated which focused upon the selection, design, and synthesis of functional monomers. The affinity generated by the molecular imprinting was evaluated in a chromatographic or batch mode. To better evaluate the utility of molecularly imprinted polymers, the induced affinity should be examined in practical applications. Here we tested the atrazine-imprinted polymer as a sorbent for solid phase extraction (SPE) of a triazine herbicide [10].

Herbicides are commonly analyzed by gas and liquid chromatography. SPE is often performed as a sample preparation technique prior to the chromatographic analysis. Although affinity-type SPE sorbents would be useful for the specific enrichment of target compounds, traditional affinity SPE sorbents, in which biomolecules are immobilized as molecular recognition elements, are not suitable when the samples are lipophilic and organic solvents are used. Here, the atrazine-imprinted polymer, proven to work in organic solvents, was applied for triazine herbicide-selective SPE using organic solvents. In this newly developed procedure, two different sequential retention modes were involved.

Since spherical particles of a regular size are advantageous for column use, a triazine herbicide-selective polymer was prepared by suspension polymerization. The polymer beads obtained were tested, prior to use in SPE, as a LC stationary phase to examine whether the atrazine structure was properly imprinted in the polymer network. Although water in the suspension may interfere with complex formation of atrazine with MAA during the imprinting process and the reaction temperature was higher than that in bulk polymerization, a clear imprinting effect was evident in the suspension polymerization system. Thus, the atrazine-imprinted polymer exhibited a capacity factor of 31.2, while the blank polymer prepared in the absence of atrazine gave a capacity factor of 2.5 (Table I). Furthermore, the retention was selective for the template atrazine among the triazine herbicides tested. Other agrochemicals without structural analogy (Figure 9) were eluted quickly. Although the atrazine selectivity vs. other triazine herbicides appeared to be lower than that for the bulk-polymerized atrazine receptor, the fact that the polymer exhibited group selectivity for triazine herbicides suggests that it can be used for recognition of triazine herbicides.

Table I. Capacity Factors[a] (k') for the Agrochemicals on the Imprinted and Non-imprinted (Blank) Polymers.

Sample	Polymer[b]	
	Imprinted	Blank
triazine herbicides		
atrazine	31.2	2.5
propazine	30.3	3.0
simazine	25.4	2.3
others		
asulam	0.3	0.2
thiram	0.0	0.0
propyzamide	0.0	0.0
iprodione	0.1	0.0

[a]Acetonitrile-chloroform (2:3, v/v) was the eluent. [b]The polymers were prepared by suspension polymerization, where: i) atrazine (1.57 g), methacrylic acid (6.29 g), ethylene glycol dimethacrylate (41.0 g), and 2,2'azobisdimethylvaleronitrile (0.94 g) were dissolved in chloroform (41.0 g); ii) the organic phase was added to an aqueous solution of poly(vinyl alcohol) (1.6 g) in water (219 mL); and iii) the mixture was stirred (300 rpm) at 50 °C for 6 hours. The blank polymer was prepared similarly, but in the absence of atrazine during the polymerization process.

As a sample herbicide for molecularly imprinted solid phase extraction (MI-SPE), we employed simazine, an analog of atrazine, which is commonly used as a herbicide for golf courses in Japan and has attracted considerable concern with respect to environmental analysis. Simazine was also strongly bound by the imprinted polymer. A model mixture was prepared as 500 mL of distilled water containing 0.1 ppm of the target simazine, as well as other golf course-related agrochemicals, such as asulam, thiram, propyzamide, and iprodione (Figure 9). The procedure for the MI-SPE is shown in Figure 10. In this method, a two-mode operation of the molecularly imprinted polymer was utilized to retain simazine selectively -- a reversed phase mode followed by an affinity mode. During the sample application step (Figure 10a), the polymer acted as a reversed-phase sorbent while the hydrogen bonding-based affinity is turned "off" due to water interference. In this stage, all of the agrochemicals were retained except asulam (70 % of the loaded asulam was not sorbed). Subsequent affinity-mode operation was performed with dichloromethane as the wash solvent (Figure 10b). Because hydrogen bonds with carboxyl groups can be dominant interactions in this less polar organic solvent, specific retention of simazine can be expected during the washing step. The washing waste contained the impurities to a high degree (> 90%), while simazine was not detected. Thus the simazine was still retained on the column. Finally, the simazine was extracted three times with 3.0 mL of methanol (Figure 10c). The recovery rate was more than 90%. Although optimized conditions have not yet been determined, the affinity of the molecularly imprinted polymer is demonstrated to be potentially useful in a practical application.

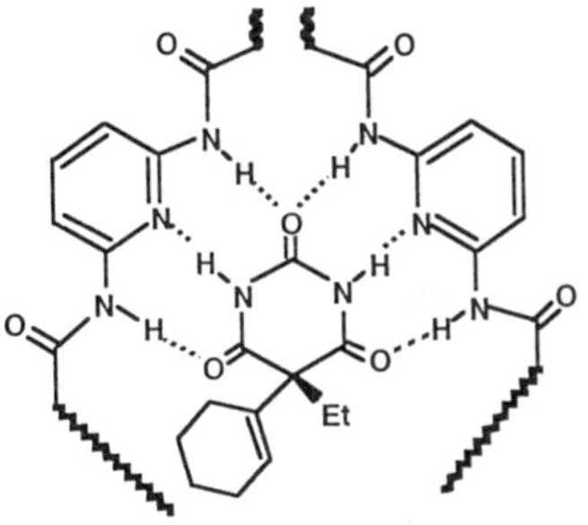

Figure 8. Possible structure of the binding site for cyclobarbital constructed by molecular imprinting with 2,6-bis(acrylamido)pyridine as a multiple hydrogen bond-formable functional monomer.

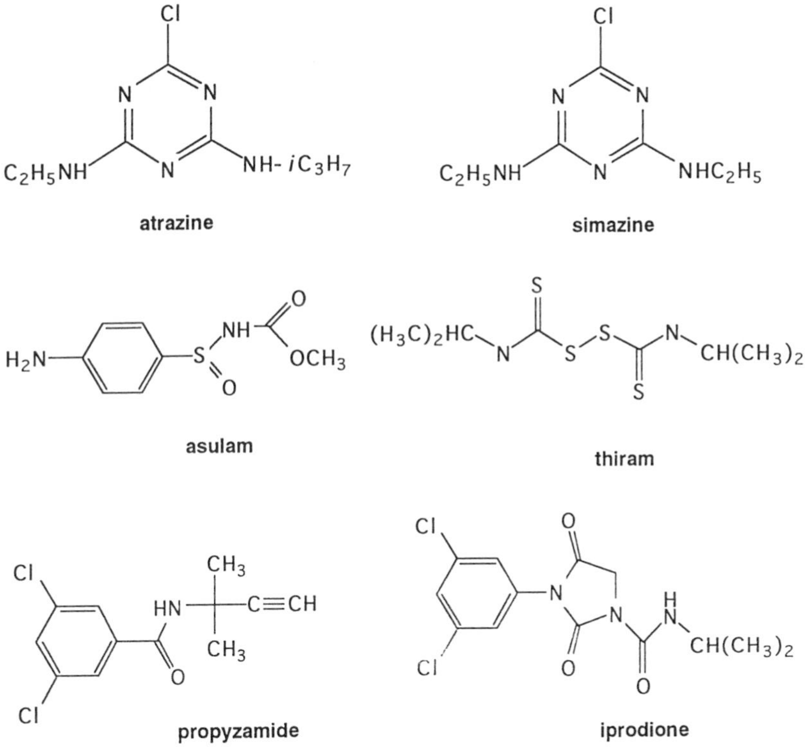

atrazine

simazine

asulam

thiram

propyzamide

iprodione

Figure 9. Structures of atrazine (template molecule), simazine (analytical target), and other golf course-related chemicals.

In Situ Molecular Imprinting

Although suspension polymerization was conducted for the preparation of the MI-SPE sorbent described above, a more common procedure for molecular imprinting involves bulk polymerization. In that case, the resultant polymer blocks have to be crushed, ground, and sieved to provide molecularly imprinted polymer particles for use as stationary phases in liquid chromatography. These procedures are tedious and time-consuming. Furthermore, all of the polymer particles prepared cannot be used because the fines are inappropriate for chromatographic use. To circumvent such problems, we have developed *in situ* molecular imprinting to prepare "tailor-made" affinity media which may be used directly in liquid chromatography. The *in situ* molecular imprinting technique utilizes a two-step procedure for the preparation of ready-to-use chromatography columns (Figure 11). All of the reagents necessary for molecular imprinting are mixed and poured into a column, then crosslinking is performed by heating. The resultant column which is filled with a macroporous, molecularly imprinted polymer rod can be used directly for chromatography. With xanthine derivatives [*11*] and optical isomers of amino acid derivatives [*12*] as template molecules, *in situ* molecular imprinting has been demonstrated to be a potentially useful method for preparing imprinted stationary phases. Enhanced retention of the template molecule was exhibited by the molecularly imprinted rods, as compared to the non-imprinted blank rods. Selectivity was also induced by imprinting as shown by the retention behaviors of the columns (Table II). A typical chromatogram for the theophylline-imprinted polymer rod is shown in Figure 12.

phenylalanine anilide (PheAn)   theophylline

Table II. Retention Characteristics of Imprinted Polymer Rods

polymer (template)	crosslinker	L-PheAn	D-PhenAm	separation factor
P(L-PheAn)	EGDMA	4.3	2.5	1.7
P(D-PheAn)	EGDMA	2.7	3.8	1.4

		theophylline	theobromine	caffeine
P(theophylline)	EGDMA	1.2	1.0	0.16
P(none)	EGDMA	0.47	0.61	0.12
P(theophylline)	ST-DVB	1.9	1.3	0.37
P(none)	ST-DVB	1.4	1.3	0.36

A nicotine-imprinted polymer rod was also prepared by the *in situ* method and evaluated in a chromatographic mode [*13*]. Although enhanced retention ability in comparison to a non-imprinted blank polymer rod was observed (Table III), the resulting capacity factor for nicotine was found to be smaller than that of a conventional nicotine-imprinted polymer. Because the enhancement of capacity factors, *i.e.* the imprinting effect, was also smaller, it was suggested that the porogenic agents employed (1-dodecanol and cyclohexanol) were detrimental to the imprinting process. Because stable complexation of the template molecule and functional monomer(s) by hydrogen bonding or electrostatic interaction is essential during the imprinting process, such protic solvents could be unfavorable because they interfere with the complexation. Crosslinking by heating (45 °C) may be also a

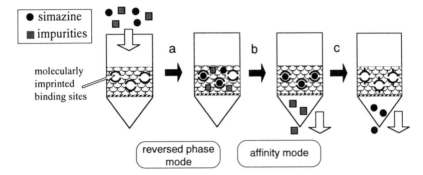

Figure 10. Solid phase extraction of simazine from water. The polymer beads (2.0 g) were packed into a 6.0 mL polypropylene cartridge and washed with methanol and water for conditioning. (a) The model sample solution was passed through the cartridge at an approximate flow rate of 30 mL/min. (b) The cartridge was dried and treated with 10 mL of dichloromethane. (c) Extraction was conducted three times with 3.0 mL of methanol.

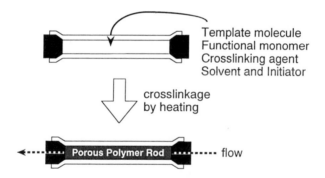

Figure 11. Illustration of the procedure for *in situ* molecular imprinting. A typical procedure for nicotine-imprinting follows: a stainless steel tube (50 mm x 4.6 mm ID) was filled with a mixture of nicotine (0.2 mmol), a functional monomer (1.2 mmol), ethylene glycol dimethacrylate (850 mg), 1-dodecanol (400 mg) and cyclohexanol (1.3 g) as porogenic agents, and 2,2'azobis-(dimethylvaleronitrile) (85 mg). The column was heated at 45 °C for 6 hours, then was carefully washed with methanol-acetic acid (8:2, v/v) at a flow rate of 0.01-0.1 mL/min.

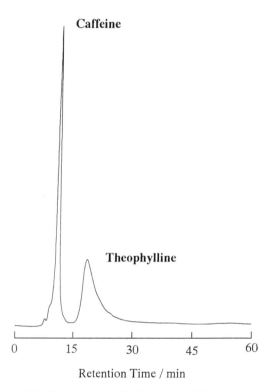

Caffeine

Theophylline

Retention Time / min

Figure 12. A typical chromatogram for the theophylline-imprinted polymer rod. The column size was 150 mm x 4.6 mm ID with acetonitrile as eluent at a flow rate of 0.25 mL/min.

detrimental factor for complexation, since the conventional bulk polymerization was photoinitiated at 3 °C. The results lead us to the conclusion that it is particularly important to use a powerful functional monomer for the *in situ* technique, since conditions are not optimal for template-functional monomer interaction. TFMAA, that was shown to be superior in bulk-type, nicotine-imprinted polymer preparations in chloroform as described above, is again expected to strongly interact with nicotine which possesses two proton accepting groups because it is a better proton donor than MAA due to the electron-withdrawing effect of the trifluoromethyl substituent.

Table III. Capacity Factors for Nicotine in the Imprinted and Non-imprinted (Blank) Polymer Rods[a]

rod	template	functional monomer	capacity factor
rod 1	nicotine	MAA	2.4
rod 2	none	MAA	1.1
rod 3	nicotine	TFMAA	28.4
rod 4	none	TFMAA	5.1

[a]The eluent was acetonitrile-acetic acid (98:2, v/v) at a flow rate of 2.5 mL/min.

On this basis, nicotine-imprinted rods were prepared using TFMAA as the functional monomer [13]. As shown in Table III, a larger imprinting effect was observed in the TFMAA-based, *in situ* imprinting system. Thus, the imprinted rods exhibited 5.6-times larger capacity factors than the non-imprinted blank rod, whereas it was only 2.2-times larger in the MAA-based system. Consequently, the imprinted polymer rod prepared with TFMAA displayed a 12-times larger capacity factor, as compared to the MAA-based, nicotine-imprinted rod. Furthermore, the selectivity against cotinine, a major metabolite of nicotine, was also improved. The ratio of capacity factors, k'(nicotine) / k'(cotinine), was 19.6 in the TFMAA-based, imprinted polymer rod and 1.5 in the MAA-based imprinted polymer rod. Accordingly, it appears that TFMAA-based imprinting can also produce high-affinity specific binding sites selective for nicotine in the *in situ* method.

It should be noted that the chromatographic tests were performed with acetonitrile-acetic acid (98:2, v/v) as the eluent. This protic and highly hydrophilic eluent system was required to allow for direct comparison of the retention abilities of the imprinted and non-imprinted polymer rods, since in the absence of acetic acid acetonitrile did not elute nicotine from the TFMAA-based imprinted rod. The results again showed high affinity performance of the nicotine-imprinted polymer rods prepared with TFMAA.

Polymer rods for diastereoseparation were also prepared using TFMAA as the functional monomer. Antimalarial drugs of cinchona alkaloids, cinchonidine and its antipod cinchonine, were used as templates [14]. Both cinchonidine-imprinted (PF1) and cinchonine-imprinted (PF2) polymers exhibited stronger retention than the non-imprinted blank polymer rod (FB) prepared in the absence of the corresponding template cinchona alkaloids (Table IV). Furthermore, diastereoselectivity was observed which is presumably induced by the imprinting effect because both polymers appeared to be selective for the original template species. We also tested MAA-based cinchonidine-imprinted (PM1) and cinchonine-imprinted (PM2) polymer rods. We expected MAA would be a better functional monomer for imprinting cinchonidine because the cinchonidine molecule possesses a quinuclidine nitrogen and a hydroxyl group that would be expected to form a double hydrogen bond with MAA in the manner described in the previous section, while cinchonine may be unfavorable for accepting such a double hydrogen bonding due to steric hindrance. As expected, a larger separation factor was exhibited for cinchonidine with MAA, while the use of MAA resulted in a reduced separation factor in the case of the cinchonine-imprinted polymer.

Table IV. Disasterioseparation of Cinchona Alkaloids by Molecularly Imprinted Polymer Rods

analyte	PF1		PF2		FB
	k'	α	k'	α	k'
cinchonidine	11.5		6.6		2.0
		2.0		5.3	
cinchonine	5.8		34.7		2.0

analyte	PM1		PM2		MB
	k'	α	k'	α	k'
cinchonidine	3.9		1.5		0.59
		2.6		3.1	
cinchonine	1.5		4.7		0.50

Thus, the *in situ* method is demonstrated to be an easy and potentially very useful technique for the preparation of molecularly imprinted affinity media for liquid chromatography. The results shown here for the comparison of MAA and TFMAA allow us to draw a general conclusion that the selection of appropriate functional monomers is usually important for achieving high affinity in molecularly imprinted materials and is especially significant when polymerization conditions are not optimal for template-functional monomer complex formation, as is the case for the *in situ* method. Our goal is to develop a large library of functional monomers for use in molecular imprinting. This is a versatile and powerful approach to the preparation of "tailor-made" materials which are capable of molecular recognition for use in analytical, bio- and organic chemical fields.

References

1. Takeuchi, T.; Matsui, J. *Acta Polymer* **1996**, *47*, 471.
2. Kugimiya, A.; Matsui, J.; Abe, H.; Aburatani, M.; Takeuchi, T. *Anal. Chim. Acta*, in press.
3. Matsui, J.; Miyoshi, Y.; Doblhoff-Dier, O.; Takeuchi, T. *Anal. Chem.* **1995**, *67*, 4404.
4. Siemann, M.; Andersson, L.I.; Mosbach, K. *J. Agric. Food Chem.* **1996**, *44*, 141.
5. Muldoon, M.T.; Stanker, L.H.; *J. Agric. Food Chem.* **1995**, *43*, 1424.
6. Welhouse, G. J.; Bleam, W. F. *Environ. Sci. Technol.* **1993**, *27*, 500.
7. Matsui, J.; Doblhoff-Dier, O.; Takeuchi, T. *Anal. Chim. Acta*, **1997**, *343*, 1.
8. Matsui, J.; Miyoshi, Y.; Takeuchi, T. *Chem. Lett.* **1995**, 1007.
9. Tanabe, K.; Takeuchi, T.; Matsui, J.; Ikebukuro, K.; Yano, K.; Karube, I. *J. Chem. Soc., Chem. Commun.* **1995**, 2303.
10. Matsui, J.; Okada, M.; Tsuruoka, M., Takeuchi, T. *Anal. Commun.* **1997**, *34*, 85.
11. Matsui, J.; Miyoshi, Y.; Matsui, R.; Takeuchi, T. *Anal. Sci.* **1995**, *11*, 1017.
12. Matsui, J.; Kato. T.; Takeuchi, T.; Suzuki, M.; Yokoyama K.; Tamiya, E.; Karube, I. *Anal. Chem.* **1993**, *65*, 2223.
13. Matsui, J.; Takeuchi, T. *Anal. Commun.* **1997**, *34*, 199.
14. Matsui, J.; Nicholls, I. A.; Takeuchi, T. *Anal. Chim. Acta*, in press.

Chapter 9

Noncovalent Molecular Imprinting of a Synthetic Polymer with the Herbicide 2,4-Dichlorophenoxyacetic Acid in the Presence of Polar Protic Solvents

Karsten Haupt

Department of Pure and Applied Biochemistry Chemical Center, Lund University, P.O. Box 124, S-22100 Lund, Sweden

Non-covalent molecular imprinting of a synthetic polymer with the herbicide 2,4-dichlorophenoxyacetic acid has been achieved in the presence of the polar solvents methanol and water. Formation of the prearranged complex results from hydrophobic and ionic interactions between the template and the functional monomer 4-vinylpyridine. The polymer obtained binds the original template with an appreciable selectivity over structurally related compounds. The potential use of micrometer-sized imprinted polymer particles as the recognition element in a radioligand binding assay for 2,4-dichlorophenoxyacetic acid is demonstrated.

Molecular imprinting is becoming increasingly recognized as a powerful technique for the preparation of synthetic polymers containing tailor-made recognition sites for certain target molecules (*1*). The imprinting process is performed by copolymerizing functional and crosslinking monomers in the presence of a molecular template. After elution of the template, complementary binding sites are revealed within the polymer network that allow rebinding of the template with a high specificity, sometimes comparable to that of antibodies (*2, 3*). The artificial receptors so obtained may be used in applications that demand specific ligand binding, such as the analytical and preparative separations of closely related compounds (*4, 5*), solid phase extraction (*6*), directed synthesis and catalysis (*7-9*), as recognition elements in sensors (*10, 11*) and immunoassay-type binding assays (*2, 12*).

There exist two conceptionally different approaches for molecular imprinting in synthetic polymers. In the "covalent approach" that has been developed by Wulff and others, the functional monomers are covalently coupled to the template prior to polymerization (*13*). These bonds have to be cleaved to liberate the template and reveal the binding sites, and are subsequently reformed during rebinding of the target molecule. The second approach, which has primarily been developed by the Mosbach group (*1*), is generally called the "non-covalent approach". It relies on a pre-arrangement of functional monomers with the template prior to polymerization via non-covalent bonds. In some cases, a "hybrid approach" has been adopted wherein the polymerization is performed with the functional monomer(s) covalently bound to the template and, after cleavage of the template from the polymer, rebinding takes place by non-covalent interactions. An interesting example of this approach has recently been reported by Whitcombe et al. (14). Also of note is another imprinting protocol which is somewhat different from the covalent and non-covalent approaches in that it involves the use of metal coordination (15).

Among these different methods for the preparation of molecularly imprinted polymers, the non-covalent approach is probably the most flexible in terms of the choice of functional monomers and possible template molecules and has therefore been most widely adopted. However, it does have some limitations. The non-covalent bonds formed during pre-arrangement, e.g. hydrogen bonds or other electrostatic interactions, are relatively weak. Therefore conditions must be chosen to shift the equilibrium towards complex formation. As a result non-covalent imprinting has been mostly performed in apolar organic solvents, since in the presence of polar solvents, and especially water, the prearranged complex is destabilized. On the other hand, hydrophobic interactions are strong in water, and ionic bonds might also add to the stability of the complex in an aqueous environment. The aim of the present work was therefore to investigate whether specific non-covalent molecular imprints can be obtained in the presence of high concentrations of water using a combination of the hydrophobic effect and ionic interactions. The herbicide 2,4-dichlorophenoxyacetic acid (2,4-D) was selected as the model template owing to its hydrophobic aromatic ring and its ionizable carboxyl group. Moreover, molecularly imprinted polymers specific for phenoxyacid herbicides could be attractive as artificial receptors in environmental analysis.

Experimental

Ethyleneglycol dimethacrylate (EDMA) and 4-vinylpyridine (4-VP) were from Merck (Darmstadt, Germany). 2,4-Dichlorophenoxyacetic acid (2,4-D), 2,4-dichlorophenoxybutyric acid (2,4-DB), 2,4-dichlorophenoxyacetic acid methyl ester (2,4-D-OMe), 4-chlorophenoxyacetic acid (CPOAc), 4-chlorophenylacetic acid (CPAc), phenoxyacetic acid (POAc), phenoxyethanol (POEtOH) and naphthoxyacetic acid (NOAc) were from Sigma (St. Louis, MO, USA). 2,2'-azobis(2,4-dimethylvaleronitrile) (ABDV) was from Wako (Osaka, Japan). All other chemicals were of analytical grade and solvents were of HPLC quality.

Preparation of Polymers. Compositions of the polymerization mixtures are summarized in Table I. The ratio of crosslinker : functional monomer : template was 20:4:1. Template, monomers, and polymerization initiator (ABDV) were weighed into glass test tubes and mixed with the solvent. The solutions were then sonicated, sparged with nitrogen for 2 minutes and placed in a thermostated water bath at 45°C for 4 hours, followed by 2 h at 60°C. The resultant hard bulk polymers were ground in a mechanical mortar and wet-sieved in acetone through a 25-μm sieve. The particles were washed by incubation in methanol/acetic acid (7:3) (2x), acetonitrile/acetic acid (9:1) (2x), acetonitrile (1x), methanol (2x) for 2 hours each time, followed by centrifugation. The particles were then resuspended in acetone and allowed to settle for 4 hours. The ones that remained in suspension (fines) were collected and the procedure was repeated 4 times. The solvent was removed by centrifugation and the particles were dried in vacuo. The fine particles obtained in this way had an average diameter of 1 μm and were used in all further experiments.

Radioligand Binding Assays. The polymer particles were suspended in the incubation solvent and appropriate volumes were added into 1.5-ml polypropylene test tubes, followed by the radioligand ^{14}C-2,4-D (0.26 nmol, specific activity 15.7 mCi/mmol), varying amounts of a solution of a competing ligand if appropriate, and solvent to give a total volume of 1.0 ml. The samples were incubated on a rocking table for 2 hours. After centrifugation, 700 μl of supernatant was withdrawn and measured by liquid scintillation counting.

Table I. Composition of the different imprinted and control polymers

Polymer	EDMA	4-VP	Template	Solvent	ABDV
1 (imprinted)	20 mmol	4 mmol	1 mmol 2,4-D	4 ml methanol + 1 ml H$_2$O	0.31 mmol
2 (control)	20 mmol	4 mmol	0	4 ml methanol + 1 ml H$_2$O	0.31 mmol
3 (control)	20 mmol	4 mmol	1 mmol toluene + 1 mmol acetic acid	4 ml methanol + 1 ml H$_2$O	0.31 mmol

Results

Preparation of the Polymers. The imprinted polymer (1, Table I) was prepared by copolymerization of a crosslinking monomer (EDMA) with a functional monomer (4-VP) in the presence of 2,4-D as the template. Given that the goal was to prepare and use the imprinted polymers in water-containing solvents, complex formation had to rely on hydrophobic and ionic interactions which, unlike hydrogen bonding, are not, or at least are to a lesser extent, disturbed in the presence of water. 4-VP was chosen as the functional monomer to allow for ionic interaction with the carboxyl group of the template as well as for the hydrophobicity of its aromatic ring.

The imprinting was performed in methanol/H$_2$O (4:1), as the crosslinking monomer EDMA and 2,4-D were only poorly soluble in pure water. After polymerization, a faint blue color was observed in the bulk polymer. The non-imprinted control polymer (2) was uncolored. However, a control polymer prepared in presence of toluene and acetic acid (3) was also slightly blue. The color disappeared in all cases during washing of the polymers.

In contrast to other reports on binding assays with imprinted polymers (2, 7, 11), in the present work we used fine particles with a diameter of about one micrometer. These particles are normally discarded. We found that micrometer-sized particles and smaller had the same binding characteristics for the target molecule as the 25 µm particles normally used. Furthermore, not only were the incubation times reduced due to shorter diffusion distances, but fines were found to be more practical for binding assays as they stayed in suspension longer and were easier to pipette. Moreover, if the template is expensive, it is undesirable to lose a considerable amount of polymer by discarding the fines.

Rebinding of [14]**C-2,4-D.** Initially, all three polymers were tested for rebinding of ^{14}C-2,4-D in the original imprinting solvent. The imprinted polymer (1) could rebind the radiolabelled template in methanol/H$_2$O (4:1). Only 200 µg polymer was needed to adsorb *ca.* 50% of the added radioligand. The control polymers 2 and 3 showed only very low binding of the radioligand.

As the polymers were intended for use in aqueous buffer, the conditions for aqueous binding assays had to be optimized. There have been some reports in the literature where polymers imprinted in organic solvents have been used in aqueous buffer for binding assays (12, 16). Normally low percentages of a water-miscible organic solvent, such as ethanol or acetonitrile, are added to the buffer to increase the wettability of the polymer and to prevent adsorption of hydrophobic templates on test tubes, pipette tips, *etc.* In initial binding assays, we found that the addition of small amounts of a non-ionic surfactant (0.02-0.1% Triton-X-100) had the same effect, but also improved the mixing of the polymer particles in the test tubes. Therefore, we employed 0.1% Triton X-100 instead of ethanol or acetonitrile in all our binding assays.

Also of importance is the pH of the buffer used. Figure 1 shows binding of ^{14}C-2,4-D to polymers 1 and 2 in different buffers from pH 3 to pH 9. Binding to polymer 1 was approximately constant between pH 3 and 7, and then decreased with further increase in pH. The local minimum at pH 5 can be attributed to the fact that at this pH both the pyridine and the 2,4-D are partially charged, which renders them less hydrophobic. Binding should now be due mostly to ionic interactions where the buffer ions can effectively act as competitors. The non-imprinted control polymer 2 showed highest binding at pH 3 which rapidly decreased to remain at a constant low level above pH 6. All subsequent experiments were performed in phosphate buffer at pH 7 where binding to the control polymer was minimized.

As can be seen from Figure 2a, polymer 1 was able to rebind radiolabelled 2,4-D in 20 mM phosphate buffer at pH 7 and 150 µg/ml of polymer was needed to bind 50% of the added radioligand. At this polymer concentration, very low binding to the corresponding blank polymer was observed.

It has to be added that only with polymers imprinted in methanol/H_2O (4:1) and having 4-VP as the functional monomer was appreciable rebinding capacity observed. Polymers prepared with methacrylic acid as the functional monomer did not specifically rebind ^{14}C-2,4-D, either in buffer at different pH or in toluene or acetonitrile. Polymers prepared with 4-VP as the functional monomer but in dry toluene as the imprinting solvent could specifically rebind ^{14}C-2,4-D, but with a 13 times lower capacity than polymer 1. The corresponding control polymer exhibited a significant amount of non-specific binding (not illustrated).

Competitive Binding Assays. Figure 2b shows the competition of ^{14}C-2,4-D binding to polymers 1-3 by unlabelled 2,4-D. A typical sigmoid competition curve similar to those observed in competitive immunoassays was obtained for the imprinted polymer 1. The useful concentration range for detection of 2,4-D is from 30 ng/ml (135 nM) to 10 µg/ml (45 µM).

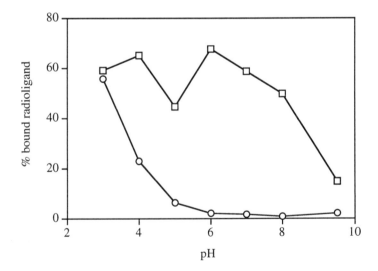

Figure 1. Binding of ^{14}C-2,4-D to polymers 1 (□) and 2 (O) as a function of pH. Conditions: 150 µg/1 ml assay; 20 mM buffer, 0.1% Triton X-100.

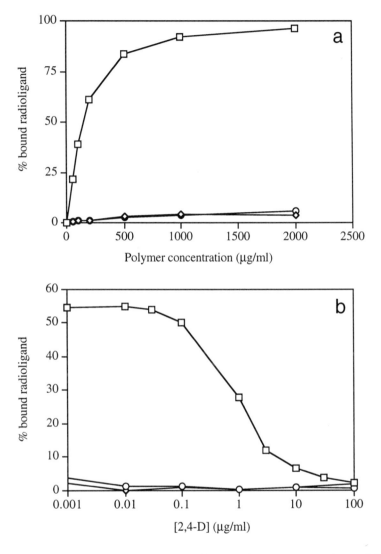

Figure 2. (a) Binding of radioligand relative to polymer concentration and (b) radioligand displacement curves with unlabelled 2,4-D as competitor at 150 µg polymer/1 ml assay for polymer 1 (□), polymer 2 (○) and polymer 3 (◇). Conditions: 20 mM sodium phosphate buffer pH 7, 0.1% Triton X-100.

Cross-reactivity Studies. To assess the specificity of the imprinted polymer, competition of ^{14}C-2,4-D binding by structurally related compounds was studied. By comparing the concentration that yields 50% inhibition of ^{14}C-2,4-D binding (IC$_{50}$ value) for the different competitors to that of 2,4-D, the cross-reactivities of the related compounds can be estimated. The structures of the different competitors and the corresponding cross-reactivities are shown in Figure 3. All evaluated compounds exhibited lower binding to the polymer than the original template. The lowest cross-reactivities were obtained with compounds not having a charged group (2,4-D-OMe, POEtOH), whereas 2,4-dichlorophenoxybutyric acid showed almost the same binding as 2,4-D.

To assess whether the added surfactant Triton X-100 had an influence on the binding of the different compounds to the polymer, binding was also studied in the same buffer containing 10% ethanol instead of the surfactant. The cross-reactivities obtained were very similar to or identical with those measured in the presence of surfactant.

Discussion

Molecular imprinting of polymers using only non-covalent interactions between the template and the functional monomers has until now been limited to apolar environments. In some cases however, the use in aqueous buffer of polymers initially imprinted in non-polar solvents has been demonstrated to be feasible. It has been suggested that the interaction between the molecule of interest and the polymer is governed by different molecular forces in non-polar solvents and in aqueous buffer (*12*). This would mean that the interactions that are predominant in the latter case, mainly hydrophobic interactions and ionic bonds, are strong enough to allow for complex formation in a polar environment. The herbicide 2,4-dichlorophenoxyacetic acid seemed to be a good model compound if the imprinting itself should also be done in presence of water, as it contains an aromatic ring and an ionizable carboxyl group. On the other hand, we presumed that there would only be a small chance of success for imprinting of 2,4-D in non-polar solvents such as toluene, as 2,4-D has too few possible electrostatic interaction points. The use of the solvent system methanol/water instead of pure water in our experiments was dictated by the low solubility of the crosslinker and the template in pure water. We believe, however, that other templates, used in conjunction with water-soluble crosslinkers may work even better at much higher water concentrations.

From a comparison of polymer 1 (imprinted with 2,4-D) and polymer 2 (control polymer, no template) in rebinding of the template, it can be concluded that polymer 1 is clearly templated. The observation of a faint blue color during polymer formation in the case of the imprinted polymer could be an indication of a charge transfer interaction between the aromatic rings of 2,4-D and 4-vinylpyridine. The formation of a π-complex (*17*) between the two aromatic rings may also be possible. This may contribute to the stability of the prearranged complex mainly formed through ionic interactions and the hydrophobic effect in the polar environment.

Hydrophobic interactions are strong in water and should thus make the biggest contribution to adsorption. However, they are generally non-specific. Therefore it was of interest to investigate the specificity of polymer 1 by use of related compounds as competitors for ^{14}C-2,4-D-binding in radioligand binding assays. Given the conditions during imprinting and rebinding as well as the structure of the template, the cross-reactivities obtained between the different competitors and 2,4-D are surprisingly low. Moreover, these data provide additional information about the contribution of different forces to the interaction. Comparing 2,4-D and 2,4-D-OMe, it is obvious that cancelling out the effect of the negative charge greatly reduces binding (7% cross-reactivity, Figure 3). Removing one (CPOAc) or two (POAc) chlorine atoms from the ring also progressively weakens the affinity (24 and 2 % cross-reactivity, respectively). In the case of POEtOH which has no chlorine and no charged group, binding is completely suppressed. On the other hand, NOAc which

Compound	Cross-reactivity (%)
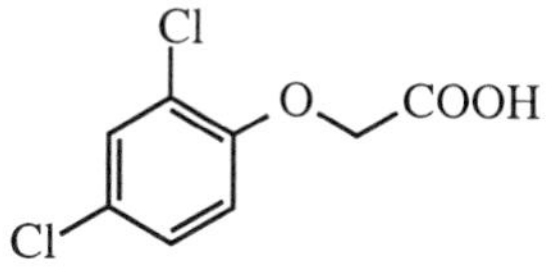 2,4-D	100
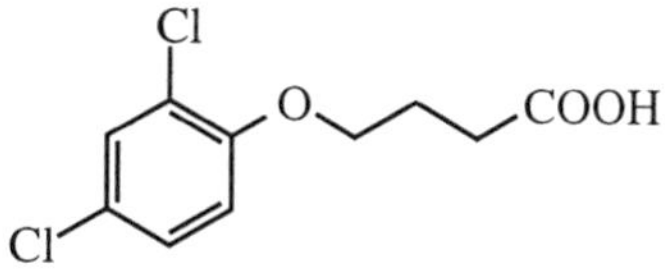 2,4-DB	95
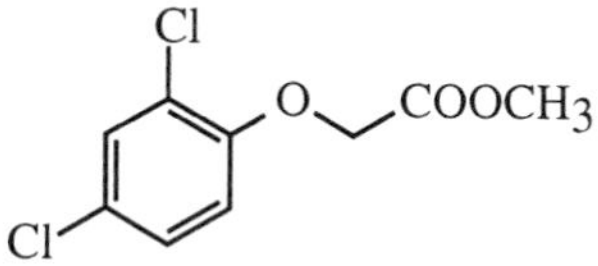 2,4-D-OMe	7
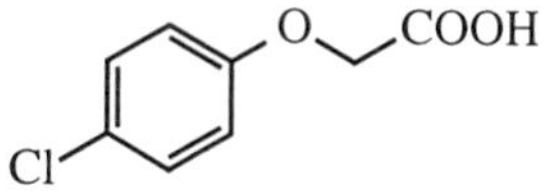 CPOAc	24
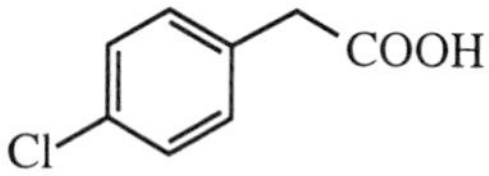 CPAc	10
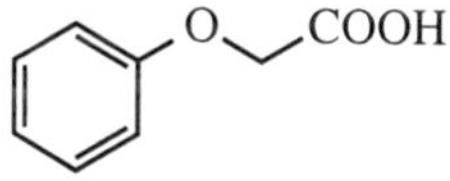 POAc	2
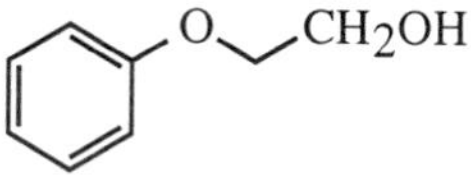 POEtOH	<0.1
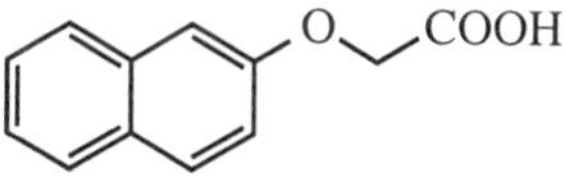 NOAc	44

Figure 3. Structures of the different compounds used as competitors in the radioligand binding assays and corresponding cross-reactivities. Cross-reactivities were calculated by dividing the IC_{50} value for 2,4-D by that of the other compounds. IC_{50} values were obtained from the competition curves by non-linear regression.

has a naphthoxy instead of the phenoxy structure nevertheless shows 44 % cross-reactivity and 2,4-DB binds to the polymer nearly as well as the original template. This indicates that the specificity of the polymer is due to specific interactions between the monomers and the template rather than to the formation of specifically shaped cavities.

To confirm the templating effect, a second control polymer was synthesized where 2,4-D was replaced by one molar equivalent each of toluene and acetic acid (polymer <u>3</u>). The fact that this polymer also did not bind 2,4-D, even though a similar bluish color appeared during polymerization, indicates that an ionizable group has been positioned close to a hydrophobic pocket in polymer <u>1</u> by templating with 2,4-D.

Conclusion

In the present work, a polymer has been imprinted in 20% water in methanol. This shows that molecular imprinting in the presence of polar protic solvents using only non-covalent interactions is possible, depending on the nature of the template molecule and the functional monomers. It was also demonstrated that an appreciable binding specificity for the original template can be obtained. These findings extend the potential applicability of non-covalent molecular imprinting particularly to cases where either the use of polar solvents and especially water may be required or the target molecule may lack the functionalities required for imprinting in non-polar solvents. The potential use of this imprinted polymer as the recognition element instead of antibodies in binding assays for 2,4-D was demonstrated, where particles with an average diameter of one micrometer were advantageous over the larger particles usually employed.

Acknowledgments

The author wishes to thank Dr. Peter Cormack for linguistic advise. Financial support of Karsten Haupt by the EU *Human Capital and Mobility* program is gratefully acknowledged.

Literature Cited

1. Mosbach, K.; Ramström, O. *Bio/Technology* **1996**, *14*, 163.
2. Vlatakis, G.; Andersson, L.I.; Müller, R.; Mosbach, K. *Nature* **1993**, *361*, 645.
3. Ramström, O.; Ye, L.; Mosbach, K. *Chem. Biol.* **1996**, *3*, 471.
4. Fischer, L.; Müller, R.; Ekberg, B.; Mosbach, K. *J. Am. Chem. Soc.* **1991**, *113*, 9358.
5. Schweitz, L.; Andersson, L. I.; Nilsson, S. *Anal. Chem.* **1997**, *69*, 1179.
6. Sellergren, B. *Anal. Chem.* **1994**, *66*, 1578.
7. Beach, JV. and Shea, KJ. *J. Am. Chem. Soc.* **1994**, *116*, 379.
8. Ohkubo, K.; Urata, Y.; Hirota, S.; Honda, Y.; Fujishita, Y.; Sagawa, T. *J. Mol. Catal.* **1994**, *93*, 189.
9. Matsui, J.; Nicholls, I.A.; Karube, I.; Mosbach, K. *J. Org. Chem.* **1996**, *61*, 5414.
10. Kriz, D.; Ramström, O.; Svensson, A.; Mosbach, K. *Anal. Chem.* **1995**, *67*, 2142.
11. Piletsky, S. A.; Piletskaya, E. V.; Elgersma, A. V.; Yano, K.; Karube, I.; Parhometz, Y. P.; El'skaya, A. V. *Biosens. Bioelectron.* **1995**, *10*, 959.
12. Andersson, L.I. *Anal. Chem.*, **1996**, *68*, 111.
13. Wulff, G. *Angew. Chem., Int. Ed. Engl.* **1995**, *34*, 1812.
14. Whitcombe, M.J.; Rodriguez, M.E.; Villar, P.; Vulfson, Y.N. *J. Am. Chem. Soc.* **1995**, *117*, 7105.
15. Dhal, P.K.; Arnold, F.H. *Macromolecules* **1992**, *25*, 7051.
16. Siemann, M.; Andersson, L.I.; Mosbach, K. *J. Agric. Food Chem.* **1996**, *44*, 141.
17. Schneider, H.J. *Angew. Chem., Int. Ed. Engl.* **1991**, *30*, 1417.

Chapter 10

Development of Uniform-Sized, Molecular-Imprinted Stationary Phases for HPLC

Ken Hosoya and Nubuo Tanaka

Department of Polymer Science, Kyoto Institute of Technology, Matsugasaki, Sakyo-ku, Kyoto 606, Japan

Non-covalently molecular-imprinted, uniform-sized, polymer-based stationary phases are prepared by a two-step swelling technique using isomers of diaminonaphthalene or chiral amides derived from (S)-α-methylbenzylamine as template molecules. Methacrylic acid is an effective host molecule for the diaminonaphthalene templates and produces a molecular recognition ability almost equivalent to that of molecular-imprinted stationary phases prepared by traditional bulk polymerization. Unexpectedly, an imprinted, crosslinked, polymer-based stationary phase without this type of strong host functionality also showed moderate molecular recognition which suggests that the crosslinked polymer network itself can memorize the shape of template in some fashion. For chiral separations of the amide derivatives, the crosslinked polymer network also memorized the shape of the chiral template to provide chiral resolution. In addition, chiral host monomers with similar functionality as the chiral amide template were found to markedly enhance the molecular recognition. Further investigation suggests that this increase in chiral recognition is due to favorable structural interactions within specific recognition sites. Also, uniform-sized, polymer-based, HPLC stationary phases with multi-chiral selectors are prepared by a combination of molecular imprinting and *in situ* surface modification. Simple additivity of the chiral recognition abilities of the two types of chiral selectors is observed. Size exclusion chromatography reveals a possible causitive factor for the selectivity.

Molecular imprinting is an effective strategy for the preparation of chromatographic stationary phases with specific molecular recognition (*1-3*). In this technique, the template molecule is mixed with monomers and polymerized to afford polymeric separation media. The template molecule usually possesses relatively polar functional groups (*4,5*), such as carboxylic acid, hydroxyl, or amino functions, and/or aromatic groups. Appropriate host monomers with functional groups which may interact with the functional groups of the template molecule through intermolecular interactions, such as hydrogen bonding and electrostatic attraction, are often utilized in the construction of an effective imprinted recognition site (*6*).

Usually various kinds of nonaqueous bulk polymerization methods (*4*), including a preparative technique for the formation of continuous rod-type separation media (*7*), have been utilized to obtain molecular-imprinted separation media for HPLC. However, the molecular-imprinted polymers prepared by bulk polymerization must be ground and then sieved to provide suitable packing materials (8). Customary oil-in-water suspension polymerization which forms spherical polymer beads requires an aqueous suspension medium. Water is thought to weaken interaction between a template and its host due to the high polarity of the aqueous medium adjacent to the oil droplets of the monomer mixture. Recently, an imprinted material with high selectivity was prepared using an *in situ* aqueous porogen (9). In that paper, the author also mentioned that weakly or moderately complexed templates require a low polymerization temperature and exclusion of water from the porogen during imprinting to produce a recognition effect.

Independently, we have reported the preparation of uniform-sized, polymer-based stationary phases for HPLC (10). Such uniform-sized stationary phases exhibit excellent column efficiency with good stability. We now report the development and chromatographic properties of uniform-sized, molecular-imprinted, polymer-based stationary phases for HPLC through investigation of the effects of the monofunctional host molecule on the separation selectivity of the base polymer and of chiral host monomers on the molecular recognition ability. In addition, an *in situ* surface modification technique is utilized to further improve the uniform-sized, molecular-imprinted stationary phases.

Experimental Section

Materials. Ethylene dimethacrylate (EDMA) was purchased from Tokyo Chemical Industry Company, Ltd. (Tokyo, Japan), while methacrylic acid (MA) was obtained from Nacalai Tesque (Kyoto Japan). Both monomers were purified by distillation *in vacuo* to remove the polymerization inhibitor. The radical initiators of benzoyl peroxide and α,α'-azobisisobutyronitrile were purchased from Nacalai Tesque (Kyoto, Japan) and utilized as received.

The 1,5- and 1,8-diaminonaphthalene and (S)-(-)-α-methylbenzylamine were purchased from Nacalai Tesque (Kyoto, Japan), while (S)-(+)-N-(3,5-dinitrobenzoyl)-α-methylbenzylamine (DNB), its antipode, and (S)-1-naphthylethylamine, as well as the other chiral solutes were obtained from Aldrich Chemical Company. The S-(+)-α-phenylglycinol and R-(-)-α-phenylglycinol were purchased from Fluka Chemika, Japan (Tokyo, Japan).

Chiral crosslinking agents and a chiral host monomer were prepared from chiral phenylglycinols and (S)-1-naphthylethylamine, respectively, by condensation with methacryloyl chloride and purified by column chromatography on silica gel with hexane-ethyl acetate (3:1, v/v) as eluent.

A Two-Step Swelling and Polymerization Method. Uniform-sized polystyrene seed particles to be utilized as a shape template were prepared by emulsifier-free emulsion polymerization and purified by the previously reported method (11). The size of the seed particles was *ca.* 1 μm in diameter.

Preparation of uniform-sized, macroporous, molecular-imprinted polymer beads, as well as non-imprinted polymer beads, by a two-step swelling and polymerization process was carried out as follows: An aqueous dispersion of the uniform-sized polystyrene seed particles (1.4 mL, 9.5×10^{-2} g/mL) was mixed with a microemulsion prepared by sonication of 0.95 mL of dibutyl phthalate as the activating solvent (12), 85 mg of benzoyl peroxide or α,α'-azobisisobutyronitrile, 4 mg of sodium dodecylsulfate, and 10 mL of distilled water. This first-step swelling was carried out at room temperature with stirring at 125 rpm. Completion of the first-step swelling was determined with an optical microscope as the vanishing point of the micro oil droplets in the microemulsion.

A dispersion of 10 mL of the crosslinking agent (or 95:5 by weight mixture of ethylene dimethacrylate and the host or additive) and 10 mL of toluene or cyclohexanol as the porogenic solvent into 90 mL of water containing 1.92 g of polyvinylalcohol (dp = 500, saponification value = 86.5 to 89 mol%) as the dispersion stabilizer was added to the dispersion of swollen particles from the first step. The second-step swelling was carried out at room temperature for 12 hours with stirring at 125 rpm. If a template molecule was to be added, 1.0 gram of the template was mixed with the monomers which formed the dispersion in the second-step swelling.

After the second-step swelling was completed, polymerization was performed at 70 °C under an argon atmosphere with slow stirring. After 24 hours, the dispersion of polymerized beads was poured into 250 mL of water to remove the suspension stabilizer (polyvinylalcohol) and the supernatant was discarded after sedimentation of the beads. The polymer beads were redispersed in methanol and the supernatant was again discarded after sedimentation. This procedure was repeated three times in methanol and twice in tetrahydrofuran (THF). Then the polymer beads were filtered on a membrane filter and washed with THF and acetone and dried at room temperature. The final bead size was 5.6 μm.

The yields of particles prepared by the polymerization were higher than 88%. Complete removal of the template molecules by the washing procedure was verified by the absence of nitrogen in combustion analysis of the polymers.

The prepared beads were packed into stainless steel columns (4.6 mm ID X 150 mm) by a slurry technique with aqueous acetonitrile as the slurry medium for evaluation of their chromatographic characteristics. The void markers employed were acetone in an acetonitrile mobile phase and both di-*tert*-butylbenzene and n-pentylbenzene in a normal phase mode.

Chromatography. All of the chromatographic solvents were purchased from Nacalai Tesque and used as received. HPLC was performed with a Jasco, 880-PU Intelligent HPLC Pump equipped with a Rheodyne 7125 valve loop injector and a Waters Model 440 UV detector set at 254 nm. Chromatography was performed at 30 °C and a Shimadzu C-R4A recorder was utilized.

Results and Discussion

Diaminonaphthalenes as Templates. As we reported previously (*13*), the isomer of diaminonaphthalene (1,5- or 1,8-diaminonaphthalene) which is utilized as the template elutes after the other isomer on each of the imprinted stationary phases with 100% acetonitrile as the mobile phase as shown in Figure 1. In these stationary phases, methacrylic acid is the host monomer for the diaminonaphthalene template. In both chromatograms, the more hydrophobic molecule naphthalene was found to be eluted much faster than both 1,5- and 1,8-diaminonaphthalene even in a reversed-phase mode. Thus the carboxylic acid groups in the imprinted polymer interact with the amino groups to retard elution of the diaminonaphthalenes compared with naphthalene itself.

The separation factors, α, observed were 1.55 for the 1,5-diaminonaphthalene-imprinted, uniform-sized stationary phase and 1.30 for the 1,8-diaminonaphthalene-imprinted, uniform-sized stationary phase. Since the values previously reported for molecular-imprinted, polymer-based stationary phases prepared by bulk polymer-ization were 1.45 and 1.50 for 1,5- and 1,8-diaminonaphthalene-imprinted stationary phases, respectively (*14*), the molecular recognition ability for the template observed for the uniform-sized, molecular-imprinted stationary phases is comparable to that of traditional imprinted stationary phases. The separation factors are proposed to arise from hydrogen bonding interactions between the diaminonaphthalene template and the carboxylic acid groups derived from the host monomer, methacrylic acid. Apparently the interaction is sufficiently to strong to allow for molecular imprinting even when water is present in the polymerization system.

The role of the carboxylic acid functionality in retention of compounds with amino groups was probed for other aromatic amines (Table I). Aniline elutes after benzene with a broader peak shape. Pyridine affords the largest k' value for all the solutes tested. Addition of 0.1% of triethylamine to the acetonitrile mobile phase did not change the k' value of benzene, whereas other amine compounds including the diaminonaphthalenes were eluted much faster than in the absence of triethylamine. Since triethylamine should competitively block the methacrylic acid groups of the stationary phase, these results support the role of carboxylic acid functionality derived from the methacrylic acid monomer on retention. The results also confirm earlier results which were obtained with buffered mobile phases (15).

Table I. Effect of Triethylamine on Separation and Retention [a]

Template [b]	Mobile phase [c]	k' (1,5-) [d]	k' (1,8-) [e]	α (k'$_{1.5-}$ / k'$_{1.8-}$)	k' (pyridine)	k' (aniline)	k' (benzene)
1,5-	AN	1.46	0.94	1.55	1.51	0.40	0.11
1,5-	AN + TEA	0.99	0.72	1.38	0.79	0.30	0.12
1,8-	AN	0.93	1.18	0.79	1.23	0.34	0.11
1,8-	AN + TEA	0.69	0.82	0.84	0.70	0.28	0.12

[a]Chromatographic conditions: flow rate, 1 mL/min: detection, UV 254 nm.
[b]The diaminonaphthalene used as the template: 1,5- = 1,5-diaminonaphthalene; and 1,8- = 1,8-diaminonaphthalene.
[c]AN = 100% acetonitrile; AN + TEA = 99.9% acetonitrile and 0.1% triethylamine.
[d]k' for 1,5-diaminonaphthalene.
[e]k' for 1,8-diaminonaphthalene.

Methacrylic acid is found to be one of the important factors for the molecular imprinting technique which allows for separation by retaining the template longer than its isomer. However, other solutes, such as pyridine, with the same functionality as the template are also affected which results in a longer retention time with broadened peaks. This may cause peak overlapping with the template molecule in the analysis of mixed solutes on uniform-sized, molecular-imprinted stationary phases.

Although the retention contributed by the carboxylic acid groups derived from the methacrylic acid host is blocked by the presence of triethylamine in the mobile phase, each diaminonaphthalene isomer used as template is still retained longer than its isomer on the corresponding imprinted stationary phase. Since the base stationary phase prepared without the template molecule does not separate the two diamino-naphthalenes, the moderate molecular recognition which we observe can be only explained based on a contribution of recognition sites formed in the crosslinked polymer network.

To certify these findings concerning the contribution of the crosslinked polymer network, we prepared an imprinted stationary phase in the absence of methacrylic acid. The results for a stationary phase with 1,8-diaminonaphthalene as the template as well as for a non-imprinted stationary phase are summarized in Table II. Retention times of both diaminonaphthalenes are found to be shorter on the stationary phase prepared without the host compared with the stationary phase prepared with the host, meth-acrylic acid; whereas for both the k' values are larger than those obtained with the base

Table II. Retention on Non-imprinted Polymers and Imprinted
Polymers Prepared with and without a Host Monomer [a]

Template	Host	k' (1,8-)[b]	k' (1,5-) [c]	α (k' 1,8- / k' 1,5-)	k' (Pyridine)
None	None	0.46	0.46	1.00	0.15
1,8- [d]	None	0.66	0.60	1.10	0.14
1,8- [d]	Methacrylic Acid	1.18	0.93	1.27	1.23
None	None	0.47 [e]	0.50 [e]	0.94	0.15 [d]

[a]Chromatographic conditions: mobile phase, 100% acetonitrile; flow rate 1.0
mL/min; detection, UV 254 nm.
[b]k' for 1,8-diaminonaphthalene.
[c]k' for 1,5-diaminonaphthalene.
[d]1,8-Diaminonaphthalene.
[e]The mobile phase was 99.9% acetonitrile and 0.1% triethylamine.

non-imprinted stationary phase prepared without the host monomer (*i.e.* neither
template nor host were present in the polymerization mixture.)

Although the base stationary phase (first entry in Table II) cannot separate the
isomers of diaminonaphthalene, the imprinted stationary phase prepared without the
host monomer, methacrylic acid (second entry) can separate the isomers with the
template isomer being retained longer. Interestingly, pyridine is eluted much faster on
both of the base stationary phase and the imprinted stationary phase in this case. The
differences in k' values for pyridine support a contribution of carboxylic acid groups
from the methacrylic acid host to the retention of compounds with amino
functionality.

Addition of triethylamine to the mobile phase for the non-imprinted base stationary
phase (fourth entry) did not appreciably affect the k' values for any of the solutes
tested. This finding means interaction between the functional groups derived from the
ethylene dimethacrylate and the amino functionality is relatively weak and the observed
molecular recognition on the imprinted stationary phase prepared without the host
monomer is due to recognition of the shape of the template by the crosslinked polymer
network.

(S)-(+)-N-(3,5-Dinitrobenzoyl)-α-methylbenzylamine as a Template.
Chiral resolution is a good target for the molecular imprinting technique. Many chiral
solutes, such as drugs, have been reported to be recognized or separated with
molecularly imprinted separation media (*16-18*). One of the templates which we
selected for this study is (S)-(+)-N-(3,5-dinitrobenzoyl)-α-methylbenzylamine, (S)-
DNB (Figure 2). Although underivatized α-methylbenzylamine could be used as the
template and should be separable by a molecular-imprinted stationary phase using
methacrylic acid as a host molecule, here the amide derived from α-
methylbenzylamine is employed to determine the structural contribution of the
crosslinked polymer network as discussed in the previous section using
diaminonaphthalene as the template.

As shown by the data in Table III, the non-imprinted base stationary phase
prepared with only ethylene dimethacrylate as the crosslinking agent (first entry) could
not resolve the enantiomers for three kinds of chiral amides. When (S)-DNB was
utilized as the template (second entry), molecular recognition of the template was
found to afford chiral resolution with longer retention times for the template
enantiomer. Although the k' values of the template and its antipode become larger by
73 and 34%, respectively, those of N-(4-nitrobenzoyl)-α-methylbenzylamine (NB),
which has a structure similar to that of the template, is increased by only 11%,
while that of N-benzoyl-α-methylbenzylamine (B) is only 7% larger without
resolution of the enantiomers in either case.

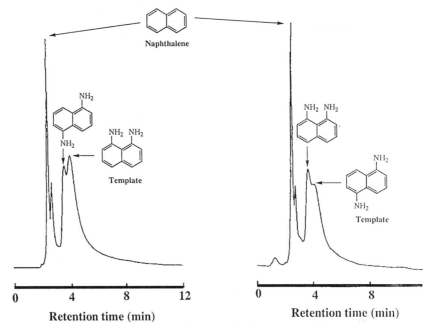

Figure 1. Chromatographic separation of diaminonaphthalene isomers on uniform-sized, molecular-imprinted stationary phases. (Chromatographic conditions: mobile phase, acetonitrile; flow rate, 1 mL/min, detection, UV at 254 nm.)

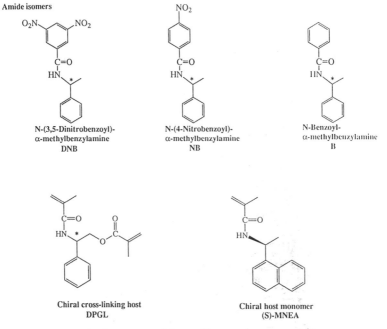

Figure 2. Structures of the amide compounds studied.

Table III. Separation of Chiral Amide Derivatives [a]

Template	Additive	(R)-DNB k'	(S)-DNB k'	α	(R)-NB k'	(S)-NB k'	α	(R)-B k'	(S)-B k'	α
None	None	1.73	1.73	1.01	1.63	1.63	1.02	1.21	1.21	1.00
(S)-DNB	None	2.32	3.00	1.29	1.82	1.82	1.02	1.30	1.30	1.01
(S)-DNB	(R)-DPGL	2.26	4.88	2.16	1.77	1.94	1.09	1.25	1.31	1.05
(S)-DNB	(±)-DPGL	2.55	3.69	1.44	1.86	1.91	1.02	1.31	1.31	1.00
(S)-DNB	(S)-DPGL	2.41	3.00	1.24	1.75	1.70	0.97	1.19	1.18	0.99

[a] Abbreviations refer to Figure 2.

As found in the previous section, the imprinted stationary phase prepared without methacrylic acid did afford larger k' values for the template molecule and its isomer compared with the non-imprinted base stationary phase. Here the imprinted stationary phase with the crosslinking agent ethylene dimethacrylate again affords chiral resolution with larger k' values for the template and the antipode in comparison with those for NB or B. These results suggest that the crosslinked polymer network can memorize specifically the shape of the chiral template.

Effect of a Chiral Crosslinking Agent as the Host Monomer. To enhance the specific molecular recognition within the imprinted polymer network, a chiral crosslinking agent, N,O-dimethacryloyl phenylglycinol (DPGL, Figure 2) was examined. The reason why we selected this chiral crosslinking agent is that it was easy to prepare and possesses amide functionality which may produce a favorable molecular interaction through amide-amide interactions with the template. In addition, as we reported in the previous paper dealing with uniform-sized, polymer-based chiral separation media with methacrylamide as the chiral selector (19), this chiral crosslinking agent did not induce any chiral resolution when the usual bulk copolymerization technique was utilized to prepare the chiral stationary phase with the crosslinking agent. Thus the chiral crosslinking agent DPGL does not function effectively as a chiral selector in normal chiral stationary phases prepared by the bulk copolymerization technique. Similar poor recognition ability was also reported previously for a related chiral crosslinking agent (20). Therefore, we investigated here (R)-DPGL and its antipode (S)-DPGL, as well as racemic DPGL, as co-crosslinking agents.

(R)-DPGL which has the same chirality as the template is found to enhance the specific molecular recognition and a larger α value of 2.16 was observed for the chiral template (third entry in Table III). Interestingly, the k' value for the solute used as the template becomes much larger than that for an imprinted stationary phase prepared without the chiral cross-linking agent, while the k' values for the antipode as well as another solutes are almost the same (compare the second and third entries in Table III). Although chiral resolution of DNB is markedly affected by the addition of (R)-DPGL, almost no chiral separation is found for the related solutes NB and B.

If chiral selectivity of the added chiral crosslinking agent is the dominant factor for enhancement of molecular recognition for the template molecule, addition of racemic DPGL should diminish the molecular recognition by the imprinted stationary phase. However, the imprinted stationary phase (fourth entry) shows intermediate molecular recognition between the stationary phases prepared with and without (R)-DPGL. In addition, the stationary phase prepared with (S)-DPGL (fifth entry), the antipode of (R)-DPGL, is found to afford quite similar molecular recognition to the imprinted stationary phase prepared without DPGL (second entry). This means that (S)-DPGL does not disturb the chiral recognition observed with the imprinted stationary phase prepared without any chiral crosslinking agents (DPGL).

The addition of racemic or (S)-DPGL does not appreciably change the k' values for the other solutes observed for the imprinted stationary phase with (R)-DPGL. A very small inversion of chiral recognition is observed with the stationary phase prepared with (S)-DPGL. This is probably due to the inverse chirality of the crosslinking agent. However as mentioned before, the chiral recognition ability of DPGL is not high *(18)* , so the inversion of chiral recognition is almost negligible.

These findings strongly suggest that favorable interaction between solutes and the chiral crosslinking agent incorporated into the imprinted site enhances specific molecular recognition toward the solute which was used as the template with specific enhancement of the retention to the template. These results also suggest that relatively weak intermolecular interaction can be enhanced within a specific recognition site where the template molecule fits well.

Another Chiral Host Monomer. Next we utilized another chiral host monomer, (S)-N-methacryloyl-1-naphthylethylamine [(S)-MNEA, Figure 2]. Although the host monomer in this case is not a crosslinking type, the naphthyl group may enhance $\pi-\pi$ interactions between the template with a dinitrophenyl group and the host. As described when DPGL was the host monomer, the chiral host monomer (S)-MNEA is also a poor chiral selector for the amide compounds employed in this study if it is utilized in a typical bulk co-polymerization method. However, as shown in Figure 3 (left), a larger α value for the template isomer was obtained with quite good molecular recognition ability among the amide compounds.

Interestingly, a contribution of the crosslinked polymer network is also found for this uniform-sized, molecular-imprinted stationary phase. When (R)-DNB was utilized as the template instead of (S)-DNB, (R)-DNB was found to be retained longer than (S)-DNB which was preferably retained when the chiral host monomer (S)-MNEA was used [Figure 3 (right)]. The α value observed in this case is 1.69, while a stationary phase prepared with (S)-DNB as the template and no host monomer gave an α value of 1.29 for (S)-DNB. These data are also comparable to those observed for a stationary phase prepared with DPGL as the chiral host monomer.

Effect of Preparation Method. Using (R)-DPGL in the same amount as the other monomer and porogen, we prepared molecular-imprinted polymers by traditional bulk polymerization followed by grinding and sieving to a 22 μm mesh. SEM pictures of the uniform-sized, molecular-imprinted stationary phase and that prepared by bulk polymerization are shown in Figure 4. Although size classification was performed for the latter, some fine particles are still evident.

Chromatograms for the separation of DNB isomers on both stationary phases are shown in Figure 5. Larger α values are noted for the stationary phase prepared by bulk polymerization. This finding can be explained based on relatively weak molecular interaction between the host monomer and the template (*i.e.* hydrogen bonding and/or $\pi-\pi$ interactions) which may be weaken further when the polymerization takes place in an aqueous medium. On the other hand, the peak shapes for the stationary phase prepared by bulk polymerization are quite poor, especially for

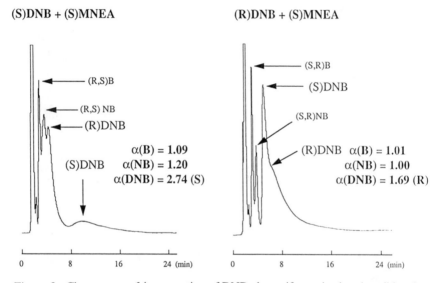

(S)DNB + (S)MNEA

(R,S)B

(R,S) NB

(R)DNB

(S)DNB

$\alpha(B) = 1.09$
$\alpha(NB) = 1.20$
$\alpha(DNB) = 2.74$ (S)

0   8   16   24 (min)

(R)DNB + (S)MNEA

(S,R)B

(S)DNB

(S,R)NB

(R)DNB $\alpha(B) = 1.01$
$\alpha(NB) = 1.00$
$\alpha(DNB) = 1.69$ (R)

0   8   16   24 (min)

Figure 3. Chromatographic separation of DNBs by uniform-sized and traditional molecular-imprinted stationary phases. (Chromatographic contitions: mobile phase, hexane-ethyl acetate, 1:1, v/v; flow rate, 1 mL/min; detection, UV at 254 nm.)

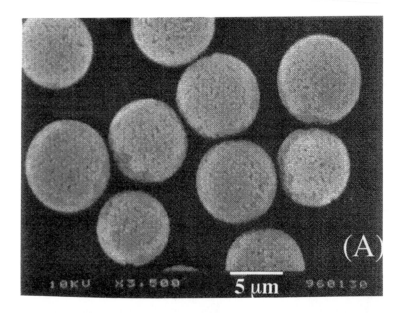

Figure 4. Scanning electron micrographs of the uniform-sized and traditional molecular-imprinted stationary phases. (A) Uniform-sized, molecular-imprinted stationary phases; (B) Molecular-imprinted polymer particles prepared by a traditional bulk polymerization techniques followed by grinding and sieving to 22 μm mesh.

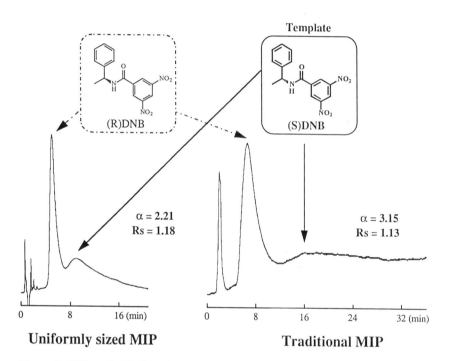

Figure 5. Effect of chirality of the template on the chromatogrpahic separation of DNB. (The chromatographic conditions were the same as those listed in the caption for Figure 3.)

the template. Thus, the Rs value is lower than that for the uniform-sized, molecular-imprinted stationary phase for which much better peak shape is obtained even for the template.

An *In Situ* Surface Modification Method. For the uniform-sized polymer beads, another advantage is the possibility of *in situ* surface modification. In this section, we explore the potential for combining the molecular imprinting technique with *in situ* surface modification using a chiral monomer. This combination has a potential for the preparation of a uniform-sized, molecular-imprinted stationary phase with an additional chiral selector, namely multi-chiral selectors.

We chose chiral molecule **1** and chiral monomer **2** for use in the molecular imprinting and *in situ* surface modification, respectively (Figure 6). The uniform-sized, molecular-imprinted stationary phase was prepared from a polystyrene seed polymer by the typical two-step swelling and polymerization method as reported in the Experimental Section (*11*) In this case, no host functional monomer was used to make the role of the additional chiral selector clear.

The chiral monomer **2** was added during the polymerization step for formation of the uniform-sized, molecular-imprinted polymer particles from **1**. This *in situ* surface modification technique is reported to afford a polymeric chiral surface functionality which displays good chiral recognition ability toward axis chirality (*19*). It should be noted that the preparative method is a one-pot procedure.

The resultant polymeric stationary phase was washed with methanol, tetrahydrofuran, and acetone to remove the template molecule and any other unbound impurities. Size uniformity of the polymer particles was excellent with a CV value of around 3%. The yield was quantitative.

For reference, three other stationary phases were also synthesized: a) the basic stationary phase prepared without either the template or the additional monomer; b) a stationary phase imprinted with **1** only; and c) a surface-modified stationary phase prepared by the *in situ* surface modification method with **2**.

We utilized the three kinds of solutes shown in Figure 7 for evaluation of the these stationary phases by HPLC. The chiral recognition abilities of the prepared stationary phases are for these solutes summarized in Table IV.

Table IV. Chiral Recognition Ability[a]

Entry	Template **1**	Monomer **2**	DNB (α)	B (α)	BN (α)
1	No	No	1.00	1.00	1.00
2	No	Yes	1.04	1.04	1.28
3	Yes	No	1.06	1.17	1.04
4	Yes	Yes	1.09	1.18	1.31

[a]Chromatographic condition: mobile phase, hexane-ethyl acetate (1:1, v/v).

The base stationary phase (entry 1) did not show any chiral recognition ability (α = 1.00) due to the absence of any chiral selectors. The surface-modified stationary phase (entry 2) exhibited a chiral recognition ability for binaphthol (BN) which has an axis chirality, but with relatively low chiral recognition abilities for the amide solutes DNB and B. The observed chiral recognition ability is comparable to the results reported in our previous work dealing with chiral stationary phases which utilized polymethacrylamide as a chiral selector (*19*).

The stationary phase which was molecular-imprinted with the template molecule (*S*)-B (entry 3) showed a relatively large separation factor for the enantiomers of B. Since no host functional monomer was used in the preparation of this stationary phase, the observed α value is not very high. However, this is clearly a molecular imprinting effect, since DNB, the solute with the larger retention, showed a smaller α value.

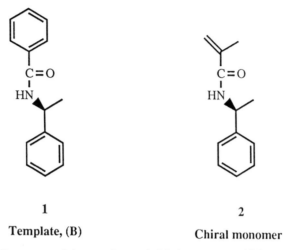

1

Template, (B)

2

Chiral monomer

Figure 6. Structures of the template and chiral monomer which were utilized for the *in situ* modification method.

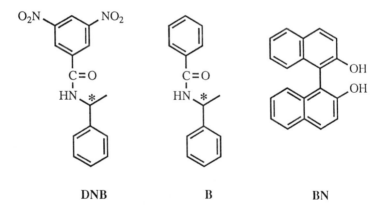

DNB   **B**   **BN**

Figure 7. Structures of solutes used to evaluate the multi-chiral selectors.

The stationary phase prepared from both of the template **1** and the additional monomer **2** (entry 4) showed mixed chiral recognition abilities. The α values for stationary phase 4 are thought to be the combination of those obtained for stationary phases **2** and **3**. These facts suggest that the molecular-imprinted sites and the additional polymeric chiral selectors function independently.

Pore Volume of the Prepared Stationary Phases. Size exclusion chromatography was utilized to determine the pore volumes of the prepared stationary phases (Table V). The total pore volume of the base stationary phase 1 is 0.95 mL/g; whereas the molecularly imprinted stationary phase 3 has a pore volume of 0.94 mL/g which is almost equal to that of the base stationary phase. On the other hand, the stationary phase prepared using the additional monomer **2** gave a pore volume of 0.79 mL/g. Clearly this is a reduced pore volume compared with that of the base stationary phase (21).

From a more detailed examination of the data it is seen that the reduction of pore volume for the surface-modified stationary phase 2 results from a reduced pore volume in the relatively large pore-size region (EVc), while the pore volume in the small pore-size region is altered for the molecularly imprinted stationary phase 3. These observations suggest that the additional chiral monomer forms bulky polymeric chiral selectors in the macropore region. On the other hand, the molecular imprinting technique affords imprinted sites in the micropore region that is formed by the highly crosslinked polymer network. This may be the reason why stationary phase 4 in Table IV shows the mixed chiral recognition abilities of stationary phases 2 and 3.

Table V. Pore Volume of the Prepared Stationary Phases

Entry	EVa/mL[a]	EVb/mL[b]	EVc/mL[c]	EVtot/mL[d]
1	0.13	0.17	0.65	0.95
2	0.13	0.19	0.47	0.79
3	0.20	0.18	0.56	0.94

[a]Elution volume (benzene) - elution volume (hexylbenzene).
[b]Elution volume (hexylbenzene) - elution volume (polystyrene MW = 760).
[c]Elution volume (polystyrene MW = 760) - elution volume (polystyrene MW = 20,000,000).
[d]EVtot = EVa + EVb = EVc.

Traditional molecular imprinting techniques involve bulk polymerization for which additional modification would be difficult. In contrast, the method reported here can afford uniform-sized stationary phases with not only molecular-imprinted sites but also an additional chromatographic selector. Although we are reporting the combination of a molecular-imprinted chiral selector and an additional polymeric chiral selector, which we term "multi-chiral selectors", other combinations are possible, such as a surface hydrophilic stationary phase with molecular-imprinted recognition sites (22). We have obtained preliminary results for such a modified, uniform-sized stationary phase which can be used for the direct analysis of serum samples that contain a target drug. These results will be reported in the near future.

Conclusions

Uniform-sized, molecular-imprinted stationary phases for HPLC have been prepared by a two-step swelling and polymerization method. The molecular recognition exhibited by these stationary phases is observed to be nearly equivalent to that obtained with traditional molecular-imprinted stationary phases prepared by bulk polymerization. However, the column efficiency is found much better on the uniform-sized, molecular-imprinted stationary phase. Therefore, the practical application of uniform-sized, molecular-imprinted stationary phases as HPLC stationary phases with specific molecular recognition ability is established.

However, this work still involves some speculative explanations since molecular-imprinted stationary phases prepared without a host monomer capable of strong interaction with the template molecule also afford moderate molecular recognition behavior which is attributed to memorization of the template shape by the crosslinked polymer network in some fashion. The host monomer plays an important role in affording specific recognition between isomers of the template molecule. However, other solutes with functionality similar to that of the template are also affected by the host functionality which results in non-specific enhancement of retention by the uniform-sized, molecular-imprinted stationary phase.

On the other hand, crosslinked, molecular-imprinted stationary phases prepared in the absence of a strong host monomer can enhance only the retention of a solute which was the template because of relatively weak intermolecular interactions. Thus specificity within solutes which having similar functionality is also obtained. Addition of an appropriate chiral crosslinking agent and a chiral host monomer contribute to the enhancement of this specific molecular recognition without a loss of solute specificity.

Moreover, an *in situ* surface modification method can afford modified, uniformed-sized, molecular-templated stationary phases very easily. In this study, multi-chiral selectors are obtained by a combination of molecular imprinting and *in situ* surface modification. Interestingly, both chiral recognition sites (selectors) appear to function independently.

Acknowledgment

This research was funded in part by the Japanese Ministry of Education (grant number 09640726).

Literature Cited

1. Sellergren, B.; Ekberg, B.; Mosbach, K. *J. Chromatogr.* **1985**, *347*, 1.
2. Shea, K. J.; Sasaki, D. Y. *J. Am. Chem. Soc.* **1991**, *113*, 4109.
3. Wulff, G. *Am. Chem. Soc. Symp. Series* **1986**, *308*, 186.
4. Wulff, G.; Sarhan, A. *Angew. Chem.* **1972**, *84*, 364.
5. Andersson, L. I.; Sellergren, B.; Mosbach, K. *Tetrahedron Lett.* **1984**, *25*, 5211.
6. Ramström, O.; Andersson, L. I.; Mosbach, K. *J. Org. Chem.* **1993**, *58*, 7562.
7. Svec, F.; Fréchet, J. M. J. *Anal. Chem.* **1992**, *64*, 820.
8. Wulff, G.; Sharhan, A.; Zabrocki, K. *Tetrahedron Lett.* **1973**, *14*, 4329.
9. Sellergren, B. *J. Chromatogr., A* **1994**, *673*, 133.
10. Hosoya, K; Fréchet, J. M. J. *J. Polym. Sci., Part A, Polym. Chem.* **1993**, *31*, 2129.
11. Smigol, V.; Svec, F.; Hosoya, K.; Wang, Q.; Fréchet, J. M. J. *Angew. Makromol. Chem.* **1992,** *195*, 151.
12. Ugelstad, J.; Kaggerud, K. H.; Hansen, F. K.; Berge, A. *Makromol. Chem.* **1979**, *180*, 737.
13. Hosoya, K.; Yoshizako, K.; Tanaka, N.; Kimata, K.; Araki, T.; Haginaka, J. *Chem. Lett.* **1994**, 1437.
14. Matsui, J.; Kato, T.; Takeuchi, T.; Suzuki, M.; Yokoyama, K.; Tamiya, E.; Karube, I. *Anal. Chem.* **1993**, *65*, 2223.
15. Sellergren, B.; Shea, K. J. *J. Chromatogr.* **1993**, *654*, 17.
16. Vlatakis, G.; Andersson, L. I.; Müller, R.; Mosbach, K. *Nature* **1993**, *361*, 645.

17. Fischer, L.; Müller, R.; Ekberg, B.; Mosbach, K. *J. Am. Chem. Soc.* **1991** *113*, 9358.
18. Kempe, M.; Mosbach, K. *J. Chromatogr. A* **1994**, *664*, 276.
19. Hosoya, K.; Yoshizako, K.; Tanaka, N.; Kimata, K.; Araki, T.; Fréchet, J. M. J. *J. Chromatogr. A* **1994**, *666*, 449.
20. Andersson, L.; Ekberg, B.; Mosbach, K. *Tetrahedron Lett.* **1985**, *26*, 3623.
21. Hosoya, K.; Maruya, S.; Kimata, K.; Kinoshita, K.; Araki, T.; Tanaka, N.; *J. Chromatogr.* **1992**, *625*, 121.
22. Hosoya, K; Kishii, Y; Kimata, K; Araki, T.; Tanaka, N.; Svec, F. Fréchet, J. M. J. *J. Chromatogr. A* **1995**, *690*, 21.

Chapter 11

Designing Metal Complexes in Porous Organic Hosts

John F. Krebs[1] and A. S. Borovik[2,3]

[1]Department of Chemistry, Kansas State University, Manhattan, KS 66506
[2]Department of Chemistry, University of Kansas, Lawrence, KS 66045

Copolymerization of metal complexes into organic hosts is an effective way of fabricating new materials as is demonstrated by the reversible binding of CO to immobilized Cu(I) complexes in porous methacrylate network polymers. The assembly of the molecular species prior to polymerization is advantageous because of the greater control of the structure and amount of metal complex incorporated into the polymer, the possibility of regulating their microenvironments and the use of these systems as mimics for metalloproteins. The functional properties of these materials can be modified by metal ion substitution as is illustrated by changes in CO binding characteristics of a network polymer that occurs by exchanging Cu(I) ions for Ag(I) ions in the immobilized metal ion sites.

The function of metal ions in biomolecules is controlled by two interrelated structural features: (1) the structure of the metal ion coordination sphere which includes the geometric relationship of metal-bound ligands and (2) the molecular architecture of the metal binding site that controls the secondary coordination sphere (or microenvironment) about the metal ion. While the role of component one is obvious in directing the activity of metalloproteins, the importance of component two cannot be overlooked. Microenvironments about the metal ion active sites, which are induced by the protein structure, regulate several properties including the hydrophobicity, polarity, electrostatics, solvation, and dielectric constant. In addition, the morphology of the metal active site in metalloproteins can govern the accessibility of substrates by the metal ions. Protein-created microenvironments thus have a significant role in controlling the reactivity of the metal ions.

The effects of the microenvironment on the function of metal ions in proteins is clearly illustrated by the diverse activity of heme-containing proteins (*1,2*). In hemoglobin and myoglobin, the steric constraints and hydrogen bonding capacity of the distal side of the heme pocket has a significant effect on oxygen binding properties of these proteins (*3*). In the oxygenases and peroxidases, the functions of enzymes are affected greatly by the various protein environments that house the catalytic iron heme moieties. For example, cytochrome P_{450} (a monooxygenase) and chloroperox-

[3]Corresponding author.

idase (which halogenates substrates) have identical heme active sites with axially bound thiolates, yet their functions are vastly different (*1*).

Protein structure also controls other necessary properties for metal ions to function in biomolecules. In most cases, the active sites are located within the interior of the proteins, isolated from each other to prevent undesirable interactions. In human hemoglobin for example (Figure 1), the four heme dioxygen binding sites are isolated from each other by the globin: the closest distance between heme sites is 25 Å (*4.*). This is imperative for reversible O_2 binding because if the heme sites were allowed physical contact, by either intra or intermolecular pathways, the four-electron auto-oxidation of O_2 would occur and lead to thermodynamically stable μ-oxo bridge iron species (Figure 2). In hemoglobin, like many metal ion containing proteins, access by external ligands to the metal sites are provided by channels that connect the active sites to the surface of the proteins. The channel structure, while providing a means of entry into the active sites, can also aid in orienting substrates as they approach the metal ion or assist in the selection of substrates.

In the last forty years, there has been great interest in developing synthetic systems that mimic the structural, physical, and functional properties of the metal ion sites found in proteins (*5,6*). One approach to examine the role of microenvironments in the functions of metal ions within proteins is to simulate various architectural features in low molecular weight systems (*6*). Design features found in proteins have been incorporated into organic ligand systems to help direct the chemistry at the metal centers in solution. The reversible binding of O_2 to synthetic iron porphyrin is one example where the exquisite design of organic ligands can dictate the reaction chemistry at a distant metal site (*3,7*). The picket-fence iron porphyrin of Collman was the first synthetic heme to reversibly bind O_2 in solution at room temperature by preventing the intermolecular iron oxygen interactions that lead to μ-oxo bridge iron species (*8*). A variety of other porphyrins and non-porphyrin ligands have since been designed containing cavity motifs that when metallated with iron are capable of forming $Fe-O_2$ adducts (*3,7,9,10*). In addition, other notable examples where ligand design has aided in mimicking biological function in synthetic systems include the specific recognition of metal ions (*11,12,13*), acceleration of the rates of chemical reactions (*14*) and artificial receptors that show strong and selective binding of organic substrates (*15,16,17,18*). In most cases, these molecular systems use a combination of morphological control of a binding cavity and weak bonding interactions to guide the recognition process.

Another approach simulating the site isolation properties of metalloproteins is to attach synthetic metal complexes onto the surface of solid supports. As in the low-molecular solution studies discussed above, there has been extensive work on developing reversible O_2 systems. Wang indicated in a 1958 report that imbedding the diethyl ester of heme in a hydrophobic matrix of polystyrene and 1-(2-phenylethyl)-imidazole permits the Fe(II) sites of the heme to reversibly bind O_2 (*19,20*). Collman and Reed showed in 1973 that crosslinked polystyrene containing attached imidazole ligands can coordinate Fe(II) tetraphenylporphyrin ($Fe^{II}TPP$) (*21*). This matrix was found to be too flexible (or the sites were not sufficiently dispersed throughout the matrix) to prevent the formation of $[Fe^{III}TPP]_2O$. In a related system, Basolo and coworkers attached $Fe^{II}TPP$ to a rigid silica gel support that was modified with 3-imidazolylpropyl groups (*22*). Reversible O_2 binding to the Fe sites was observed but the binding was weak. At -127°C the binding is irreversible and a $P_{1/2}(O_2)$ of 230 torr was measured at 0°C. For comparison, the estimated $P_{1/2}(O_2)$ for hemoglobin at 0°C is 0.14 torr.

Other types of matrices have been used to immobilize metal complexes for the purpose of reversible binding of O_2. These include the encapsulation of iron porphyrins in dendrimer cages (*23,24*) and membranes (*25*), and the immobilization of cobalt Schiff base complexes in zeolite cages (*26,27,28*). The dendrimer porphyrins show great promise in their ability to stabilize $Fe-O_2$ adducts as illustrated by the recent work of Aida and Collman and Diederich. The zeolite systems use a "ship in the bottle" protocol to assemble the Co(II) Schiff base complexes inside zeolite cages.

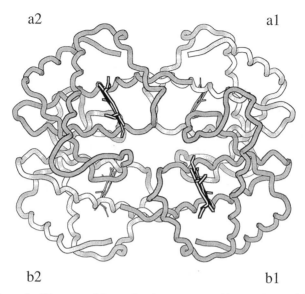

Figure 1. Diagram of the molecular structure of human hemoglobin.

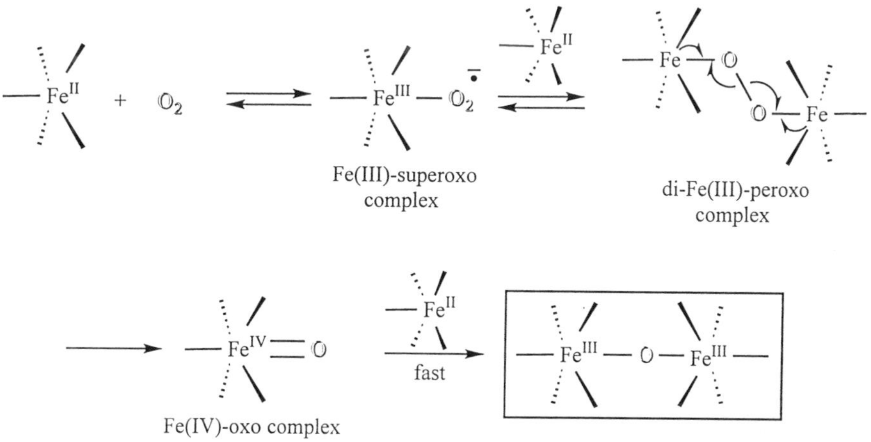

Figure 2. The auto-oxidation of O_2 by five-coordinate Fe(II) complexes.

The limitations of this technique are the difficulty in matching the size and shape of the metal complex to that of the preformed cage and the inability of external gases to access the metallated sites within the interior of the zeolite. Reversible binding has been observed in these systems; however, the amount of Co sites involved in the binding is low (< 25%) (25).

Design Considerations

As discussed above, despite some promising results, many of these systems do not sufficiently isolate the metal sites from each other or utilize only a small percentage of their immobilized metal sites in the binding of dioxygen. We have been investigating template copolymerization techniques to immobilize metal complexes within porous organic hosts (Figure 3). The copolymerization techniques we have employed, as outlined in Figure 4, are modeled after those used to make molecular imprinted polymers (29,30,31). The protocol utilizes template complexes that are synthesized prior to polymerization and copolymerized with a large excess of an organic crosslinker in the presence of a porogenic agent. The assembly of the template prior to polymerization is advantageous because of the greater control of the structure and amount of species incorporated into the porous organic host, in addition to providing control of the microenvironments about immobilized metal complexes. Thus the porous organic matrices in our systems are designed to have the same function as the protein structure of metalloproteins, that is, they can isolate metal sites while allowing external reagents to access the metals *via* its porous network structure. Moreover, materials fabricated in this manner should be able to function under conditions (e.g., high temperatures and pressure) where most biomolecules are unstable and are not active.

In the design of these polymers, the metal complexes immobilized in the porous organic host can be utilized in two general ways (Figure 4). In one case, the template complexes can serve as reaction sites in the porous host. For this to happen, the template complexes need to be coordinatively unsaturated or possess non-polymerizable ligands. A second possibility is that the metal ions used to form the monomeric template can be removed from the complex after copolymerization. This would position the endogenous ligands (*i.e.*, those covalently attached to the organic host) in a fixed-spatial arrangement within the metal ion binding sites which can then act as new chelating ligands to rebind different metal ions than initially used in synthesizing the polymers. Therefore, rebinding a different metal ion in the microcavity may yield coordination chemistry not observed for a previous metal ion.

Reversible CO Binding to an Immobilized Cu(I) Complex (32)

Our objective in this study is to verify that (1) the copolymerization process can be used to synthesize metal ion sites in a polymer matrix that can reversibly bind low-molecular weight molecules, and (2) the removal of the metal ion leaves the site with ligands arranged to rebind other metal ions not present during the polymerization process. Thus, the chemical and physical properties of the metal sites, and those of the polymer, can be changed through metal ion substitution.

The choice of a Cu(I) system is based on its ability to reversibly bind CO in solution (33) (equation 1). We reasoned that a similar reaction could occur within a network polymer, if the CO-free form has a coordinatively unsaturated three-coordinate Cu(I) center. (The reversible CO binding to Cu(I) complexes attached to

$$-Cu^+- \quad \xrightarrow[\text{-CO, 298 K}]{\text{+CO, 298 K}} \quad \text{Cu}^+ \quad (1)$$

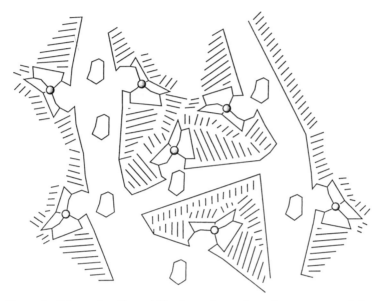

Figure 3. Schematic of immobilized metal complexes in porous organic hosts.

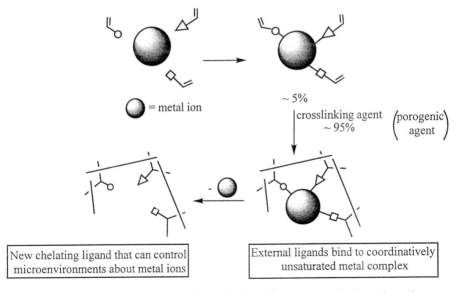

~ 5%
crosslinking agent ⎛porogenic⎞
~ 95% ⎝ agent ⎠

New chelating ligand that can control
microenvironments about metal ions

External ligands bind to coordinatively
unsaturated metal complex

Figure 4. Schematic illustrating the method used for copolymerization of metal
complexes in porous organic hosts.

the surface of a solid support has been reported by Drago (*34*).) The ligand used in our metallo-monomer is derived from bis(2-pyridylethyl)amine (*35*) through addition of a styrene group by treating the amine with 4-vinylbenzyl chloride to afford **1** (50%). Allowing **1** to react with [Cu(CH$_3$CN)$_4$]PF$_6$ in acetonitrile produced Cu**1**PF$_6$ in 85% yield (*36*).

The Cu**1** network polymer (**P-1Cu**) was synthesized using the protocol outlined in Figure 5 to afford insoluble network polymer **P-1Cu** (average surface area of 100 m^2/g). The solid-state binding of CO to **P-1Cu** was monitored using diffuse reflectance infrared Fourier transform (DRIFT) spectroscopy. When **P-1Cu** (20 mass % in KBr) is exposed to CO, a new band at v = 2085 cm^{-1} appears in the DRIFT spectrum within 5 minutes (Figure 6b). This band is indicative of carbon monoxide coordination to a Cu(I) center (*37*). Reversion of **P-1Cu(CO)** to **P-1Cu** is accomplished by applying a vacuum or by flushing the polymer with a stream of N$_2$ for 15 min; this conversion is nearly quantitative with > 85% of **P-1Cu** recovered (Figure 6c). Solid-state rebinding of CO to **P-1Cu** is also essentially quantitative, with > 95% of the original signal for **P-1Cu(CO)** obtained after re-exposure of the polymer to CO (Figure 6d). This process has been repeated for ten cycles without any measurable loss in CO binding.

We have also synthesized Ag**1** and its corresponding network polymer **P-1Ag**. This silver(I) polymer serves as a structural but not functional mimic for the Cu(I) sites in **P-1Cu**. The polymeric metallo-sites formed by Ag**1** and Cu**1** complexes should have analogous architecture since Ag(I) and Cu(I) complexes with nitrogenous ligands have similar structures (*38*). However, CO binding to Ag(I) complexes is rare (*39*), and is unlikely to occur in the coordination environment provided by **1**. In fact, when **P-1Ag** is exposed to CO there is no evidence in the DRIFT spectrum for CO binding to the Ag(I) immobilized sites, even when **P-1Ag** is exposed to 1 atm of CO for 30 min (Figure 8).

P-1Ag can be converted into a polymer which binds CO by replacing Ag(I) with Cu(I) ions in the metal binding sites. Silver ions were removed from the polymer by treating **P-1Ag** with an aqueous solution of EDTA, shaking for 6 days, washing with acetonitrile, and drying under vacuum (Figure 7). This process removes ~66% of the Ag(I) from the polymer producing an apopolymer (**P-1**) that has binding sites (the immobilized tridentate ligand **1**) available to chelate metal ions. The reconstituted Cu(I) polymer **P-1[Ag → Cu]** is made by allowing **P-1** to react with an acetonitrile solution of [Cu(CH$_3$CN)$_4$]$^+$ under dinitrogen. Repeated washing of **P-1[Ag → Cu]** with acetonitrile (to remove unbound Cu(I) ions) yielded a polymer containing copper in 52% of the available sites (not optimized). When **P-1[Ag → Cu]** is exposed to CO (1 atmosphere, 15 minutes) a peak at v = 2085 cm^{-1} is observed in the DRIFT spectrum (Figure 8), showing that this polymer is now competent to bind CO at its Cu(I) sites. Moreover, this CO binding process is again reversible, making **P-1[Ag → Cu]** functionally similar to **P-1Cu**, the original network polymer containing Cu(I) immobilized sites.

The luminescent properties of **P-1Cu** are also affected by binding of CO to the immobilized copper(I) sites. Figure 9 shows data on the quenching of the emission signal from **P-1Cu** when CO is introduced to the polymer suspended in 2-toluene at room temperature. The emission at 530 nm for **P-1Cu** is from a MLCT excited state (Cu → py) (*40*); when CO is bound the excitation band associated with this excited state is shifted to higher energy which is no longer assessible at λ_{ex} = 345 nm (*41*). These results suggest that it is possible to use the photophysical properties of the Cu(I) centers as an additional mode for monitoring the binding of CO to the immobilized sites.

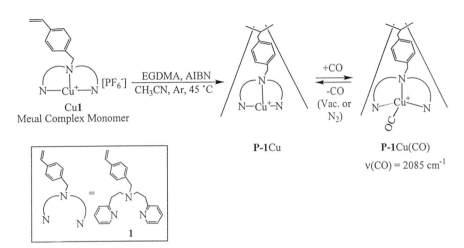

Figure 5. Synthetic scheme outlining the synthesis of the Cu(I) immobilized complexes.

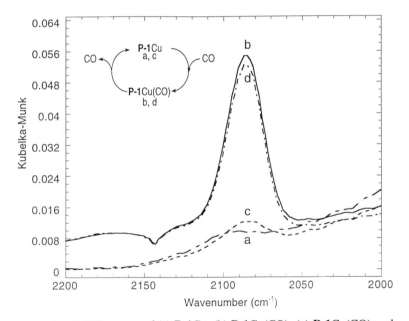

Figure 6. DRIFT spectra of (a) **P-1Cu**; (b) **P-1Cu(CO)**; (c) **P-1Cu(CO)** under vacuum (10 min, 293K); (d) Rebinding of CO to the polymer formed in spectrum c. The carbonyl polymers were formed by treating the **P-1Cu** with 5 mL of CO. Spectra of **P-1Cu(CO)** were recorded after a 15 s vacuum was applied to remove unbonded CO.

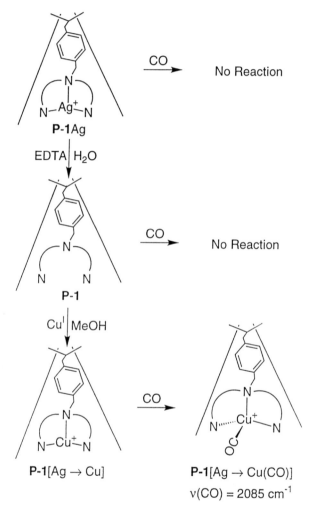

Figure 7. Scheme for the synthesis of **P-1**Ag and **P-1**[Ag → Cu(CO)]

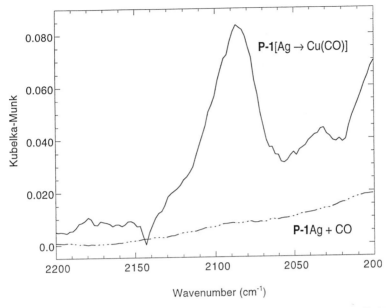

Figure 8. DRIFT spectra of **P-1**Ag under 1 atm of CO (— ·· —) and **P-1**[Ag → Cu(CO)] (—).

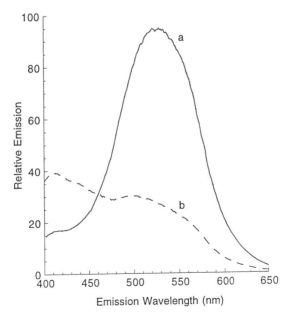

Figure 9. Emission spectra of (a) **P-1**Cu and (b) after treating **P-1**Cu with CO. The polymer was suspended in toluene during data collection.

Summary

These results demonstrate the reversible solid-state binding of CO to Cu(I) sites in a network polymer. They contrast with those results obtained recently for another solid state metal-based system which irreversibly binds CO (*42*). The copolymerization technique that we have used allows for the convenient incorporation of metal complexes of known structure into a polymeric matrix and low-molecular weight external molecules can reversibly access the immobilized metal center. This is important because reversible binding is a prerequisite for using these polymeric systems as biomimics. Moreover, metal substitution within templated sites shows that templating with one metal ion and reconstituting with a different metal can change the chemical properties of the immobilized sites.

These findings are important first steps in using this technique of copolymerizing molecular assemblies as a way of modeling metalloprotein function. Clearly, additional systems need to be developed that can further test whether the organic host in these synthetic systems can mimic properties of protein structure. For example, we have recently examined dioxygen binding to immobilized Co(II) complexes to ascertain the site isolation capabilities of the hosts (Krebs and Borovik, submitted for publication). Significant stabilization of Co-O_2 adducts at room temperature are obtained for the immobilized sites (50-80%). By comparison, the monomeric metal template complexes show < 10 % of the Co-O_2 adducts in solution because of intermolecular interactions that occur between the unhindered metal complexes. The organic host, with its ability to prevent these undesirable metal-metal interactions, allows for new chemistry to be uncovered that does not occur in solution.

Acknowledgment is made to the donors of Petroleum Research Fund administered by the American Chemical Society (26743-G3), NSF (OSR-9255223) and ONR (N00014-96-1-1216).

Literature Cited.

1. Dawson, J. H. *Science* **1988**, *240*, 433.
2. Ortiz de Montellano, P.R. *Acc. Chem. Res.* **1987**, *20*, 289.
3. Suslick, K.S.; Reinert, T.J. *J. Chem. Educ.* **1985**, *62*, 974.
4. Perutz, M.F.; Fermi, G.; Luisi, B.; Shaanan, B.; Liddington, R.C. *Acc. Chem. Res.* **1987**, *20*, 309.
5. Ibers, J. A.; Holm, R. H. *Science* **1980**, *209*, 223.
6. Karlin, K. D. *Science* **1993**, *261*, 701.
7. Momenteau, M.; Reed, C. A. *Chem. Rev.* **1994**, *94*, 659.
8. Collman, J. P. *Acc. Chem. Res.* **1977**, *10*, 265.
9. Jones, R. D.; Summerville, D. A.; Basolo, F. *Chem. Rev.* **1979**, *79*, 139.
10. Busch, D. H.; Alcock, N. W. *Chem. Rev.* **1994**, *94*, 585.
11. Cram, D. J. *Angew. Chem., Int. Ed. Engl.* **1988**, *27*, 1009.
12. Lehn, J. M. *Angew. Chem., Int. Ed. Engl.* **1988**, *27*, 89.
13. Pedersen, C. J. *Angew. Chem., Int. Ed. Engl.* **1988**, *27*, 1021.
14. Breslow, R. *Science* **1982**, *218*, 532.
15. Hamilton, A. *J. Chem. Educ.* **1990**, *67*, 821.
16. Diederich, F. *J. Chem. Educ.* **1990**, *67*, 813.
17. Tjivikua, T.; Ballester, P.; Rebek, J., Jr. *J. Am. Chem. Soc.* **1990**, *112*, 1250.
18. Webb, T.H.; Hongsuk, S.; Wilcox, C.S. *J. Am. Chem. Soc.* **1991**, *113*, 8554.
19. Wang, J. H. *J. Am. Chem.Soc.* **1958**, *80*, 3168.
20. Wang, J. H. *Acc. Chem. Res.* **1970**, *3*, 90.
21. Collman, J. P.; Reed, C. A. *J. Am. Chem. Soc.* **1973**, *95*, 2048.

22. Leal, O.; Anderson, D. L.; Bowman, R. G.; Basolo, F.; Burwell, R. L., Jr. *J. Am. Chem. Soc.* **1975**, *97*, 5152.
23. Aida, T.; Jiang, D.-L. *J. Chem. Soc., Chem. Commun.* **1996**, *1*, 1523.
24. Collman, J. P; Fu, L.; Zingg, A.; Diederich, F. *J. Chem. Soc., Chem. Commun.* **1997**, *2*, 193.
25. Tsuchida, E.; Komatsu, K.; Arai, K.; Nishide, H. *J. Chem. Soc., Dalton Trans.* **1993**, 2465.
26. Howe, R. F.; Lunsford, J. H. *J. Phys. Chem.* **1975**, *75*, 1836.
27. Herron, N *Inorg. Chem.* **1986**, 4714.
28. De Vos, D. E.; Feijen, E. J. P.; Schoonheydt, R. A.; Jacobs, P. A. *J. Am. Chem. Soc.* **1994** *116*, 4746.
29. Wulff, G., *Angew. Chem., Int. Ed. Engl.* **1995**, *34*, 1812 and references therein.
30. Shea, K. J. *Trends Poly. Sci.* **1994**, *A32*, 166.
31. Dhal, P. K.; Arnold, F. H. *Macromolecules* **1992**, *25*, 7051.
32. Krebs, J. F.; Borovik, A. S. *J. Am. Chem. Soc.* **1995**, *117*, 10593
33. Sorrell, T. N.; Jameson, D. L. *J. Am. Chem. Soc.* **1982**, *104*, 2053.
34. Balkus, Jr., K. J.; Kortz, A.; Drago, R. S. *Inorg. Chem.* **1988**, *27*, 2955.
35. Brady, L. E.; Freifelder, M.; Stone, G. R. *J. Org. Chem.* **1961**, *26*, 4757.
36. Blackburn, N. J.; Karlin, K. D.; Concannon, M.; Hayes, J. C.; Gultneh, Y.; Zubieta, J. *J. Chem. Soc., Chem. Commun.* **1984**, 939.
37. Villacorta, G. M.; Lippard, S. J. *Inorg. Chem.* **1987**, *26*, 3672.
38. F. A. Cotton; Wilkinson, G. Advanced Inorganic Chemistry; 5th Edition; Wiley: New York, NY, 1988.
39. Hurlburt, P. K.; Anderson, O. P.; Strauss, S. H. *J. Am. Chem. Soc.* **1991**, *113*, 6277.
40. McMillan, D. R.; Kirchhoff, J. R.; Goodwin, K. V. *Coord. Chem. Rev.* **1985**, *64*, 83.
41. Sorrell, T. N.; Borovik, A. S. *Inorg. Chem.* **1987**, *26*, 1957.
42. Mirkin, C. A.; Wrighton, M. S. *J. Am. Chem. Soc.* **1990**, *112*, 8596.

Chapter 12

Molecularly Imprinted Polymeric Membranes for Optical Resolution

Masakazu Yoshikawa

Department of Polymer Science and Engineering, Kyoto Institute of Technology, Matsugasaki, Kyoto 606-0962, Japan

An alternative molecular imprinting technique is used to obtain polymeric membranes for enantioselective separations. In this technique, the "molecular memory" of the print molecule is introduced into the membrane at the same time that the polymeric membrane is being cast from its polymer solution. We have found that a membrane which contains an oligopeptide residue from an L-amino acid and is imprinted by a L-amino acid derivative recognizes the L-isomer in preference to the corresponding D-isomer, and *vice versa*. By application of electrodialysis, the amino acid isomer which is preferentially adsorbed in the membrane may be transported across the membrane.

Developing novel membrane materials for optical resolution is an important objective with applications in the pharmaceutical industry, food preparation, agricultural chemicals, and so forth. Conventionally optical resolution has been performed by fractional crystallization, microbiological methods, kinetic enzymatic resolution technology, and high-performance liquid chromatography. Optical resolution with a permselective membranes is an attractive technique for separating optically active compounds since it offers continuous operation, as well as simplicity and energy efficiency compared with the conventional optical resolution methods mentioned above.

Optical resolution with synthetic membranes was first studied in liquid membranes containing chiral crown ethers (1-4). Due to a lack of durability for liquid membrane systems, attention shifted to the use of polymeric membranes for optical resolution. The pioneering studies of polymeric membrane materials for performing optical resolution include: plasma-polymerized membranes of *d*-camphor (5), *l*-menthol (5,6), and terpenes (7); polymeric membranes having cyclodextrin moieties (8); polymeric chiral crown ethers (9,10); poly(amino acid) membranes having amphiphilic side chains (11,12); enantioselective ultrafiltration membranes bearing amino acid condensate (13), (+)-poly{1-[dimethyl(10-pinanyl)silyl]-1-propyne} membranes (14); polysulfone membranes with immobilized bovine serum albumin (15,16); sericine membranes (17); cellulose tris(3,5-dimethylphenylcarbamate) membranes (18); poly{γ-[3-(pentamethyl-disiloxyl)-propyl]-L-glutamate) membranes (19); poly(methyl methacrylate) membranes containing (-)-oligo{methyl(10-pinanyl)siloxane} (20); and (+)-poly{1-[dimethyl](10-pinanyl)silyl]-1-propyne} membranes (21).

Optically active compounds, such as amino acids, are mirror image isomers which

170

have identical physicochemical properties, except for optical activity. Therefore, physical stereoselectivity might be an important factor for the recognition and separation of optically active compounds. All of the synthetic membranes for optical resolution listed above possess such chiral microenvironments (1-21). Oligopeptides are promising candidates for producing a chiral microenvironment in a synthetic membrane. To this end, the author's research group has prepared polymeric membranes which bear various oligopeptide residues as the molecular recognition sites and investigated their capacity for the optical resolution of racemic α-amino acids.

Strategy of Membrane Design.

Crown ethers (22-24), cyclophanes (25), cyclodextrins (26), calixarenes (27-29), molecular clefts (30-32), and ampiphilic complexes (33-35) have been investigated intensively in connection with molecular recognition. From the reported results for polymeric membranes (1-21), it can be deduced that the introduction of a chiral microenvironment at the molecular recognition site in the synthetic membrane is indispensable for the membrane to show optical resolution. However the preparation of suitable crown ethers, cyclophanes, etc. for incorporation into synthetic membranes may be synthetically challenging. Molecular imprinting (36-41) is another promising method for the introduction of chiral recognition sites into synthetic membranes. This approach has been applied to the transport of nucleic acid components (42,43). In these studies, the polymeric membranes were prepared by radical polymerization in the presence of the print molecule.

For our studies, oligopeptides, which are expected to produce a chiral micro-environment in polymeric membranes, were selected as the recognition site for optical resolution. The molecular imprinting technique described in References 42 and 43 is not applicable in the present case. This led us to employ the alternative molecular imprinting technique (44) which is depicted in Figure 1. In this process, the "molecular memory" of the substrate to be recognized or separated is introduced at the same time that the membrane is cast from the polymer solution. Step 1 shows specific interaction of the print molecule with the oligopeptide residue before and during the membrane preparation process to incorporate molecular memory into the membrane. In Step 2, the print molecule is extracted from the membrane to complete the preparation of the molecularly imprinted polymeric membrane (Step 3). When a racemic mixture of the print molecule or of print molecule analogues is contacted with the imprinted polymeric membrane (Step 4), the recognition site preferentially interacts with the print enantiomer which was utilized in forming the membrane or a print enantiomer analogue (Step 5).

In the present study, the objective is to prepare molecularly imprinted polymeric membranes for optical resolution of α-amino acids. In this case, the separation will be conducted in an aqueous solution or similar polar environment. To prevent structural deformation of the recognition sites and to retain the "molecular memory" of the membrane, the hydrophobic protecting groups of side chain groups used in the peptide synthesis are retained in the present study.

Adsorption Selectivity of Molecularly Imprinted Polymeric Membranes.

The structures of various membrane materials with pendant oligopeptide units are shown in Figure 2. These membranes were prepared using Merrifield's technique of solid phase peptide synthesis (45). Since these polymers do not form suitable membranes by themselves, an acrylonitrile-styrene copolymer (AS) was adopted as the membrane matrix (46). This article will focus upon results obtained with molecularly imprinted polymeric membranes obtained from a membrane material (DIDE-Resin) which bears a tetrapeptide residue, H-Asp(OcHex)-Leu-Asp(OcHex)-Glu(OBzl)-CH$_2$-. This is the top structure in Figure 2.

172

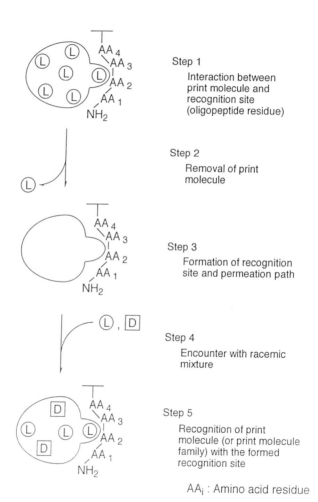

Step 1

Interaction between print molecule and recognition site (oligopeptide residue)

Step 2

Removal of print molecule

Step 3

Formation of recognition site and permeation path

Step 4

Encounter with racemic mixture

Step 5

Recognition of print molecule (or print molecule family) with the formed recognition site

AA_i : Amino acid residue

Figure 1. The Concept of an Alternative Method for Molecular Imprinting.

Asp-Ile-Asp-Glu-Resin
(DIDE-Resin)

Asp-Leu-Asp-Glu-Resin
(DLDE-Resin (L))

Phe-Phe-Phe-Resin
(FFF-Resin)

Glu-Phe-Phe-Resin
(EFF-Resin)

Glu-Glu-Glu-Resin
(EEE-Resin)

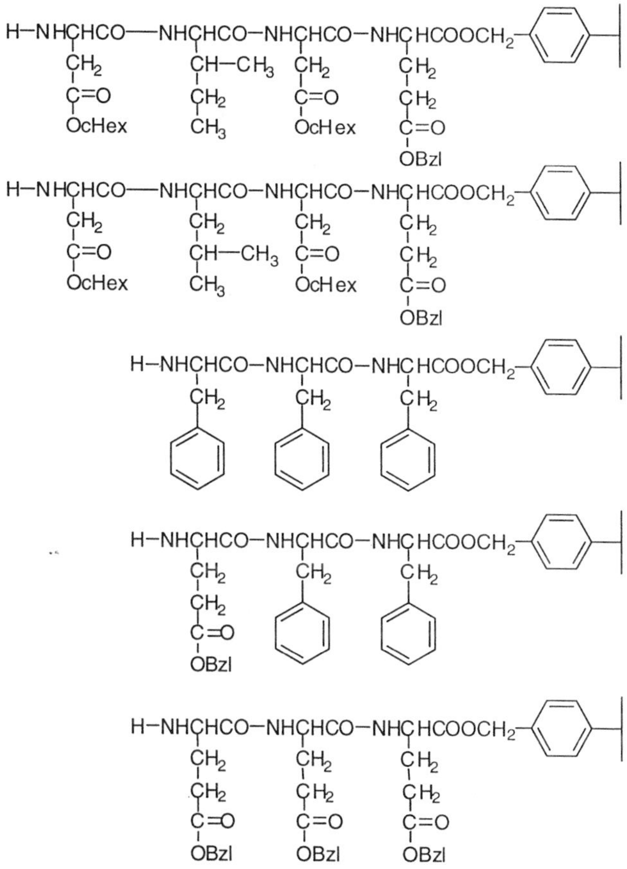

Figure 2. Structures of Various Candidates for Chiral Recognition Sites.

Figure 3 shows the effect of the membrane preparation (molecular imprinting) condition on the adsorption of racemic N-α-acetyltryptophans (Ac-Trp) (47). The mole ratio of the print molecule, Boc-L-Trp, to the tetrapeptide units was varied in the membrane preparation process: 5, 3, 1, and 0.5. The resultant imprinted polymeric membranes were then equilibrated with racemic Ac-Trp solutions. In Figure 3a, the amount of each isomer adsorbed by the membrane is plotted as a function of imprinting condition. Thus the ratio of the amount of adsorbed isomer relative to the amount of DIDE derivative in the membrane is shown in Figure 3a. The amount of each isomer adsorbed in the membrane increased linearly with the increase in the molar ratio of print molecule to DIDE units. Even though this molecular imprinting condition was varied, N-α-acetyl-L-tryptophan (Ac-L-Trp) was always adsorbed in the membrane in preference to D-isomer. The excess amount of Ac-L-Trp preferentially adsorbed by the membrane was 0.8 times that of the DIDE derivative contained in the membrane. Below a mole ratio of one for the print molecule to the tetrapeptide units, most of Ac-Trp adsorbed was the L-isomer.

The adsorption selectivity, $S_{A(L/D)}$, toward Ac-L-Trp (48) is plotted against the imprinting condition in Figure 3b. The adsorption selectivity is enhanced with a decrease in the (Boc-L-Trp)/(DIDE) ratio and reaches a value of 6 for a mole ratio of 0.5. From these results, it can be concluded that the presence of Boc-L-Trp in the membrane preparation process affects not only the formation of non-selective cavities, but also for that of molecular (chiral) recognition sites. A tentative structure for the interior of the molecularly imprinted polymeric membrane is shown in Figure 4.

Adsorption Isotherms of Ac-D-Trp and Ac-L-Trp for Boc-L-Trp Imprinted Polymeric Membranes.

It is of interest to probe the substrate specificity of the chiral recognition site which is formed by the print molecule during the membrane preparation process. To this end, the adsorption isotherms of Ac-D-Trp and Ac-L-Trp were determined. Thus, the imprinted polymeric membrane which was prepared from the DIDE-Resin and AS and imprinted with Boc-L-Trp was contacted with solutions of pure Ac-D-Trp or pure Ac-L-Trp and allowed to equilibrate. Figure 5 shows both adsorption isotherms (49).

For the adsorption isotherm of the D-isomer (Ac-D-Trp), the relationship is a straight line passing through the origin. This implies that the D-isomer is adsorbed in the membrane without any specific interaction with the membrane. On the other hand, the adsorption isotherm of Ac-L-Trp is more complicated. Above a concentration of 1 x 10^{-3} M, the relationship is a straight line which is parallel to that of D-isomer. When extrapolated to the Ac-L-Trp concentration of zero, this straight line has a positive intercept and does not pass through the origin. The adsorption isotherm of the L-isomer can be explained as follows. The isotherm is a combination of adsorption by non-specific adsorption, C_D, and by adsorption on the specific sites, C_C, which are produced by the presence of the print molecule during the membrane preparation process (Equation 1),

$$\text{Amount of Ac-L-Trp adsorbed} = C_D + C_C$$
$$= k_D[\text{Ac-L-Trp}] + \frac{(nK_C[\text{DIDE}][\text{Ac-LTrp}])}{1 + K_C[\text{Ac-L-Trp}]} \quad (1)$$

where k_D is the Henry's law dissolution constant, n is the ratio of the maximum amount of Ac-L-Trp adsorbed in the membrane to the amount of the recognition site DIDE derivative in the membrane, K_C is the affinity constant, [DIDE] is the concentration of the DIDE derivative in the membrane, and [Ac-L-Trp] is the Ac-L-Trp concentration. The adsorption isotherm of Ac-L-Trp supports the proposal that the membrane contains the membrane structure shown in Figure 4. Also it can be concluded that the L-isomer

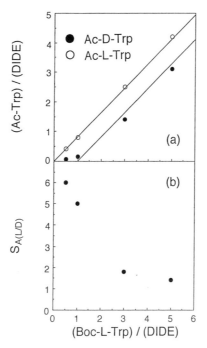

Figure 3. Effect of the Membrane Preparation Condition on a) Ac-Trp Adsorption and b) Adsorption Selectivity Toward Ac-L-Trp.

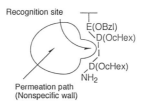

Figure 4. Proposed Structure of the Molecularly Imprinted Polymeric Membrane.

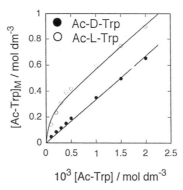

Figure 5. Effect of the Substrate Concentration on Adsorption of the Amino Acid. (The mole ratio of print molecule, Boc-L-Trp, to the DIDE derivative in the membrane preparation process was fixed at 3.0; $k_D = 3.4 \times 10^2$; $n = 1.1$; $K_C = 9.5 \times 10^3$ M; [DIDE] = 0.25 M)

was exclusively incorporated in the recognition site which was formed by the print molecule and the D-isomer was not adsorbed on the recognition site.

It is also interesting to probe the mode of interaction between the amino acid and the DIDE derivative in the membrane. Adsorption of Trp, whose α-amino group is not protected by an acetyl moiety, was studied and the results are summarized in Table I together with those for Ac-Trp (50). It can be seen that the adsorption behavior of Trp is very similar to that of Ac-Trp. This leads to the conclusion that recognition of the L-isomer is attained by an interaction between the carboxyl moiety of the amino acid and an amino group in the tetrapeptide residue and the absolute configuration of the side chain (indole moiety) in Trp.

Table I. Adsorption of Racemic Tryptophan and N-Acetyl Tryptophan by Molecularly Imprinted Membranes

(BOC-L-Trp) (DIDE)	Amino acid (AA) enantiomer	$(AA)_{mem}$ x 10^6, mol	$\frac{(AA)_{mem}}{(DIDE)}$	$S_{A(L/D)}$
For Trp				
5	D	1.12 ± 0.03	3.20	
	L	1.54 ± 0.03	4.40	1.4
3	D	0.53 ± 0.03	1.50	
	L	0.91 ± 0.04	2.60	1.7
1	D	0.07 ± 0.02	0.18	
	L	0.32 ± 0.03	0.90	5.0
0.5	D	0.03 ± 0.02	0.10	
	L	0.18 ± 0.03	0.60	6.0
For Ac-Trp				
5	D	1.09 ± 0.03	3.10	
	L	1.47 ± 0.03	4.20	1.4
3	D	0.49 ± 0.03	1.40	
	L	0.88 ± 0.05	2.50	1.8
1	D	0.06 ± 0.03	0.16	
	L	0.28 ± 0.04	0.80	5.0
0.5	D	0.02 ± 0.02	0.07	
	L	0.12 ± 0.03	0.42	6.0

Enantioselective Adsorption of Other Amino Acids.

The enantioselective adsorption of racemic Ac-Trp and Trp was described in the previous section. Table II shows the adsorption selectivity of racemic amino acids with nonpolar side chains, such as Phe and Ala, together with that for Trp (50). The adsorption selectivity of racemic amino acids with charged polar side chains is summarized in Table III (51). From the data presented in these two tables, it is clear that the molecularly imprinted polymeric membranes which were prepared from DIDE-Resin and **AS** and imprinted by Boc-L-Trp, recognized not only the print molecule family Trp or Ac-Trp, but also other racemic α-amino acids. At this time it is difficult to relate comprehensively the adsorption behavior to the structural variables, such as the net charges on the amino acids under the experimental conditions (Table IV), or the dimensions (52) and hydrophobicity (or hydrophilicity) of the side groups of these amino acids (53-56). The data in Table II also reveal another important fact. That is, the recognition site toward the L-amino acid was produced by Boc-L-Trp as the print molecule using DIDE-Resin which contained oligopeptides from L-amino acid as the membrane material. Thus it can be expected that a membrane which bears an oligopeptide

Table II. Adsorption of Racemic Amino Acids by Molecularly Imprinted Membranes[a]

Print molecule	Substrate	$(AA)_{mem}$ x 10^6, mol	$\frac{(AA)_{mem}}{(DIDE)}$	$S_{A(L/D)}$
BOC-D-Trp	D-Trp	0.57 ± 0.01	1.6	
	L-Trp	0.54 ± 0.02	1.5	1.0
BOC-L-Trp	D-Trp	3.06 ± 0.02	8.7	
	L-Trp	3.46 ± 0.02	9.9	1.1
BOC-D-Trp	D-Phe	0.46 ± 0.02	1.3	
	L-Phe	0.45 ± 0.02	1.3	1.0
BOC-L-Trp	D-Phe	2.14 ± 0.04	6.1	
	L-Phe	2.42 ± 0.03	6.9	1.1
BOC-D-Trp	D-Ala	0.40 ± 0.03	1.1	
	L-Ala	0.38 ± 0.03	1.1	1.0
BOC-L-Trp	D-Ala	1.54 ± 0.03	4.4	
	L-Ala	1.76 ± 0.03	5.0	1.1

[a]The mole ratio of print molecule (Boc-L-Trp or Boc-D-Trp) to the DIDE derivative in the membrane preparation was fixed at 10.

Table III. Adsorption of Racemic Amino Acids by Molecularly Imprinted Polymeric Membranes[a]

Substrate	$(AA)_{mem}$ x 10^6, mol	$\frac{(AA)_{mem}}{(DIDE)}$	$S_{A(L/D)}$
D-Arg	1.51 ± 0.04	4.3	
L-Arg	2.00 ± 0.04	5.7	1.3
D-Lys	1.16 ± 0.04	3.3	
L-Lys	1.30 ± 0.04	3.7	1.1
D-His	0.84 ± 0.05	2.4	
L-His	1.12 ± 0.05	3.2	1.3
D-Asp	1.47 ± 0.03	4.1	
L-Asp	1.75 ± 0.05	5.0	1.2
D-Glu	1.68 ± 0.04	4.8	
L-Glu	1.93 ± 0.04	5.5	1.2

[a]The mole ratio of the print molecule (Boc-L-Trp) to the DIDE derivative in the membrane preparation was fixed at 3. (Reproduced with permission from Reference 51.)

Table IV. pH Values and Net Charges for Amino Acids Under the Experimental Conditions

Amino acid	pH of the solution[a]	Net charge
Arg	~7.9	+0.93
Lys	~7.7	+0.95
His	~5.1	+0.89
Asp	~5.2	-0.95
Glu	~5.4	-0.93
Trp	~6.4	~0.00
Ac-Trp	~6.3	-1.00

[a]The racemic amino acid was dissolved in aqueous ethanol (1:1, v/v) to form a solution which was 1.0 M in each enantiomer.

consisting of D-amino acids and imprinted by the D-isomer will recognize D-α-amino acids in preference to the corresponding L-isomers. This expectation will be realized in the later section on Membrane Materials Which Recognize D-α-Amino Acids.

Enantioselective Permeation of Racemic Amino Acids Using a Concentration Difference as the Driving Force.

With our alternative molecular imprinting technique, molecularly imprinted polymeric membranes are easily obtained which exhibit affinities toward the print molecule family. It is of interest to explore the application of such behavior in molecularly imprinted polymeric membranes as separation agents for enantioselective permeation.

Figure 6 shows the time transport curves of D- and L-Trp through Boc-L-Trp-(Figure 6a) and Boc-D-Trp-imprinted (Figure 6b) membranes which were prepared from DIDE-Resin and AS with the mole ratio of print molecule to the DIDE derivative fixed at 10. For permeation through the Boc-D-Trp imprinted membrane (Figure 6b), only very slight permselectivity is noted. On the other hand, D-Trp was transported in preference to L-Trp through Boc-L-Trp-imprinted polymeric membranes and the separation factor (57) toward the D-isomer ($a_{D/L}$) at steady state was determined to be 1.4 (46,50). When the driving force for membrane permeation is a concentration gradient, the permselectivity is found to be opposite to the adsorption selectivity as described in the previous section. This might be due to suppression of the L-Trp permeability by its relatively high affinity toward the membrane.

The data in Table V show the effect of membrane materials (Figure 7) on the enantioselective permeation of racemic amino acids. For a membrane prepared from DIDE-Resin and AS in the absence of a print molecule, permeation of Trp was barely detectable after 500 hours. On the contrary, D- and L-Trp were transported through other membranes which were prepared in the presence of Boc-L-Trp or Boc-D-Trp as a print molecule.

From these results, it can be concluded that the presence of Boc-L-Trp or Boc-D-Trp in the membrane preparation process plays an important role, at least as a porogen to form permeation pathways for the permeant. Optical resolution was observed only in the case of the Boc-L-Trp-imprinted membrane from DIDE-Resin and AS as shown in Figure 6a. In this case, the presence of Boc-L-Trp during the membrane preparation process not only forms permeation pathways but also introduces sites for molecular recognition. In the present case, it is concluded that Boc-L-Trp functions as a porogen to form the permeation pathways and as a print molecule to shape the molecular recognition site.

Enantioselective Electrodialysis of Racemic Amino Acids.

It is important to transport the selectively adsorbed isomer across the membrane. Electrodialysis was adopted as a membrane transport system to realize such perm-

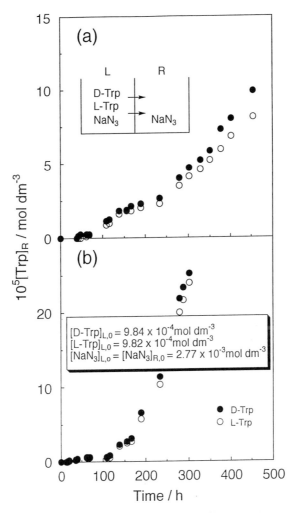

Figure 6. Time-Transport Curves of D- and L-Trp through a) Boc-L-Trp- and b) Boc-D-Trp-Imprinted Membranes Prepared from DIDE-Resin and AS at 40 °C. (The mole ratio of print molecules to DIDE derivatives was fixed at 10.)

Table V. Permeation Results for Imprinted and Nonimprinted Polymeric Membranes[a]

Resin/template/membrane matrix[b]	Permeant	$J \times 10^6$, mol cm^{-2} h^{-1}	$P \times 10^4$, cm^2 h^{-1}	$\alpha_{D/L}$
DIDE-Resin/Boc-D-Trp/AS	L-Trp	29.7	4.29	1.0
	D-Trp	30.4	4.39	
DIDE-Resin/Boc-L-Trp/AS	L-Trp	3.8	0.55	1.4
	D-Trp	5.1	0.74	
Boc-DIDE-Resin/Boc-D-Trp/AS	L-Trp	18.0	2.72	1.0
	D-Trp	17.2	2.60	
Boc-DIDE-Resin/Boc-L-Trp/AS	L-Trp	15.3	2.22	1.0
	D-Trp	15.3	2.22	
Cl-Resin/Boc-D-Trp/AS	L-Trp	21.0	3.06	1.0
	D-Trp	20.8	3.03	
Cl-Resin/Boc-L-Trp/AS	L-Trp	26.0	3.97	1.0
	D-Trp	25.9	3.95	
DIDE-Resin/--[c]/AS	L-Trp	--[d]	--[d]	
	D-Trp	--[d]	--[d]	

[a]The mole ratio of the print molecule (Boc-L-Trp or Boc-D-Trp) to the DIDE derivative in the membrane preparation was fixed at 10.
[b]See Figure 7 for structures of the functionalized resins.
[c]The membrane was prepared in the absence of a print molecule.
[d]Permeation of Trp was barely detectable after 500 hours.

selectivity. Amino acids with nonpolar side chains, such as Trp, and those with uncharged polar side chains, such as Ser, are not suitable substrates for electrodialysis because their net charges under the experimental conditions are calculated to be approximately zero (Table IV). Thus these amino acids are assumed to be zwitterions which are overall neutral (52). Therefore, electrodialysis of amino acids with charged polar side chains and Ac-Trp, in which the α-amino moiety is protected by an acetyl group, were investigated. In the present article, the results of enantioselective electrodialysis of Ac-Trp (47) are described as an example.

The effect of the applied potential difference on the enantioselective electrodialysis of Ac-Trp is shown in Figure 8. The total flux (J) is linearly proportional to the applied potential difference (ΔE). When ΔE exceeds 15.0 V, the enantioselective permeation is negligible. That is, the separation factor, $\alpha_{L/D}$, was determined to be unity. However, the separation factor increased with a decrease in ΔE and below 2.5 V of ΔE the separation factor toward the L-isomer (Ac-L-Trp) reached 6, which is equal to the adsorption selectivity. This suggests that electrodialysis of racemic amino acids through the molecularly imprinted polymeric membrane can attain the permselectivity reflected in its adsorption selectivity. Other amino acids with charged polar side chains gave results similar to Ac-Trp (51).

Membrane Materials Which Recognize D-α-Amino Acids.

So far, it is noted that a membrane which contain an oligopeptide residue from L-amino acids and imprinted by an L-isomer recognizes L-isomers in preference to the corresponding D-isomers. Therefore, it is expected that a membrane carrying an oligopeptide derivative consisting of D-amino acids and imprinted by a D-isomer will show D-amino acid adsorption selectivity as mentioned in the last paragraph of the section on Enantioselective Adsorption of Other Amino Acids. Novel membrane materials with the tetrapeptide derivative, H-D-Asp(OcHex)-D-Leu-D-Asp(OcHex)-D-

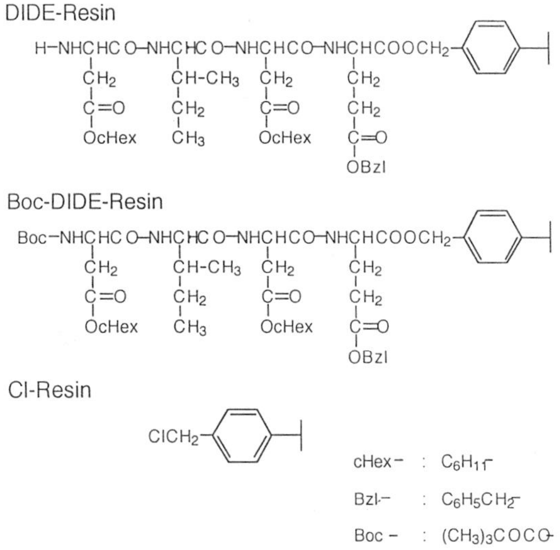

DIDE-Resin

H—NHCHC O—NHCHC O—NHCHCO—NHCHCOOCH$_2$—⟨benzene⟩—|
　　 | 　　　　 | 　　　　 | 　　　 |
　　CH$_2$　　 CH-CH$_3$　 CH$_2$　　 CH$_2$
　　 | 　　　　 | 　　　　 | 　　　 |
　　C=O　　 CH$_2$　　 C=O　　 CH$_2$
　　 | 　　　　 | 　　　　 | 　　　 |
　　OcHex　 CH$_3$　　 OcHex　 C=O
　　　　　　　　　　　　　　　　 |
　　　　　　　　　　　　　　　 OBzl

Boc-DIDE-Resin

Boc—NHCHC O—NHCHC O—NHCHCO—NHCHCOOCH$_2$—⟨benzene⟩—|
　　　 | 　　　　 | 　　　　 | 　　　 |
　　　CH$_2$　　 CH-CH$_3$　 CH$_2$　　 CH$_2$
　　　 | 　　　　 | 　　　　 | 　　　 |
　　　C=O　　 CH$_2$　　 C=O　　 CH$_2$
　　　 | 　　　　 | 　　　　 | 　　　 |
　　　OcHex　 CH$_3$　　 OcHex　 C=O
　　　　　　　　　　　　　　　　　 |
　　　　　　　　　　　　　　　　 OBzl

Cl-Resin

ClCH$_2$—⟨benzene⟩—|

cHex—　:　C$_6$H$_{11}$—

Bzl—　:　C$_6$H$_5$CH$_2$—

Boc —　:　(CH$_3$)$_3$COCO—

Figure 7. Structures of Side Groups in the Resins

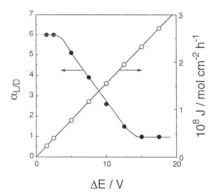

$\Delta E \; / \; V$

Figure 8. Influence of the Difference in Applied Potential on Enantioselective Electrodialysis and the Total Flux. (The mole ratio of print molecule, Boc-L-Trp, to DIDE derivative in the membrane preparation process was fixed at 0.5. The concentration of enantiomer in the racemic Ac-Trp was fixed at 1.0 x 10^{-3} M. Reproduced with permission from Reference 47.)

Glu(OBzl)-CH$_2$-, were prepared from D-amino acids and their potential for chiral recognition was investigated.

Figure 9 shows the dependence on imprinting conditions for the adsorption of Ac-D- and Ac-L-Trp from racemic mixtures by Boc-D-Trp-imprinted membranes (*58*). As expected, the D-isomer (Ac-D-Trp) was always incorporated in the membrane in preference to Ac-L-Trp. On the other hand, the membranes imprinted by Boc-L-Trp exhibited only slight adsorption selectivity. The present results suggest that the membranes bearing an oligopeptide consisting of D-amino acids and imprinted by a D-isomer will recognize the D-isomer in preference to corresponding L-isomer, and *vice versa*, as depicted in Figure 10.

Other Membrane Materials for Alternative Molecularly Imprinted Polymeric Membranes.

In the previous sections, imprinted membranes prepared from polymeric materials containing oligopeptide units were found to produce some optical resolution. It is of interest to try to prepare alternative molecularly imprinted polymeric membranes with a potential for optical resolution from other polymeric materials, such as natural polymers and synthetic polymers.

Figure 11 shows the dependence of D-/L-Glu adsorption on the imprinting conditions (*59*). For the experiments summarized in this figure, the membranes were prepared from cellulose acetate (CA) with an acetyl content of 40 % and imprinted by N-α-Boc-L-glutamic acid (Boc-L-Glu). As can be seen in the figure, L-Glu was always incorporated in the membrane in preference to D-Glu. The adsorption selectivity toward the L-isomer increased with a decrease in the molecular imprinting ratio, as previously described, and reached a value of 2.3 at the mole ratio of 0.5.

By applying electrodialysis, the molecularly imprinted CA membranes show permselectivity toward the L-isomer. The membrane with an imprinting ratio was 0.5 gave a separation factor of 2.3, which is equal to adsorption selectivity. It was recently established that CA membranes imprinted by a D-isomer shows D-isomer adsorption selectivity, and *vice versa* (*60*).

Molecularly imprinted polymeric membranes from modified polysulfones (synthetic polymers) also exhibited a potential for optical resolution. The membranes imprinted by an L-isomer gave L-isomer adsorption selectivity, and *vice versa* (*61*).

In this article, the utility of an alternative molecular imprinting technique, which was first proposed by the author's research group, to produce membranes with enantioselective permeation of amino acids is demonstrated. The usefulness of this alternative molecular imprinting technique has also been demonstrated by Kobayashi *et al.* (*62,63*) and by Ohya *et al.* (*64*).

Conclusions.

Molecularly imprinted polymeric membranes, bearing tetrapeptide derivatives, were prepared by the presence of a print molecule during the membrane preparation (casting) process. Membranes which contain oligopeptide residues from L-amino acids and imprinted by a L-amino acid derivative recognize the L-isomer in preference to the corresponding D-isomer, and *vice versa*. In electrodialysis of racemic amino acids, the permselectivity reflects the adsorption selectivity. It was also established that the alternative molecular imprinting technique can be applied to any membrane materials which bear oligopeptide units, such as derivatives of natural polymers and synthetic polymers. The present study suggests that the molecularly imprinted polymeric membranes have a potential to attain the optical resolution of amino acids.

Based on the results obtained so far, it seems that the alternative molecular imprinting technique, as was first proposed by the author's research group, is a viable method for the construction of artificial receptors, which can be applicable in membrane

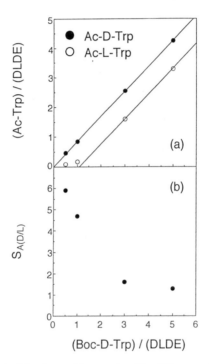

Figure 9. Effect of the Membrane Preparation Condition on a) Ac-Trp Adsorption and b) the Adsorption Selectivity toward Ac-D-Trp.

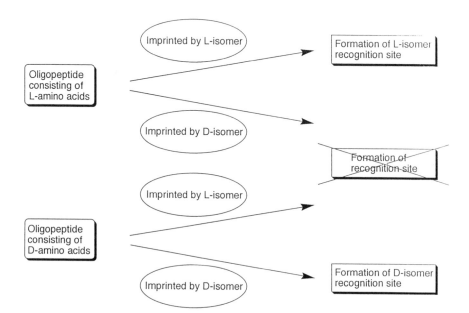

Figure 10. Summary of Alternative Molecularly Imprinted Polymeric Membranes Bearing an Oligopeptide as the Chiral Recognition Site.

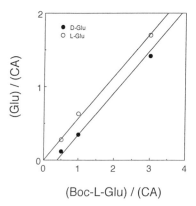

(Boc-L-Glu) / (CA)

Figure 11. Influence of the Amount of Print Molecule on Glu Adsorption in Alternative Molecularly Imprinted Membranes. (Reproduced with permission from Reference 59.)

separations, chromatography, sensors, catalysts, *etc.*in connection with various industries, such as the pharmaceutical industry, food production, agricultural chemicals, *etc.*

Acknowledgment.

The author wishes to express his appreciation to Dr. Takeo Shimidzu of the Kansai Research Institute and Professor Emeritus of Kyoto University for his continuous encouragement.

Literature Cited and Footnotes.

1. Newcomb, M.; Helgeson, R. C.; Cram, D. J. *J. Am. Chem. Soc.* **1974**, *96*, 7367.
2. Newcomb, M.; Toner, J. L.; Helgeson, R. C.; Cram, D. J. *J. Am. Chem. Soc.* **1979**, *101*, 4941.
3   Yamaguchi, T.; Nishimura, K.; Shinbo, T.; Sugiura, M. *Chem. Lett.* **1985**, 1549.
4. Kikuchi, H.; Hattori, J.; Mori, Y.; Kajiyama, T. *Kagaku Kogaku Ronbunshu* **1989**, *15*, 617
5. Osada, Y.; Ohta, F.; Mizumoto, A.; Takase, M.; Kurimura, Y. *Nippon Kagaku Kaishi* **1986**, 866.
6. Tone, S.; Masawaki, T.; Hamada, T. *J. Membr. Sci.* **1995**, *103*, 57.
7. Tone, S.; Masawaki, T.; Eguchi, K. *J, Membr. Sci.* **1996**, *118*, 31.
8. Ishihara, K.; Suzuki, N.; Matsui, K. *Nippon Kagaku Kaishi* **1987**, 446.
9. Kakuchi, T.; Takaoka, T.; Yokota, K. *Polymer J.* **1990**, *22*, 199.
10. Shinbo, T.; Kanamori, T.; Ogasawara, K.; Yamasaki, A.; lwatsubo, T., Masuoka, T.; Yamaguchi, T. *Sen'i Gakkaishi* **1996**, *52*, 105.
11. Maruyama, A.; Adachi, N.; Takatsuki, T.; Tom, M.; Sanui, K.; Ogata, N. *Macromolecules* **1990**, *23*, 2748.
12. Maruyama, A.; Adachi, N.; Takatsuki, T.; Sanui, K.; Ogata, N. *Nippon Kagaku Kaishi* **1990**, 1178.
13. Masawaki, T.; Sasai, M.; Tone, S. *J. Chem. Eng. Jpn.* **1992**, *25*, 33.
14. Aoki, T. Shinohara, K.; Oikawa, E. *Makromol. Chem., Rapid Commun.* **1992**, *13*, 565.
15. Higuchi, A.; Ishida, Y.; Nakagawa, T. *Desalination* **1993**, *90*, 127.
16. Higuchi, A.; Hara, M.; Horiuchi, T.; Nakagawa, T. *J. Membr. Sci.* **1994**, *93*, 157.
17. Yamada, H.; Fuwa, N.; Nomura, M.; Yoshikawa, M.; Kunugi, S. *Maku (Membrane)* **1993**, *18*, 301.
18. Yashima, E.; Noguchi, J.; Okamoto, Y. *J. Appl. Polym. Sci.* **1994**, *54*, 1087.
19. Aoki, T.; Tomizawa, S.; Oikawa, E. *J. Membr. Sci.* **1995**, *99*, 117.
20. Aoki, T; Maruyama, A.; Shinohara, K.; Oikawa, E. *Polymer J.* **1995**, *27*, 547.
21. Shinohara, K.; Aoki, T.; Oikawa, E. *Polymer* **1995**, *36*, 2403.
22. Cram, D. J. *Angew. Chem., Int. Ed. Engl.* **1988**, *27*,1009.
23. Lehn, J.-M., *Angew. Chem., Int. Ed. Engl.* **1988**, *27*, 89.
24. Pedersen, C. J. *Angew. Chem., Int. Ed. Engl.*, **1988**, *27*, 1021.
25. Diederich, F. *Angew. Chem., Int. Ed. Engl.* **1988**, *17*, 362.
26. Breslow, R. In *Advances in Enzymology and Related Areas of Molecular Biology*, Vol. 58; Meister, A., Ed.; Wiley: New York, 1986; pp. 1-60.
27. *Calixarenes*; Gutsche, C. D., Ed.; Royal Society of Chemistry: Cambridge, 1989.

186

28. Izatt, R. M.; Bradshaw, H. S.; Pawlak, K.; Bruening R. L.; Tarbet, B. J. *Chem. Rev.* **1992**, *92*,1261.
29. Takeshita, M.; Shinkai, S. *Bull. Chem. Soc. Jpn.* **1995**, *68*, 1088.
30. Rebek, Jr., J.; Askew, B.; Nemeth, D.; Parris, K. *J. Am. Chem. Soc.* **1987**, *109*, 2432.
31. Rebek, Jr., J. *Science* **1987**, *235*, 1478.
32. Rebek, Jr., J. *Angew. Chem., Int. Ed. Engl.* **1990**, *29*, 245.
33. Sasaki, R. Y.; Kurihara, K.; Kunitake, T. *J. Am. Chem. Soc.* **1991**, *113*, 9685.
34. Nowick, J. S.; Chen, J. S.; Noronha, G. *J. Am. Chem. Soc.* **1993**, *115*, 7636.
35. Kimizuka, N.; Kawasaki, T.; Kunitake, T. *J. Am. Chem. Soc.* **1993**, *115*, 4387.
36. Wulff, G. In *Polymeric Reagents and Catalysts*; Ford, W. T., Ed; ACS Symposium Series 308, American Chemical Society: Washington, DC, 1986, Chapter 9.
37. Ekberg, B.; Mosbach, K. *Trends Biotechnol.* **1989**, *7*, 92.
38. Flam, F. *Science* **1994**, *263*, 1221.
39. Mosbach, K. *TIBS* **1994**, *19*, 9.
40. Shea, K. J. *Trends Polym. Sci.* **1994**, *2*, 166.
41. Wulff, G. *Angew. Chem., Int. Ed. Engl.* **1995**, *34*, 1812.
42. Piletskii, S. A.; Dubei, 1. Ya.; Fedoryak, D. M.; Kukhar, V, P. *Biopolim. Kletka* **1990**, *6*, 55.
43. Mathew-Krotz, J.; Shea, K. J. *J. Am. Chem. Soc.* **1996**, *118*, 8154.
44. Yoshikawa, M.; Izumi, J.; Kitao, T.; Koya, S.; Sakamoto, S. presented at The 16th Annual Meeting of the Membrane Society of Japan, **1994**, A1-1-1.
45. Bodanszky, M.; Bodansiky, A., *The Practice of Peptide Synthesis*, 2nd edn.; Springer-Verlag: Berlin, 1994.
46. Yoshikawa, M.; Izumi, J.; Kitao, T.; Koya, S.; Sakamoto, S. *J. Membr. Sci.* **1995**, *108*, 171.
47. Yoshikawa, M.; Izumi, J.; Kitao, T. *Chem. Lett.* **1996**, 611.
48. Adsorption selectivity toward the L-isomer ($S_{A(L/D)}$) is defined as $S_{A(L/D)}$ = $[(AA)_L / (AA)_D] / (C_L/C_D)$. In this equation, $(AA)_i$ and C_i are the amount of amino acid in the membrane and the concentration in the solution after equilibrium is reached, respectively.
49. Yoshikawa, M.; Izumi, J.; Fujisawa, T,; Kitao, T., *Proceedings of the International Membrane Science and Technology Conference*, Fane, A. G., Ed.; Unesco Center for Membrane Science and Technology, Sydney, 1996, D.5.
50. Yoshikawa, M.; Izumi, J.; Kitao, T.; Sakamoto, S. *Macromolecules* **1996**, *29*, 8197.
51. Yoshikawa, M.; Izunii, J.; Kitao, T. *Polymer J.* **1997**, *29*, 205.
52. Voet, D.; Voet, J. G. *Biochemistry* ; Wiley: New York, 1990.
53. Tanford, C. *J. Am. Chem. Soc.* **1962**, *84*, 4240.
54. Nozaki, Y.; Tanford, C. *J. Biol. Chem.* **1971**, *246*, 2211.
55. Fauchere, J.-L.; Pliska, V. *Eur. J. Med. Chem.-Chim. Ther.* **1983**, *18*, 369.
56. Chen, C. C.; Zhu, Y.; King, J. A.; Evans, L. B. *Biopolymers* **1992**, *32*, 1375.
57. The separation factor $a_{j/i}$ is defined as the flux ratio J_i/J_j divided by the concentration ratio C_i/C_j in the feed side.
58. Yoshikawa, M.; Izumi, J.; Kitao, T. presented at The Membrane Symposium of the Membrane Society of Japan, **1996**, 107.
59. Izumi, J.; Yoshikawa, M.; Kitao, T. *Maku (Membrane)* **1997**, *22*, 149.
60. Yoshikawa, M.; Ooi, T.; Izumi, J.; Kitao, T. presented at The 19th Annual Meeting of the Membrane Society of Japan, **1997**, D-3-2.

61. Yoshikawa, M.; Soda, K.; Ooi, T,; Izumi, J,; Kitao, T.; Guiver M.D.; Robertson, G. P. presented at The 13th Symposium on Membrane Science and Technology, **1997**, 3DPO7.
62. Kobayashi, T.; Wang, H. Y.; Fujii, N. *Chem. Lett.* **1995**, 927.
63. Kobayashi, T.; Fujii, N.in this Symposium Series volume.
64.   Ohya, Y.; Miyaoka, J.; Ouchi, T. *Macromol. Chem., Rapid Commun.,* **1996**, *17*, 871

Chapter 13

Molecular Imprinted Membranes Prepared by Phase Inversion of Polyacrylonitrile Copolymers Containing Carboxylic Acid Groups

Takaomi Kobayashi, Hong Ying Wang, Takahiro Fukaya, and Nobuyuki Fujii

Department of Chemistry, Nagaoka University of Technology, Kamitomioka, Nagaoka, Niigata 940-21, Japan

Phase inversion of polyacrylonitrile copolymers containing carboxylic acid groups is utilized to prepare molecular imprinted membranes for theophylline (THO). The copolymer cast solution of dimethyl-sulfoxide (DMSO) with the THO template is immersed in water which causes the polymer including the template to coagulate in the nonsolvent. After removal of the template from the solidified copolymer membrane, high selectivity for rebinding of THO is observed compared to binding of the structurally related molecule caffeine. The molecular imprinting process was affected by the template concentration in the DMSO cast solution, the coagulation conditions, and the carboxylic acid content in the copolymer. High binding selectivity of the THO solute by the THO-imprinted copolymer membrane arises from the formation of imprinted sites in which free OH groups of the carboxylic acid units form hydrogen bonds with the THO.

The molecular imprinting technique has been found to be an effective means for encoding molecular scale information in bulk polymeric materials (1-3). The procedure involves incorporation of small amounts of a template molecule in the polymerization medium. Then an extraction procedure removes the template to leave functionalized sites for the template in the polymer network (3). Most of molecular recognition polymers prepared by this method are for applications in separations of chiral compounds (4) and amino acid derivatives (5) and in drug assays (6). More recently, a phase inversion polymerization process has been used to prepare molecular imprinted membranes of poly(acrylonitrile-co-acrylic acid) (P(AN-co-AA) (7,8). In this technique, coagulation of the polyacrylonitrile copolymer is utilized to encode information for the template molecule theophylline (THO) in porous polymeric membranes.

The Phase Inversion Process for the Preparation of Molecular Imprinted Membranes

Porous membranes have an advantage of low resistance to mass transfer of solute in solution. Therefore, membrane separation processes possess a potential for treating

188

$$\left(CH_2-\underset{\underset{CN}{|}}{CH}\right)_x\left(CH_2-\underset{\underset{COOH}{|}}{CH}\right)_y$$

P(AN–co–AA)

R: H = Theophylline
CH$_3$ = Caffeine

large amounts of solution per unit time with substantial energy savings (9-11). Most of the separation membranes typically used in ultrafiltration, microfiltration, and reverse osmosis filtration are prepared by immersion precipitation processes (9). Generally, homogeneous polymer solutions are used and phase inversion is achieved in an excess of nonsolvent (12-16). Phase inversion involves transformation from a liquid phase polymer solution to a solid state (Figure 1). For polyacrylonitrile (PAN), DMSO and water are suitable for the cast solvent and the coagulation medium, respectively (17). To control the permeability of the resulting PAN membranes, especially for ultrafiltration uses, a water-soluble additive, such as lithium nitrate (18,19), poly(vinyl alcohol) (20,21), or poly(ethylene glycol) (22) is added to the polymer cast solution. Solidification takes place with the additive in the nonsolvent medium. After the additive is washed away with water, the resulting polymeric membrane contains small pores corresponding to the size of the additive. This approaches allows us to prepare imprinted membranes by a phase inversion technique.

For a molecular imprinting process of PAN copolymer containing COOH groups, the functional groups which interacts with the template (8), solidification in water takes place in the presence of the template (Figure 2). Here the PAN segments in the copolymer coagulate strongly in the aqueous nonsolvent medium and envelope the THO template molecules. After coagulation, the template molecules are removed from the coagulated-membrane by washing with aqueous acetic acid (AcOH). The resultant polymeric membrane has memorized the dimensions of the template.

Membrane Preparation by Phase Inversion of P(AN-co-AA). Copolymerization of acrylonitrile (AN) and acrylic acid (AA) was performed in DMSO solution as previously reported (23). Preparation of P(AN-co-AA) membranes by the phase inversion method was conducted by the following procedure (7). The THO template was dissolved in the DMSO cast solution in the range of 0-4.7 weight percent. At concentrations higher than 4.7 weight percent, the DMSO cast solution becomes saturated with the THO template. The DMSO cast solution containing 10 weight percent of P(AN-co-AA) copolymer and various weight percentages of the template were well mixed at 50 °C for 20 hours. The solutions were cast on a glass plate which had been warmed to 50 °C and were then coagulated in water. The resultant membrane was rinsed with a large volume of water to remove the DMSO. In a control experiment, the cast solution contained 10 weight percent of P(AN-co-AA), but no THO template. Also, a PAN homopolymer membrane was obtained for comparison.

To provide information about the interaction between the copolymer and THO molecules in the DMSO cast solution, the viscosity of the cast solution containing various concentrations of THO was measured at 50 °C with a B-type viscometer at different rotation rates (rpm) (8). P(AN-co-AA) containing AA segments with y = 0.145 mole fraction (x + y = 1) was used for the experiments. For 10 weight percent of P(AN-co-AA) and PAN solutions containing no THO, the viscosities were 5200 and 1200 (cp), respectively, at 6 rpm (Figure 3a). Also, there was a slight difference in the viscosities measured for P(AN-co-AA) at a high rotation rate. This indicates that the

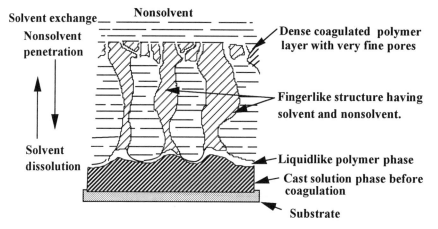

Figure 1. Cross sections of coagulated polymer membranes formed by phase inversion.

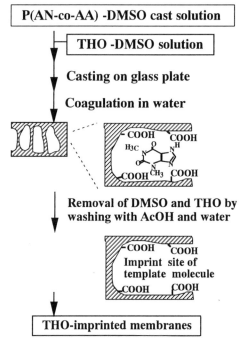

Figure 2. Imprinting processes of THO in PAN copolymer membranes by the phase inversion method.

AA segments of the copolymer interact with each other to produce a high viscosity. The viscosity value was found to decrease upon addition of the THO template to the copolymer cast solutions. The decrease of viscosity by THO addition implies that the template in the cast copolymer solution disrupts the hydrogen bonding interaction between the AA segments in the P(AN-co-AA) membrane. For the PAN cast solutions, the viscosity values are independent of the template concentration. The data for PAN also suggest no interaction of THO with the AN segments. For dilute solutions of the copolymer and PAN homopolymer, the reduced viscosities (η_{sp}/C) were measured with an Ubbelohde viscometer to compare the molecular weights of the resultant polymers (Figure 3b). The intrinsic viscosities, $[\eta]$, of P(AN-co-AA) and PAN were determined to be 1.6 and 1.7 (cm^3/g), respectively, by extrapolating the linear plots. This indicates that the molecular weights of both polymers are nearly the same. It is noted that the slope of the η_{sp}/C vs. C plot for P(AN-co-AA) is larger than that for PAN. This also suggests interactions between the AA segments in the copolymer.

Photographs of the cross sections of the P(AN-co-AA) membranes were obtained by scanning electron microscopy (SEM) to evaluate the effect of the coagulation condition on the copolymer (Figure 4). The concentration ratio [THO]/[COOH] for the THO and the carboxylic acid groups in the cast solution was varied. It is obvious from the SEM photographs that the copolymer membranes have an asymmetric structure, which consists of a dense top layer supported by large porous sublayer with a finger-like structure for a total thickness of about 70 μm. The copolymer membrane prepared from a cast solution which contained no template has a similar cross section to that of PAN (16). Comparison of the SEM photographs for the P(AN-co-AA) membranes suggests that the membrane morphology depends on the template concentration in the cast solution. These membranes have very thin dense top layer (<1 μm) on a finger-like support layer which is typical of ultrafiltration membranes (9,13,14). When the template concentration in the cast solution was increased to [THO]/[COOH] = 1, the morphology of the top layer for the resultant membrane looks like a thick sponge, with a thickness of about 20 μm. As the template ratio is increased even higher, the top layer of the membrane becomes very thin. The differing morphologies observed in the SEM photographs reveals that the template concentration in the cast solution influences the coagulation process.

In the phase inversion process for PAN, solidification is initiated first by a transition from the liquid state of the DMSO cast solution to two liquid phases of the cast solvent and the coagulate solution (liquid-liquid demixing) (14). In the solvent exchange (Figure 1), DMSO solvent dissolution occurs as water penetrates into the polymer cast solution. At the same time, phase separation of the polymer takes place at the interface between the water and polymer cast solution. Finally, a solid polymer membrane is obtained which has an asymmetric structure. The support layer with finger-like cavities appears when the nonsolvent and the cast solvent show high affinity. Namely, finger-like macrovoid formation in this type of asymmetric membrane is believed to arise from instantaneous demixing of the cast solvent-water medium (24). The SEM photographs for the DMSO-water system imply that coagulation of the PAN segments in the copolymer arises due to a large exothermic heat of mixing (25). Several types of experimental evidence show that the disappearance of the macrovoids is due to slow rates of polymer coagulation in the medium (16). The results for P(AN-co-AA) membranes suggest that the affinity between the solvent and the nonsolvent is diminished and solidification of the copolymer proceeds slowly when [THO]/[COOH] = 1. At higher ratios, the morphology change in the copolymer membranes indicates that water penetration occurs immediately and the copolymer instantly coagulates upon immersion in the aqueous medium.

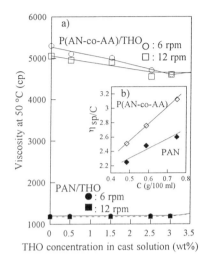

Figure 3. a) Viscosities of 10 weight percent of DMSO cast solutions containing varying amounts of THO for P(AN-co-AA) and PAN. b) Reduced viscosity of P(AN-co-AA) and PAN measured at different concentrations in DMSO.

a) [THO]/[COOH]=0   b) [THO]/[COOH]=1   c) [THO]/[COOH]=1.9

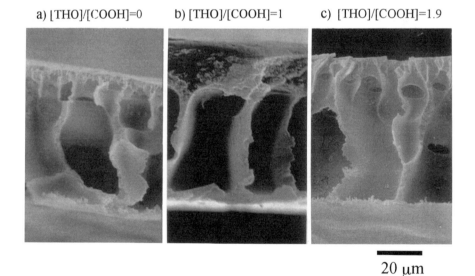

20 μm

Figure 4. Morphology of the cross section of P(AN-co-AA) membranes prepared from DMSO cast solutions with a) [THO]/[COOH]=O, b) 1.0 and c) 1.9.

Before studying the permeation of THO across the copolymer membranes, FT-IR spectra of membranes with different mole fractions of AA groups for y = 0, 0.04 and 0.21 were compared (Figure 5). For the FT-IR measurements, copolymer membranes of 3-5 μm thickness were used (8). FT-IR spectra of the coagulated membranes have characteristic peaks for CN stretching at 2240 cm^{-1} for the PAN segments (26) and C=O stretching in the 1720-1730 cm^{-1} region for the AA segments of the copolymer (27). Peaks for CH bending and stretching appeared at 1447 and 2937 cm^{-1}, respectively. Note that the OH stretching absorption which appears in 3200-3500 cm^{-1} region has two peaks in the spectra for the P(AN-co-AA) membranes. In general, carboxylic acid groups exist as dimers due to strong intermolecular hydrogen bonding (28,29). The carboxylic acid dimer shows very broad OH stretching in the region of 2500-3300 cm^{-1}. Thus, the broad IR peaks in that region in the P(AN-co-AA) spectra imply the presence of hydrogen bonding between the COOH groups in the membrane. In addition, a free OH stretching vibration is observed near 3500 cm^{-1}. Relative to the CH stretching intensity at 2937 cm^{-1}, the OH stretching intensity increases as the molar fraction of AA segments increases. In addition, a comparison of FT-IR spectra was made for P(AN-co-AA) membranes with an AA content of y = 0.145 mole fraction (Figure 6). Here, the [THO]/[COOH] ratio in the cast solution was varied in the range of 0-1.9 and the membranes were coagulated at 23 °C. As the THO concentration in the cast solution increased, the IR peak for the OH stretching vibration at 3520 cm^{-1} increased. This indicates that when the THO template is present in the copolymer membrane, it interacts with the carboxylic acid groups in the copolymer via hydrogen bonding and disrupts the inter- and/or intramolecular hydrogen bonding between AA segments in the membrane (8).

Molecular Recognition by Molecular Imprinted Membranes

THO Uptake by P(AN-co-AA) Membranes. To estimate the permeability and molecular size exclusion effect of the resultant P(AN-co-AA) membranes, filtration experiments with THO and dextrans were carried out with an ultrafiltration cell with 2.5 kPa of applied pressure (7). The solution containing 3.6 μmol/dm^3 of THO substrate was permeated through the copolymer membranes with y = 0.145. The volume flux values for the THO feed solution varied with the template concentration in the cast solutions in the range of 5×10^{-7}-1×10^{-6} (m^3/(m^{2}s). With an increase in the THO concentration, the volume flux diminished until [THO]/[COOH] = 1 and, in the high THO concentration, the volume flux increased. These tendencies correlate with the thickness of the dense top layer as shown in the SEM photographs (Figure 4). This is expected since the dense layer property determines the transport rates of solute and solution in such asymmetric porous separation membranes (9). Permeation data for various molecular weight dextrans reveal that the P(AN-co-AA) membranes do not reject THO molecules by a molecular size exclusion effect (8), since the pore sizes of the membrane are much larger than the dimensions of the THO molecule.

For THO binding (Figure 7), the permeate solution was collected at hourly intervals and the substrate concentrations in the feed and permeate solution were analyzed (7). The filtration experiments were continued until no binding of the substrate was observed. For each permeate solution collected at i intervals, the amount of THO or CAF lost to the membrane, $[S]_i$, was calculated from the reduction in the solute concentration. The values for the total binding amount of the solute, $[S]$ (μmol/g-membrane) binding to a unit weight W (in g) of membrane, were calculated by the equation

$$[S] = \Sigma[S]_i = \Sigma(C_0 - C_{pi})V_i/W \qquad (1)$$

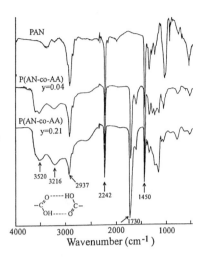

Figure 5. FT-IR spectra of P(AN-co-AA) membranes with various AA contents.

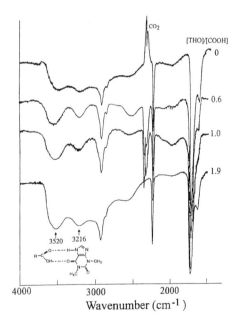

Figure 6. FT-IR spectra of P(AN-co-AA) membranes prepared with different [THO]/[COOH] ratios with an AA content of y = 0.145 mole fraction.

where C_0 and C_{pi} are concentrations (in $\mu mol/dm^3$) for the feed solution and permeate solution with volumes V_i (dm^3) collected at i intervals, respectively. The values of [S] increased with an increase in the filtration time and, then, become constant after a few hours. In the filtration time for the constant [S] region, the total amount of THO solute taken up by the copolymer membranes is defined as $[S]_{THO}$. The data reveal that the THO solutes is taken up by the imprinted sites in the copolymer membrane and saturation is reached when all of the imprinted sites are occupied. For example, the $[S]_{THO}$ value for the copolymer membrane prepared from a cast solution with [THO]/[COOH] = 1.9 is about 0.52 $\mu mol/g$. It is noted that the $[S]_{THO}$ value increases as the THO concentration in the cast solution is enhanced. This suggests that the number of THO imprinted sites increases for a high [THO]/[COOH] ratio. Interestingly, the amounts of THO taken up into the PAN membrane without AA segments are almost zero, even though the template concentration in the PAN cast solution was same as that for P(AN-co-AA) with [THO]/[COOH] = 1.9. This comparison suggests that the presence of AA segments is very important for THO to be taken up at the imprinted sites. Consequently, in copolymer membranes prepared from cast solutions with high template concentrations, the THO solute is effectively taken up by the imprinted sites in the membrane.

To test the permselective transport through the copolymer membrane, a 3.6 $\mu mol/dm^3$ solution of caffeine (CAF) was passed across the membranes (8). CAF has methyl group on 7-nitrogen atom instead of H group of THO molecules. It is known that CAF molecules are less effectively bound than THO molecule by the imprinted sites in methacrylic acid segments of a crosslinked gel matrix due to less hydrogen bonding of CAF to the COOH groups of the matrix (1,6). As was done for the THO solute permeation, the total amounts of CAF taken up by the membrane, $[S]_{CAF}$, were estimated. The values of $[S]_{THO}$ and $[S]_{CAF}$ for each membrane are summarized in relationship between $[S]_{THO}$ or $[S]_{CAF}$ and [THO]/COOH] shown in Figure 8. For the CAF taken up by the THO-imprinted membrane, less effective binding is observed in all of the [THO]/[COOH] regions. As a result, the structurally analogus CAF does not bind effectively to the THO-imprinted sites.

Influence of Coagulation Temperature on THO Uptake by THO-Imprinted Membranes. For phase inversion polymers, the resultant membrane properties depend strongly on the preparation conditions, such as the coagulation temperature (9,30). It is known that the condition of the coagulation medium influences primarily the exchange rate of between the cast solvent and the nonsolvent. Nonsolvent penetration into the polymer cast solution medium decreases at low temperatures, because of high viscosity and low solubility. Consequently, a polymer coagulates slowly at low temperature (31). For P(AN-co-AA), the temperature of the water coagulation medium was varied to examine the temperature dependence of substrate binding by THO-imprinted membranes. After extraction of the template from the membranes, THO uptake experiments were conducted (Figure 9). With CAF as the substrate, the THO-imprinted membranes exhibit very low binding. On the other hand, the THO substrate is bound efficiently. Note that the uptake of THO depends on the coagulation temperature. A decrease in the coagulation temperature produces an increase in THO uptake by the copolymer membrane. For the membranes prepared at 10 °C and 40 °C, the $[S]_{THO}$ values are 1.25 and 0.26 (mmol/g-membrane), respectively. The increase of THO binding for the resultant membrane coagulated in low temperature may be due to an increase in the number of THO-imprinted sites, but only for THO substrate.

SEM photographs for various membranes show that the thickness of the top layer increases as coagulation temperature is decreased (32). This difference strongly suggests that coagulation of P(AN-co-AA) is influenced by the temperature of the aqueous medium. For a coagulation temperature of 10 °C, the observation at a thick

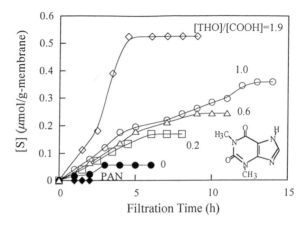

Figure 7. Amounts of THO solutes taken into the copolymer membranes in permeation experiments with 3.6 μmol/dm^3 solutions of THO.

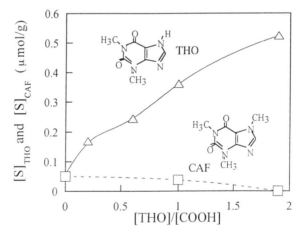

Figure 8. Relationship between the amounts of THO and CAF solutes binding to P(AN-co-AA) membranes prepared with different [THO]/[COOH] ratios.

top layer with sponge-like structure shows that liquid-liquid demixing is retarded relative to that at high temperatures. On the other hand with the higher coagulation temperature, the membrane has significant asymmetry with a very thin top layer. This indicates that the solvent exchange between DMSO and water occurs quickly at a higher temperature because of the strong coagulation of PAN segments in an excess of the aqueous nonsolvent. Since the THO template is water soluble, some of the template molecules may move into the aqueous nonsolvent medium together with the DMSO during coagulation. Such loss of template molecules may diminish the formation of the imprinted sites in the coagulated membrane.

A comparison of ¹H-NMR spectra was made for the copolymer membranes prepared without template extraction. The results reveal that the amounts of THO template dissolved into the aqueous medium is decreased with a decrease in the coagulation temperature.

IR spectra of THO-imprinted membranes coagulated in water at different temperatures were measured to provide information about the coagulation temperature dependence. IR spectra of P(AN-co-AA) membranes were taken before THO extraction by washing with aqueous AcOH (Figure 10). As mentioned above, there are two kinds of OH stretching vibrations for carboxylic acid groups in the copolymer. One is the IR peak around 2500-3300 cm^{-1} for carboxylic acid dimers of the AA segments. The other is the peak near 3520 cm^{-1} for free carboxylic acid groups. Because of the dimerization of COOH group, this kind of COOH group may not contribute to form hydrogen bonding with THO templates. In contrast, the presence of the free OH group of AA segments may be suitable for hydrogen bond formation with THO in the uptake experiments. Thus, the free COOH segments create the THO recognition sites in the copolymer membrane matrix. For FT-IR spectra of the THO-imprinted membranes obtained after THO extraction, the spectra are shown by dashed

COOH dimer   Interaction via hydrogen bonds   Free COOH

lines in Figure 10. (Extraction of the THO template was confirmed by measuring with ¹H-NMR spectrum for the membrane.) The difference in IR spectra for the membranes without and with the template removal is remarkable for the low coagulation temperature. Note that the intensity of the IR peak near 3520 cm^{-1} for OH stretching of free COOH groups increases upon removal of THO from the membrane coagulated at low temperature. With coagulation at 40 °C, the IR spectra show no significant difference for membranes prepared with and without the template extraction. These results suggest that free OH groups of COOH segments, which interact with the THO solute via hydrogen bonds, are responsible for the THO-imprinted sites.

THO Binding to P(AN-co-AA) Membranes with Varying AA Group Content. Since the carboxylic acid groups interact with THO via hydrogen bonding, the amount of THO bound will depend on the AA content of the copolymer (Figure 11). P(AN-co-AA) membranes with differing AA contents were prepared by the phase inversion method (Kobayashi, T., *Anal. Chim. Acta*, in press). For CAF, the binding amounts of the substrate are independent on the AA content of the copolymer membrane. This is due to a low level of interaction of CAF with COOH groups by hydrogen bonding. However, THO binding to the membranes increases with

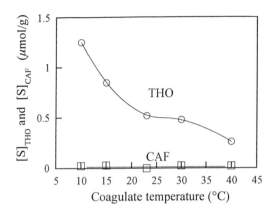

Figure 9. Amounts of THO and CAF taken up by THO-imprinted membranes coagulated at various temperatures of the water nonsolvent.

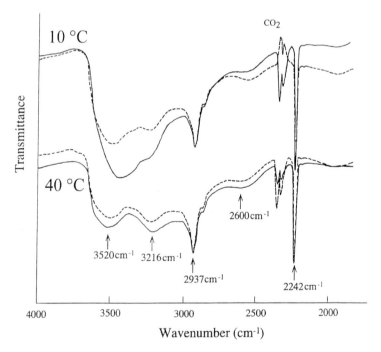

Figure 10. FT-IR spectra of the THO-imprinted P(AN-co-AA) membranes without and with THO extraction for coagulation temperatures of 10 and 40 °C.

enhancement of the AA content and becomes almost constant over an AA segment mole fraction of about 0.15. At higher than y = 0.3, it was very difficult to prepare the copolymer membranes by this phase inversion method. This is due to the very brittle nature of the resultant P(AN-co-AA) membranes with high AA contents.

Template Copolymerization of AN and AA in the Presence of THO. It is useful to compare the incorporation of THO substrate into P(AN-co-AA) copolymers prepared by radical copolymerization of AN and AA in the absence and the presence of THO template molecule. By the previously reported method (23), the copolymerization was carried out in the presence of THO template with monomer feed amounts of AN/AA/THO--O.572/0.107/0.026 (mole ratios) in DMSO. Then, membranes were prepared by phase inversion of the resultant templated copolymer (Kobayashi, T., unpublished data). THO was extracted from the resultant copolymer membranes by washing with aqueous AcOH. For the experiments of THO binding to the templated copolymer (TCP), a weighed amount of the membrane was immersed in a solution containing 1.1 μmol/dm^3 of THO at 30 °C for 24 hours (Figure 12). The THO binding per unit weight of copolymer increased with time and became constant after about 10 hours. This behavior of THO binding shows that the copolymer has imprinted sites for THO and the imprinted sites are produced by a template copolymerization of AN and AA. However, for PAN template polymerization (TP), the data show that there are no THO binding sites in the resultant homopolymer. This demonstrates the importance of the presence of AA groups in the polymeric membrane.

Figure 12 also contains data for a P(AN-co-AA) membrane (TCPM) prepared by phase inversion of the templated copolymer. After solubilization of the templated copolymer in DMSO, reprecipitation was carried out in 23 °C water to prepare the membrane. Although the amount of THO bound to the membrane is smaller than that measured before the membrane preparation, the copolymer membrane definitely shows THO binding. The values of bound THO for TCP and TCPM are higher than that for P(AN-co-AA) prepared without an imprinting treatment. Thus it is shown that polymeric template membranes can encode the volumetric size of a THO template. This approach will contribute to the development of a new type of molecular imprint membranes.

Acknowledgment. This research was supported partly by a Grant-in-Aid for Science Research (C) (09650757) from the Ministry of Education, Science, Sports and Culture of Japan.

Literature Cited

1. Mosbach, K. *Trends Biochem. Sci.* **1994**, *19*, 9.
2. Wulff, G. *Molecular Interaction in Bioseparations*, Ngo, T. Ed.; Plenum: New York, NY, 1993, p. 363.
3. Andersson, I. L.; Ekberg, B.; Mosbach, K. *Molecular Interaction in Bioseparations*, Ngo, T. Ed.; Plenum: New York, NY, 1993, p. 383.
4. Fischer, L.; Muller, R.; Ekberg, B.; Mosbach, K. *J. Am. Chem. Soc.* **1991**, *113*, 9358.
5. Sellergren, B.; Ekberg, B.; Mosbach, K. *J. Chromatogr.* **1985**, *347*, 1.
6. Viatakis, G.; Andersson, L. I.; Muller. R.; Mosbach, K. *Nature* **1993**, *361*, 645.
7. Kobayashi, T.; Wang, H. Y.; Fujii, N. *Chem. Lett.* **1995**, 927.
8. Wang, H.Y.; Kobayashi, T.; Fujii, N. *Langmuir* **1996**, *12*, 4850.
9. Mulder, M.; *Basic Principles of Membrane Technology*, Kluwer: Dordrecht, The Netherlands, 1991, pp.83-86.

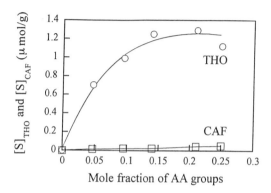

Figure 11. Amounts of THO and CAF taken up by THO-imprinted membranes of P(AN-co-AA) with different mole fractions of AA groups.

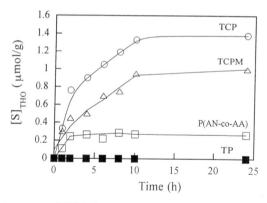

Figure 12. Amounts of THO bound to P(AN-co-AA) obtained from template polymerization without and with membranes prepared by phase inversion from 10 weight percent DMSO solutions.

10. Baker, R. W.; Cussler, E. L.; Eykamp, W.; Koros, W. J.; Riley, R. L.; Strathmann, H. *Membrane Separation Systems: Recent Developments and Future Directions*, Noyes Data Corporation: Park Ridge, NJ, 1991, pp. 9-13.
11. Noble, R. D.; Stern, S. A. *Membrane Separations Technology: Principles and Applications*, Elsevier: Amsterdam, The Netherlands, **1995**, pp. 1-43.
12. Loeb, S.; Sourirajan, S. *Adv. Chem. Ser.* **1962**, *38*, 117.
13. Cabasso, I. In *Encyclopedia of Polymer Science and Engineering*, Mark, H. F.; Bikales, N. M.; Overberger, C. G.; Menges, G.; Kroschwitz, J. I., Eds.; Wiley: New York, NY, 1987, vol. 9, p. 5O9. and references therein.
14. Munari, S.; Boffino, A.; Capanneli, G. *J. Membr. Sci.* **1983**, *16*, 181.
15. Koenhen, D. M.; Mulder, M. H. V.; Smolder, C. A. *J. Appl. Polym. Sci.* **1977**, *21*, 199.
16. Bottino, A.; Capannelli, G.; Munari, S. *J. Appl. Polym. Sci.* **1985**, *30*, 3009.
17. Matsuura, T. *Synthetic Membranes and Membrane Separation Processes*, CRC Press: Boca Raton, FL, **1994**, pp. 11-31.
18. Nakao, S.; Osada, H; Kurata, H.; Turu, T.; Kimura, S. *Desalination* **1988**, *70*, 1519.
19. Jitsuhara, I.; Kimura, S. *J. Chem. Eng. Jpn.* **1983**, *16*, 389.
20. Miyarna, H.; Tanaka, K.; Nosaka, Y.; Fujii, N. *J. App. Polym. Sci.* **1988**, *36*, 925.
21. Kobayashi, T.; Nagai, T.; Suzuki, T.; Nosaka, Y.; Fujii, N. *J. Membr. Sci.* **1994**, *86*, 47.
22. Kobayashi, T.; Miyarnoto, T.; Nagai, T.; Fujii, N. *J. Appl. Polym. Sci.* **1994**, *52*, 1519.
23. Oak, M.; Kobayashi, T.; Wang, H. Y.; Fukaya, T.; Fujii, N. *J. Membr. Sci.*, **1997**, *123*, 185.
24. Friendrich, C.; Driancourt. A.; Noel. C.; Monnerie, L. *Desalination* **1981**, *36*, 39.
25. Guillotin, M.; Lemoyne, C.; Noel, C.; Monnerie, L. *Desalination* **1977**, *21*, 165.
26. Yamadera, R. *Koubunshi Kagaku* **1964**, *21*, 362.
27. Lee, Y. M.; Oh, B. *J. Membr. Sci.* **1995**, *98*, 183.
28. Silverstein, R. M.; Bossier, G. C.; Morrill, T. C. *Spectrometric Identification of Organic Compounds*, 4th ed., Wiley: New York, NY, **1991**, p. 183.
29. Herzberg, G. *Molecular Spectra and Molecular Structure, II. Infrared and Raman Spectra of Polyatomic Molecules*, Van Nostrand Reinhold: New York, NY, p. 334.
30. Koyama, K; Nishimura, M. *Membrane* **1980**, *5*, 189.
31. Kobayashi, T; Nagai, T.; Ono, M.; Wang, H. Y.; Fujii, N. *Eur. Polym. J.* **1997**, 33, 1191.
32. Wang, H. Y.; Kobayashi, T.; Fuji, N. *Langmuir* **1997**, *13*, 5396.

Chapter 14

Imprinting of Proteins on Polymer-Coated DNA for Affinity Separation with Enhanced Selectivity

Daisuke Umeno, Masafumi Kawasaki, and Mizuo Maeda

Department of Chemical Science and Technology, Faculty of Engineering, Kyushu University, Hakozaki, Faculty of Engineering, Kyushu University, Hakozaki, Fukuoka 812-81, Japan

A temperature-responsive polymer, poly(N-isopropyl acrylamide) terminated with psoralen, has been synthesized. Photo-induced reaction between DNA and psoralen end group on the polymer resulted in the grafting of poly(N-isopropyl acrylamide) on double-stranded DNA. The conjugate was found to form a precipitate at temperatures above 31 °C and was easily collected by centrifugation. In this precipitation process, the conjugate collected DNA binders from solution. On the other hand, the binding efficiency of these molecules to the conjugate decreased significantly with an increasing level of DNA modification by the polymer. This blocking effect was utilized for protein imprinting on the DNA. Pre-incubation of the restriction endonuclease *Eco*RI with DNA results in selective preservation of its binding site from polymer modification. Bioimprinting is a potentially facile method for the preparation of selective collectors for site-specific DNA binders.

Molecular imprinting is a powerful strategy for the construction of a desired recognition site and has been widely applied to the preparation of selective separators for various target molecules and ions *(1,2)*. However, it is generally difficult to realize high separation selectivity by this method. Only a portion of the functional groups in the system actually interact with the template molecules in the polymerization step and the remaining functional groups form non-specific recognition sites. A very recent approach is to mask the functional groups in the non-specific recognition sites by chemical reaction and thereby enhance selectivity (McNiven, S., Tokyo University, personal communication, 1997).

When macromolecules of a characteristic size or shape are the target species for separation, molecular masking itself could bring about the imprinting effect. Ahrne *et al.* report the imprinting of bacteria on the surface of a polyamide latex *(3)*. The polyamide surface of the resin was coated with perfluoropolyether in the presence of bacteria. Bacteria adhering to the resin blocked the adhesion sites from coating with the polymeric perfluoropolyether. Lectin was then immobilized on the uncoated domains which had the shape of the bacterial template to give a latex with bacteria-selective binding properties.

Living organisms often display a variety of host functions simultaneously. To utilize them as functional materials, methodologies for choosing one of the functions are important. For instance, even a single molecule of DNA displays a large number

202

of recognition sites for DNA-binding proteins. Each of them is highly specific, but the host function of the entire DNA molecule is much less selective because the different recognition sites act independently. At present, no method is available to obtain an affinity ligand with perfect selectivity except by the synthesis of a small DNA fragment with the recognition sequence of the target protein (4). This is a laborious process. Furthermore this approach is ineffective for a protein whose DNA binding site is unknown.

Protein imprinting/masking offers the potential for modification of the binding selectivity of the DNA. This methodology may provide a technique for the preparation of a highly specific separator for different DNA-binding species from a single source.

We have synthesized poly(N-isopropyl acrylamide) terminated with a DNA-binding group, psoralen. DNA grafted with this polymer exhibits reduced, but significant, affinity to DNA-binding molecules. The diminished affinity due to polymer introduction was found to be controllable. Dense introduction of the polymer inhibited access of DNA-binding proteins almost completely. In this paper, we examine protein imprinting of DNA by polymer modification to produce a selective separator of DNA-binders with enhanced selectivity.

Materials and Methods

Materials. N-Isopropylacrylamide (NIPAAm) was obtained from Tokyo Kasei Kogyo and recrystallized from benzene-hexane. N,N'-Azobisisobutyronitrile (AIBN) and 3-mercaptopropanoic acid (MPA) were purchased from Kishida chemicals and Aldrich, respectively. Trioxalene (4,5',8-trimethylpsoralen) was obtained form Aldrich. pBR 322 was purchased from Fermentas. *Hind* III and *Eco*RI were obtained from Boehringer Mannheim. Other reagents and solvents were commercial products and were used as received. pSV-β-Galactosidase vector was a gift from Dr. A. Maruyama of Tokyo Institute of Technology.

Psoralen-Terminated PolyNIPAAm (PsoPNIPAAm). The synthetic route is shown in Scheme 1. Initially, semitelechelic polyNIPAAm with a carboxyl group was prepared by the method of Okano *et al.* (5). Radical polymerization of NIPAAm was performed in dimethylformamide (DMF) with AIBN as initiator and 3-mercapto-propanoic acid (MPA) as the chain transfer agent. The resultant polymer was purified by repeated reprecipitation from DMF-diethyl ether. The carboxyl group at the polymer terminus was esterified with reaction with N-hydroxysuccinimide and dicyclohexyl-carbodiimide (DCC) in dry DMF at *ca.* 25 °C for 24 hours. The reaction mixture was concentrated *in vacuo* and poured into diethyl ether to give a precipitate. Ester formation was confirmed by the UV spectrum (6) of the polymer. The activated polyNIPAAm (0.50 g, 94 μmol) and 0.07 g (235 μmol) of the amino derivative of psoralen, (4'-[[N-(2-aminoethyl)amino]methyl]-4,5',8-trimethylpsoralen synthesized by the method of Lee *et al.* (7)), were dissolved in dry DMF (10 mL in total volume). The solution was stirred at *ca.* 25 °C for 12 hours, concentrated *in vacuo*, and poured into diethyl ether. The precipitate was filtered and washed with 100 mL of diethyl ether. The resultant white powder was dried *in vacuo*. This polymer was dissolved in pure water to provide a stock solution.

The molecular weight of the polymer was determined at each step with terminal-group analysis by acid-base titration for the carboxyl terminus and UV absorption for the ester (260 nm) and psoralen (300 nm) moieties.

To determine the phase transition temperature (cloud point) of an aqueous solution of the polymer, optical transmittance was measured at 500 nm with a Hitachi U-3210 UV/vis spectrophotometer as the temperature was increased at a rate of 0.4 °C/minute.

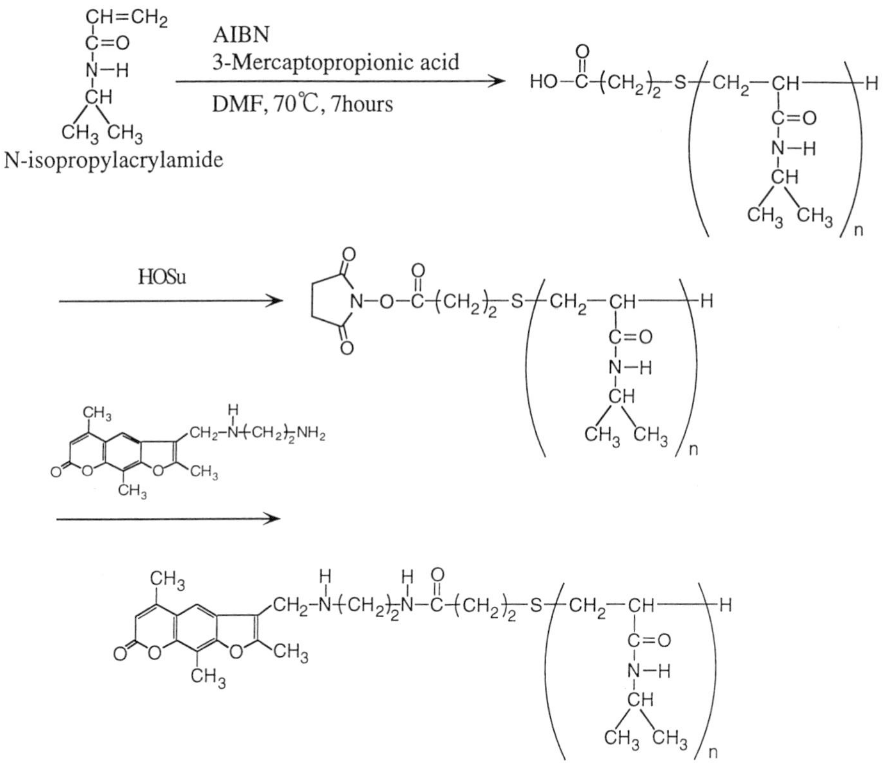

Scheme 1. Synthetic Route to Poly(N-isopropylacryamide) Terminated with Psoralen.

The temperature for 90 % light transmittance of the polymer solution was defined as the cloud point.

Photoimmobilization of PsoPNIPAAm on Plasmid pBR 322 DNA. To a solution of linearized pBR 322 (4,302 base pairs in length) in a 1.5-mL test tube (Eppendorf) was added a TE solution (10 mM Tris-HCl, 1.0 mM EDTA, pH 7.4) of PsoPNIPAAm from the stock solution (100 μM per strand). The total volume was adjusted to 100 μL. The final concentration of DNA was 58.5 μM in base pairs (bp). The concentration of PsoPNIPAAm was varied in the range of 0-50 μM (per strand). Each solution was irradiated (*ca.* 30 mW/Cm^2) in an ice bath with a 500-W ultra-high pressure Hg lamp equipped with a high-pass filter (Toshiba, UV-31) for 10 minutes.

The resultant solutions were analyzed by gel electrophoresis. A 15-μL portion of each sample solution was combined with 5 μL of a gel-loading solution consisting of glycerin and water (1:1, v/v). the mixture was loaded on a 1.0 %-agarose gel and the gel electrophoresis was performed at 7 V/cm for 1.5 hours in TBE buffer (89 mM Tris borate, 2.4 mM EDTA, pH 8.0). The gel electrophoresis was performed at 10 °C since precipation of polyNIPAAm and its conjugates from aqueous glycerin (25 %) take place at ambient temperature (*8*). After electrophoresis, the DNA in the gel was stained with ethidium bromide.

Accessibility of DNA Binders to the DNA-PolyNIPAAm Conjugate. The conjugates between the DNA and polyNIPAAm were examined for their accessibility to ethidium bromide and *Eco*RI.
Ethidium Bromide. Calf-thymus DNA (sonicated, 50-500 bp in length) and PsoPNIPAAm were conjugated by the method described above. The final concentration of the DNA was fixed at 58 μM in bp, while the concentration of PsoPNIPAAm for the photoreaction was varied in the range of 0-100 μM in strand. To the resulting conjugate was added an excess of ethidium bromide (25 μM). The fluorescence intensity (λ_{ex} = 546 nm, λ_{em} = 595 nm) due to the DNA-binding (intercalation) of ethidium ion was measured for each sample.
*Eco*RI. To the conjugate between pSV-β-galactosidase vector (6,921 bp, linearized by Hind III) and PsoPNIPAAm ([DNA] fixed at 58 μM in bp, [PsoPNIPAAm] ranged from 0 to 50 μM), *Eco*RI (60 units) and digestion buffer were added and the solution was incubated at 29 °C. After 24 hours, the solution was subjected to electrophoresis. After the gel was stained with ethidium bromide, the band corresponding to the digested fragment of the DNA-polyNIPAAm conjugate was evaluated by scanning densitometry to estimate the ratio of the conjugate retaining the binding sites for *Eco*RI.
T4 Ligase. Reactivity of T4 ligase was also examined for the conjugates between DNA and polyNIPAAm. To the conjugate between pBR 322 (linearized by *Eco*RI) and PsoPNIPAAm ([DNA] = 29 μM in bp, [PsoPNIPAAm] = 0-10 μM), T4 ligase (1,000 units) was added and the solution was incubated at 16 °C. After two hours, samples were subjected to electrophoresis. After the gel was stained with ethidium bromide, the bands corresponding to the ligated conjugate were evaluated by scanning densitometry.

Protein Imprinting of the Polymer-Coated DNA. *Eco*RI (0-100 units) was added to pBR 322 (58 μM) in the absence of Mg^{2+}. PsoPNIPAAm was added and the solution was irradiated with UV light. The total amount of PsoPNIPAAm added for the photochemical reaction ranged from 0 to 60 µM in strand. For control samples, the same procedure was performed using the same amount of *Eco*RI which had been denatured by heating at 95 °C for 30 minutes.

The accessibility of *Eco*RI to the resulting conjugates was examined. To a solution of the conjugate, *Eco*RI or denatured *Eco*RI were added so that all samples had the same total amount of *Eco*RI. Mg^{2+} was added to the samples to initiate digestion. After incubation at 28 °C for 24 hours, the sample solutions were subjected to gel electrophoresis. Accessibility of *Eco*RI to these conjugates was evaluated from the relative band intensities for the digested conjugate.

Temperature-Induced Precipitation of the DNA-PolyNIPAAm Conjugate. The thermally-induced precipitation of the resulting conjugates between pBR 322 and polyNIPAAm was examined. To the conjugate solution (100 µL), an aqueous solution of polyNIPAAm was added (1.5 wt % total). The mixture was heated to 40 °C and centrifuged (15,000 rpm for 3 minutes) at that temperarure to give a white precipitate. After centrifugation, the supernatant was collected and subjected to gel electrophoresis (0.7 % agarose, 7 V/cm for 2 hours in TBE). After the gel was stained with ethidium bromide, the DNA band due to the conjugate was evaluated by scanning densitometry to estimate the amount of unprecipitated conjugate.

Separation of DNA Binders by Affinity Precipitation. To a solution of the λDNA in a 1.5 mL test tube (Eppendorf) was added PsoPNIPAAm for the stock solution (100 µM in strand). The final concentration of the DNA was 100 µM in bp. The concentration of PsoPNIPAAm was varied between 0 and 50 µM in strand. Samples were irradiated with UV light as described above. The resulting conjugates between polyNIPAAm and λDNA were mixed with an aqueous solution of ethidium bromide (1.0 µM) and polyNIPAAm (1.0 wt % in total). The mixed solutions (60 µL in total volume) were incubated at *ca.* for 1 hour. After incubation, the mixed solutions were heated in a water bath at 40 °C for 3 minutes. After centrifugation (15,000 rpm) at 40 °C for 60 seconds, the supernatant was collected and mixed with a TE solution of sonicated calf thymus DNA (1.0 mM), which was used to enhance and saturate the fluorescence intensity of the ethidium. The amount of ethidium remaining in the supernatant was evaluated by spectrofluorimetry (Shimadzu Rf-5300PC spectrofluorimeter, $\lambda_{ex} = 546$ nm, $\lambda_{em} = 595$ nm).

Affinity separation of the restriction endonuclease *Eco*RI was performed as follows. To a solution of the linearized pBR 322 (*Bsm* I digest) in a 1.5 mL test tube (Eppendorf) was added PsoPNIPAAm from the stock solution. The final concentration of the DNA was 56 µM in bp. The concentration of PsoPNIPAAm was varied between 0 and 60 µM in strand. The samples were irradiated with UV light as described above. The resulting conjugates between polyNIPAAm and pBR 322 (*Bsm* I digest) were mixed with *EcoRI* (60 units) and polyNIPAAm (1.0 wt % in total). The mixed solutions (60 µL in total volume) were incubated at *ca.* 25 °C for 1 hour. The solutions were heated in a water bath to 40 °C and centrifuged (15,000 rpm) for 60 seconds at 40 °C. The supernatant (30 µL) was collected and mixed with reaction buffer containing Mg^{2+} (final concentration: 5.0 mM Tris-HCl, 10.0 mM NaCl, 1.0 mM $MgCl_2$, 0.10 mM dithioerthritol, pH = 7.5). Each of the solutions was combined with a substrate DNA (pBR 322/*Bsm*I digest) solution except the supernatant of a reference sample which was a simple mixture of the DNA and polyNIPAAm. (No substrate DNA was added to the reference solution since the DNA was found to remain in the supernatant even after the 'heat and precipitation' procedure, if it was not conjugated with polyNIPAAm.) After incubation for 30 minutes at 37 °C, the samples were subjected to gel electrophoresis and the band corresponding to the digested substrate was evaluated by scanning densitometry to estimate the enzymatic activity remaining in the supernatant.

Results and Discussion.

Psoralen-Terminated PolyNIPAAm. The synthetic route to psoralen-terminated poly(N-isopropylacrylamide) (PsoPNIPAAm) is shown in Scheme 1. Semitelechelic polyNIPAAm with a carboxyl end group was prepared according to the method of Takei *et al.* (*7*). The number-average molecular weight of the polyNIPAAm was determined by base titration of the terminal carboxylic acid groups to be about 5,400 (*ca.* a 50-mer), which is in good agreement with the previous report (*5*). In each step of derivatizing the polymer terminus, the end groups were evaluated and the apparent molecular weights of the polymers were estimated. The calculated values for PsoPNIPAAm were 6,700 for the ester terminus and 6,000 for the psoralen terminus. There is no significant difference between these values.

PolyNIPAAm is a well-known polymer which shows a phase transition (*9*). When an aqueous solution of polyNIPAAm is gradually heated, the polymer becomes insoluble abruptly at 31 °C and precipitates. The same transition temperature was found for the reversible precipitation process for the polyNIPAAm prepared in this study.

Photoinduced Conjugation of PsoPNIPAAm to pBR 322. The photo-reactivity of the psoralen moiety which was introduced at the terminus of polyNIPAAm was examined for pBR 322 plasmid DNA by gel electrophoresis.

PsoPNIPAAm was added to pBR 322 and the mixture was irradiated with UV light. As can be seen in lanes 6-9 (*?*) of Figure 1, the gel electrophoretic migration of the resulting conjugates revealed significant retardations depending on the amount of PsoPNIPAAm which was added before the photochemical reaction to form the conjugate. However, the presence of PsoPNIPAAm without irradiation did not produce retardation (lane 10). The drastic changes in migration behavior shown in lanes 6-9 demonstrates modification of the DNA by PsoPNIPAAm to make the DNA larger. It should be noted that irradiation of solutions of unmodified polyNIPAAm and pBR 322 (lanes 2-5) did not affect the electrophoretic migration behavior. Thus the terminal psoralen group is necessary for the DNA modification.

The dependence of retardation upon the amount of PsoPNIPAAm added before irradiation indicates that the degree of DNA modification by polyNIPAAm can be regulated by varying the amount of PsoPNIPAAm which is used in the reaction. It was also found that the degree of modification can be regulated by changing the period of UV irradiation.

Accessibility of DNA Binders to Polymer-Coated DNA. The accessibility of DNA-binding molecules to the conjugates was then investigated. For ethidium bromide, which is a well known DNA stain, the accessibility to the conjugate was assessed by its fluorescence intensity when intercalated. As shown in Figure 2A, the fluorescence intensity gradually decreased with an increasing amount of PsoPNIPAAm present during the photoreaction to form the conjugate. However, the diminution in fluorescence leveled off as the amount of PsoPNIPAAm present during irradiation increased and was never reduced to below 50 % of the initial value.

We defined the ration of the DNA conjugate which is digested by *Eco*RI as the accessibility of *Eco*RI. The profile for accessibility of *Eco*RI was found to be quite different from that for ethidium ion (Figure 2A). The *Eco*RI accessibility decreased linearly as the amount of PsoNIPAAm present during irradiation increased. This difference can be explained as follows. The polymer chain attached to the DNA

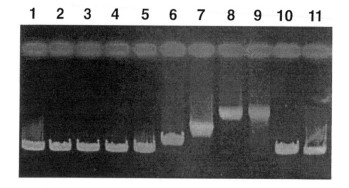

Figure 1. Gel Electrophoresis (1.2 wt %) of Linear pBR322 (58 μM in bp) Irradiated by UV Light in the Presence of Unmodified and Terminally-Modified Poly(*N*-isopropylacrylamide). Lanes 1 and 11, DNA alone (control); lanes 2-5. [PNIPAAm] = 5. 20, 50, and 200 μM; lanes 6-9. [PsoPNIPAAm] = 5, 20. 50, and 200 μM; lane 10. [PsoPNIPAAm] = 200 μM but without irradiation.

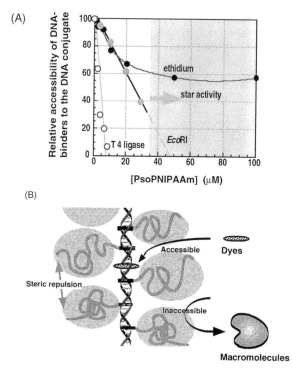

Figure 2. Accessibility of DNA Binders to the DNA-PolyNIPAAm Conjugates. (a) Accessibility of ethidium ion, the restriction endonuclease *Eco*RI, and T4 ligase with [DNA] = 58 μM in bp. (b) Selective repulsion for the macromolecular binders introduced by polymer coating.

should occupy a certain space on the DNA (Figure 2B). This situation would allow access of smaller molecules, while bulkier, potential binders would be rejected. Thus PsoPNIPAAm grafted on DNA inhibits almost all of the binding sites for macromolecules, such as enzymes and anti-DNA antibodies. This means that we can selectively mask the DNA against macromolecules.

When the modification exceeded a certain level, it became difficult to evaluate the accessibility. Significant band broadening due to star activity (*10*) was observed when more than 50 μM PsoPNIPAAm was present during the photoreaction. Under this condition, multiple sites can be attacked by *Eco*RI resulting in an increase in apparent accessibility.

In the case of T4 ligase, the enzymatic action was more severely inhibited than with *Eco*RI. This may be ascribed to the difference in their mode of action. It is reported that each intercalation of psoralen rewinds the double strands about 24 ° (*11*). This would diminish the flexibility of the DNA. In addition, introduction of the polymer should make it difficult for DNA to bend to form a circular structure. Both factors should strongly affect the enzymatic activity of T4 ligase.

Effect of Protein Imprinting on the DNA-PolyNIPAAm Conjugate. As can be seen in Figure 2, the dense introduction of polymers almost completely blocks access of DNA-binding proteins. Under this condition, the DNA is essentially 'coated' by the polymer. When protein imprinting/masking is applied to the photo-induced polymer coating, the binding selectivity of DNA as the host molecule can be modified.

The protocol is shown in Figure 3. To the plasmid pBR 322 is added the restriction endonuclease *Eco*RI. The solution was free of Mg^{2+} which is required for nuclease activity of *Eco*RI. Under this condition, the *Eco*RI strongly binds to its recognition site ($K_d \sim 10^{-10}$) without scission of the strands (*12*). PsoPNIPAAm is added and the solution is irradiated by UV light. Reference samples were subjected to the same procedure except that denatured *Eco*RI was used. A reaction buffer was added and the samples were digested at 28 °C for 24 hours. The results are shown in Figure 3B.

For the DNA conjugate prepared in the presence of denatured *Eco*RI (100 units), the accessibility decreased with an increasing degree of polymer introduction. Thus denatured *Eco*RI does not seem to bind to the proper site. On the other hand, DNA conjugates prepared in the presence of native *Eco*RI showed no decrease in accessibility to *Eco*RI. The DNA conjugates prepared in this fashion retained almost the same accessibility with native *Eco*RI. The results strongly suggest that *Eco*RI imprinting was successfully performed in the PNIPAAm coating of the DNA.

To ascertain the importance of the *Eco*RI added prior to the polymer grafting, the experiment illustrated in Figure 3A was performed under several different conditions. In this series of experiments, the DNA (58 μM in bp) and PsoPNIPAAm (60 μM in strand) present in the photoreaction were fixed, while the amount of *Eco*RI (or denatured *Eco*RI) added prior to polymer grafting was varied in the range of 0-100 units. The final amounts of *Eco*RI and denatured *Eco*RI were adjusted to 100 units for the *Eco*RI digestion test. The results are shown in Figure 3C. For the DNA conjugate prepared in the presence of native *Eco*RI, the accessible sites for *Eco*RI increased with an increasing amount of *Eco*RI fed prior to the polymer grafting. When 100 units of *Eco*RI were used, the *Eco*RI accessibility to the conjugate reached almost the same level as native DNA. On the other hand, all of the conjugates prepared in the presence of denatured *Eco*RI have almost the same level (40 %) of accessibility regardless of the amount. It is apparent that *Eco*RI is crucial to the protection of its binding site from polymer grafting: the *Eco*RI actually plays a role as a "masking agent".

(A)

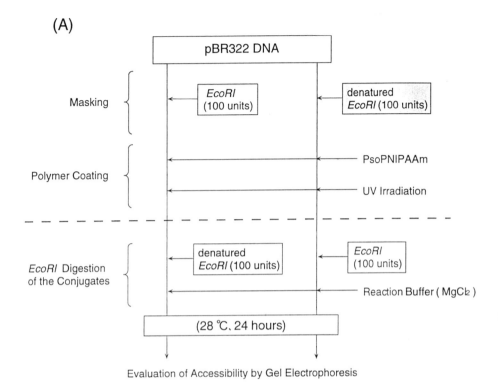

(B)

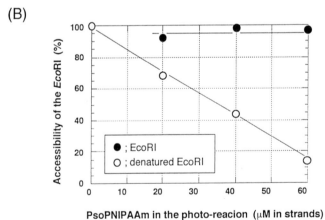

PsoPNIPAAm in the photo-reacion (μM in strands)

Figure 3. Protein Imprinting on Double-Stranded DNA. (a) The protocol: In the presence of *Eco*RI or denatured *Eco*RI, PsoPNIPAAm was photochemically immobilized on the DNA. Accessibility of *Eco*RI to the resulting conjugates was studied. (b) Imprinting effects on the accessibility to *Eco*RI. (c) The effect of the amount of *Eco*RI (masking agent) added prior to the polymer grafting to the efficiency of *Eco*RI-imprinting.

(C)

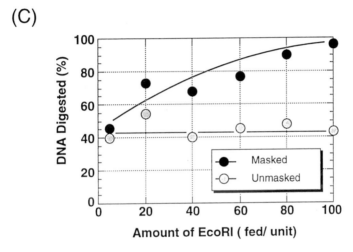

Figure 3. *Continued.*

Thus, the protein imprinting illustrated in Scheme 2 can be realized by the protocol shown in Figure 3A. It seems quite reasonable to expect that this technique will be applicable to the other site-specific DNA-binders. This technique provides a facile method to diminish or enhance the binding selectivity of DNA

Temperature-Responsive Precipitation of the Conjugates. Poly(N-isopropyl acrylamide) (polyNIPAAm) is known to exhibit temperature-responsive behavior. When attached to DNA, the polymer should induce this property in the conjugate.

For the separation by precipitation, the concentration of polyNIPAAm was adjusted to 1.5 wt % by adding the homopolymer of NIPAAm, since it was found that a certain concentration (>1.0 wt %) is required for reproducible precipitation. The conjugate solution was heated to 40 °C and centrifuged (15,000 rpm) at that temperature to give a white precipitate. The supernatant was carefully collected and subjected to gel electrophoresis. The data in Figure 4 show that the precipitation efficiency of the DNA conjugate was dependent upon the amount of PsoPNIPAAm present during the photochemical reaction. In contrast, a mixture of DNA and PsoPNIPAAm which had not been irradiated showed only slight precipitation, which is ascribed to a non-covalent modification of DNA by PsoPNIPAAm. It is presumed that a certain level of polymer modification is required for an efficient reprecipitation of the conjugates.

Separation of DNA Binding Molecules by Affinity Precipitation. Interaction of DNA-binding molecules with the conjugate should cause them to be precipitated when the temperature of the solution is raised. As a preliminary demonstration, we examined the separations of the DNA-binding species ethidium ion and *Eco*RI by affinity precipitation.

The protocol is illustrated in Figure 5. λDNA was used for the separation of ethidium ion. To a fixed amount of ethidium bromide (1.0 μM), the conjugate between λDNA and PsoPNIPAAm was added. After the addition of polyNIPAAm (1 wt %), the sample was heated to 40 °C and centrifuged at this temperature. The precipitate was pink-colored. The distribution of ethidium ion was determined by its fluorescence intensity in both the supernatant and precipitate. As can be seen in Figure 6, the separation efficiency increases as the amount of PsoPNIPAAm present during the photoreaction to produce the conjugate is increased. This suggests that a certain level of polymer incorporation is required for efficient separation of this DNA binder.

For the separation of *Eco*RI, the conjugate between pBR 322 and PsoPNIPAAm was employed. The conjugate was added to a buffer solution containing *Eco*RI. It should be noted that the *Eco*RI solution was free from Mg^{2+} at this stage to prevent DNA scission. polyNIPAAm (1 wt %) was added and the solution was heated to 40 °C and centrifuged at this temperature. The supernatant was collected and the endonuclease activity of the *Eco*RI in the supernatant was determined by adding substrate DNA and Mg^{2+}-containing reaction buffer. The results are shown in Figure 7.

When the conjugate was replaced with DNA and PsoPNIPAAm which had not been irradiated and the precipitation was performed as before, the supernatant gave complete digestion of the DNA. Thus the *Eco*RI was present in the supernatant after the precipitation of polyNIPAAm and was able to digest the substrate DNA. In contrast, the endonuclease activity of the supernatant formed by addition of *Eco*RI to the pBR322-polyNIPAAm conjugate diminished drastically after the centrifugation at 40 °C which indicates that the amount of *Eco*RI in the supernatant was decreased. The degree of diminution is found to depend upon the concentration of PsoPNIPAAm which was present during the photochemical modification of the DNA. Therefore it is

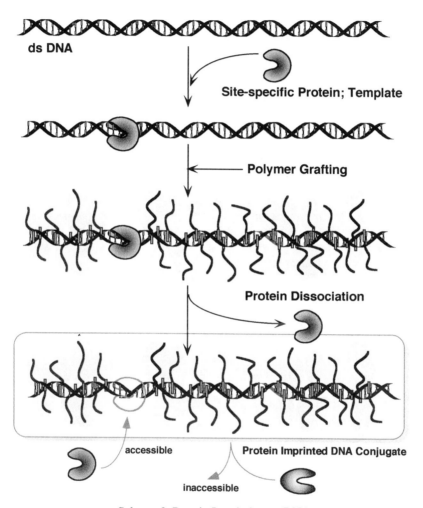

ds DNA

Site-specific Protein; Template

Polymer Grafting

Protein Dissociation

accessible

Protein Imprinted DNA Conjugate

inaccessible

Scheme 2. Protein Imprinting on DNA.

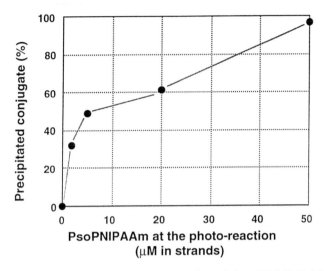

Figure 4. Temperature-Responsive Precipitation of the pBR322-PolyNIPAAm Conjugates. The conjugates were heated to 40℃ in the presence of the homopolymer polyNIPAAm and centrifuged at that temperature. The precipitation efficiency of the conjugate was evaluated by measuring the conjugates which remained in the supernatant.

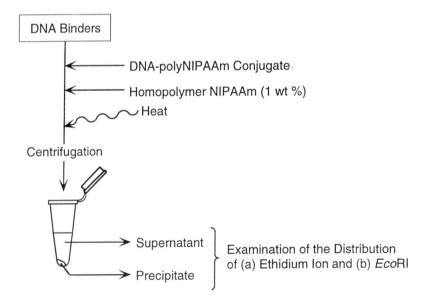

Figure 5. Procedure for the Separation of DNA-Binders by Affinity Precipitation.

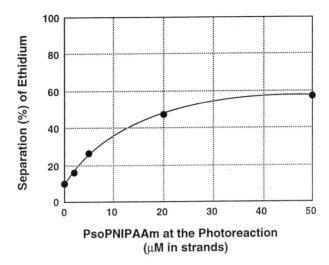

Figure 6. Separation of Ethidium Ion by Affinity Precipitation. To the solution of ethidium bromide was added the DNA-polyNIPAAm conjugate. The solution was heated to 40℃ and centrifuged. The relative amount of ethidium ion in the supernatant was determined by its fluorescence intensity. (λex= 546 nm, λem= 595 nm)

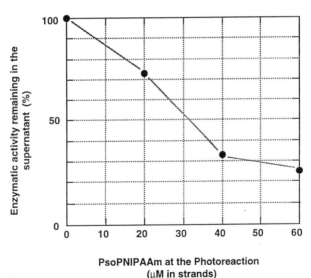

Figure 7. Separation of *Eco*RI by Affinity Precipitation. To the solution of *Eco*RI was added the DNA-polyNIPAAm conjugate. The mixture was heated to 40℃ and centrifuged at that temperature. The endonuclease activity of the resulting supernatant was evaluated by the addition of substrate DNA and Mg^{2+}. The relative amount of the substrate DNA digested (%) was calculated from the band intensity.

concluded that the DNA-polyNIPAAm conjugate is responsible for the decrease endonuclease activity of the supernatant. From these results it is apparent that *Eco*RI is captured by the conjugate, precipated from the aqueous solution, and concentrated in the polymer phase. This conjugate provides a rapid and useful separation technique since complexation of the target species takes place in homogeneous solution.

Conclusions.

By application of protein masking to a polymer coating on DNA, we have been able to enhance the binding selectivity of DNA, which exhibits low selectivity in the unmodified state. The polyNIPAAm polymer coating produces precipitation in a narrow temperature range. Photoinduced grafting of PsoPNIPAAm in the presence of a template protein may provide selective separations for various DNA-binding proteins. However we have not yet applied this method in a separation system due to difficulty in establishing complete removal of the template protein. However, we still believe that DNA conjugates prepared by protein imprinting holds promise as an easily prepared, selective collector for various DNA-binding proteins.

It should be noted that this technique differs from the general concept of molecular imprinting because the recognition domain is not introduced in the process, but already exists. However, it is related to molecular imprinting in that the selective binding property is produced by the template and the desired species may be separated from other species.

An advantage of this method is that one can obtain a wide variety of host functions by choosing a combination of DNA and "masking agents". This technique is applicable to any DNA-binder as long as it is strong and seqeuence-specific. By use of a set of several proteins, once can obtain combinationally imprinted DNA. Taking into consideration that the primary structure of DNA can be precisely designed, this technique would also provide an interesting approach for the construction of molecular architectures with precisely-regulated nano-structures.

Acknowledgment.

This work was supported in part by a Grant-in-Aid for Scientific Research from the Ministry of Education, Science, Sports and Culture of Japan. Financial support by the Kumagai Foundation for Science and Technology is also acknowledged.

Literature Cited.

1. Wulff, G. *Angew. Chem., Int. Ed. Engl.* **1995**, *34*, 1812.
2. Takeuchi, T.; Matsui, J. *Acta Polymer* **1996**, *47*, 471.
3. Aherne, A.; Alexander, C.: Payne, M. J.; Perz, N.; Vulfson, E. N. *J. Am. Chem. Soc.*, **1996**, *118*, 8771
4. Goss, T. A.; Bard, M.; Jarrett, J. W. *J. Chromatogr.* **1991**, *588, 157.*
5. Takei, Y. G.; Aoki, T.; Sanui, K.; Ogata, N.; Okano, T. Sakurai, Y. *Bioconjugate Chem.* **1993**, *4*, 42.
6. Miron, T.; Wilchek, M. *Anal. Biochem.* **1982**, *126*, 433.
7. Lee, B. L.; Murakami, A.; Blake, K. R.; Lin, S.-B.; Miller, P. S. *Biochemistry* **1988**, *27*, 3197.
8. Maeda, M.; Nishimura, C.; Inenaga, A.; Takagi, M. *Reactive Polymers* **1993**, *21*, 27.
9. Schild, H. G.*Prog. Polym. Sci.* **1992**, *17*, 163.
10. Cole, R. S. *Biochim. Biophys. Acta* **1970**, *217*, 30.
11. Polysky, B.; Greene, P.; Garfin, D. E.; McCarthy, B. J.; Gopodman, H. M.; Boyer, H. W. *Proc. Nat. Acad. Sci. USA* **1980**, *191*, 593.

METAL ION RECOGNITION
WITH ORGANIC-BASED POLYMERS

Chapter 15

Templated Polymers for the Selective Sequestering and Sensing of Metal Ions

Xiangfei Zeng, Anton Bzhelyansky, Sue Y. Bae, Amanda L. Jenkins, and
George M. Murray

Department of Chemistry and Biochemistry, University of Maryland Baltimore
County, 1000 Hilltop Circle, Baltimore, MD 21250

The template approach is applied to the synthesis of ion-templated
polymers using a variety of template metals. The resulting polymers
are used as selective sequestering resins, and are extensively tested and
characterized. Several of these polymers are also used to construct ion
selective electrodes for sensing the template metal ion. A parallel path
of research entails synthesis of the polymers with ligating monomers
that possess chromo- and fluorophores, allowing for the construction of
chemically selective optical metal ion sensors. These and further
studies are aimed at the elucidation and optimization of a general
approach for the synthesis of selective metal ion sequestering and
sensing polymers.

The application of molecular imprinting to the construction of selective polymeric
adsorbents for relatively large molecules and molecular ions is an established practice.
Much less work has been performed toward making such adsorbents for metal ions.
Up to now, success in the area have generally been limited to selective adsorbents for
metal ions that do not exhibit directed bonding. Thus, by using high levels of
crosslinking combined with macromolecular ligating monomers, selective polymers for
Ca(II) and Mg(II) were obtained (1). The following work outlines the general
principles of synthesis and discusses analytical applications of metal ion templated
polymers that exhibit directed bonding and possess an economic or environmental
importance.

Synthesis of Selective Polymeric Adsorbents

An ideal metal ion selective separation or measurement technique: 1) takes place in a
simple single-stage operation involving no elaborate apparatus; 2) is rapid; 3) needs
little or no pre-treatment of the aqueous phase (a requirement for *in situ*
measurements); 4) involves an agent that is insoluble in the aqueous phase; and

5) entails a minimal cost. An ideal complexant to be used in the construction of ion selective sequestering agents and sensors will complex only a specific ion. The following discussion concerns the development and prospective applications of metal ion complexing polymers that meet the above criteria.

Selectivity for a specific metal ion is obtained by providing polymers with cavities lined with complexing ligands so arranged as to match the charge, coordination number, coordination geometry, and size of the metal ion. These cavity-containing polymers are produced by using a specific ion which acts as a template around which monomeric complexing ligands self-assemble and later polymerize. The ligands of choice contain functional groups known to form stable complexes with the specific ion and less stable complexes with other cations. The experiments of Neckers, Wulff, Kabanov, Nishide, Kuchen, Harkins, Fish and their coworkers (2-8) detail the approaches to making templated resins for metal ions that have improved selectivity, but exhibit low capacity. We have extended and amended their methodologies by changing the order of the steps, by the inclusion of sonication, by using higher template complex loading, and by the selection of functional groups with higher complexation constants. The obtained polymers were used for selective ion extraction and fabrication of selective metal ion sensors (9).

Production and Characterization of Metal Ion Templated Polymers. The following steps constitute the essence of the research undertaken toward production and characterization of the metal ion templated polymers: 1) selection and preparation of ligand monomers; 2) synthesis of metal complexes of the monomers; 3) preparation of polymers using the monomeric metal complexes; 4) testing of the polymers for metal ion selectivity; and 5) initiating the detailed physico-chemical characterization of the successful polymers.

Ligand Monomer Selection. Initially, monomeric ethylene, propylene and styrene derivatives with substituted coordinating groups (X), both acidic and neutral, ($H_2C=CH-X$, $H_2C=CH-CH_2-X$, and $H_2C=CH-C_6H_4-X$) were evaluated as specific ion complexants. Some were purchased while others were synthesized. Preparations for those compounds not available commercially are well-known and readily available in the literature (10). Ligands were selected on the basis of the thermodynamic affinities for the specific metal ion (maximum affinity) versus the affinities for competing metal ions (maximum difference in affinities). Affinity data are readily available from the chemical literature, for example, the tables of complexation constants in the CRC Handbook, Gmelin, and other sources (11,12). Table I lists the association constants of some metal ions with carboxylic acids that were considered as possible coordinating moieties of the ligand monomers.

Based on the affinities information, 4-vinylbenzoic acid (VBA) was chosen as the first candidate ligand for lead ion complexation. Although vinyl benzoic acid is commercially available, we found that it required extensive purification before use. Therefore it was synthesized in house by the Wittig reaction (13).

Several other vinyl-substituted ligands have also found use. Methyl 3,5-divinylbenzoate (DVMB) was synthesized by the method of Shea and Stoddard (14). The divinyl compound was prepared primarily to give greater structural integrity to the coordination site by providing each ligand with two possible connections to the polymer network. It was anticipated also that the methyl-esterified compound would eliminate the pH dependence, characteristic of the VBA carboxyl functionality. Vinylsalicylaldehyde oxime was synthesized by a method adapted from Wulff and Ahmed (15) and tested as a selective agent for both lead and uranyl ions.

Table I. Log K_1 for Some Divalent Metal Ions with Carboxylic Acids

Metal Ion	Acetic Acid	Benzoic Acid	Formic Acid	Chloroacetic Acid
Pb(II)	1.39-2.70	2.0	0.74-0.78	1.52
Cd(II)	1.30-2.00	1.4	1.04-1.73	1.2
Cu(II)	1.61-2.40	1.6-1.92	1.53-2.02	0.9-1.61
Zn(II)	0.76-1.59	0.9	0.70-1.20	0.4-0.56
Ni(II)	0.41-1.81	0.9	0.46-0.67	0.23
Co(II)	0.32-1.36	0.55	0.68	0.23
Ca(II)	0.53-0.77	0.2	0.48-0.80	0.14
Mg(II)	0.51-0.82	0.1	0.34	0.23

Synthesis of Metal Ion Complexes with the Polymerizable Ligands. With a careful choice of complexing ligands, metal complex synthesis can be accomplished by mixing stoichiometric amounts of a metal salt and the complexing ligand in aqueous solution and evaporating to near dryness. Our experience with making complexes shows that water or alcohol/water mixtures of the metal and ligand in stoichiometric ratios when evaporated to dryness result in a near quantitative yield of the desired complex compound. The stoichiometry of the complex was verified by determining the mass percent of metal ion using Inductively Coupled Plasma Atomic Emission Spectroscopy (ICP-AES). Table II shows a good agreement between the calculated and experimentally determined metal content in the complexes.

Table II. Table of Compounds Verified by ICP-AES

Complex	% Metal Calculated	% Metal Found
Eu(methyl 3,5-divinylbenzoate)$_4$(NO$_3$)$_3$	49.2	46.4
Eu(methyl 3,5-divinylbenzoate)$_3$(NO$_3$)$_2$PMP	68.6	70.3
Pb(methyl 3,5-divinylbenzoate)$_2$(Br)$_2$	66.4	67.5
Pb(3,5-divinylbenzoate)$_2$	28.8	28.7
Pb(vinylbenzoate)$_2$	41.8	41.3

Synthesis of the Copolymers. Here and elsewhere in this paper crosslinking is defined as the mole % content of the crosslinking agent, divinyl benzene (DVB). The polymer matrix was initially formed at a relatively high crosslinking level to enforce the retention of the cavity "memory" upon removal of the metal ion. Experience with templated resins for metal ions that exhibit directed bonding has since revealed that relatively small amounts of crosslinking result in more selective resins. This may be a result of the bulk of the polymer possessing metal ion-to-ligand ionic and dative bonds, as well as covalent crosslinks.

Introduction of microporosity for increased reagent accessibility is achieved through the addition of a non-coordinating monomer (styrene). Solutions of the monomeric complex (metal ion complexed with polymerizable ligands), styrene, initiator (AIBN) and crosslinker (DVB) in various amounts of a suitable solvent (0-20 weight % of pyridine) are prepared. The dissolution of the complex/monomer mixture has historically been a difficult step. We have found that this step can be simplified if the copolymerization is carried out using ultrasonication. The sonication has the serendipitous effect of maintaining the temperature at *ca.* 60°C, which is the recommended temperature for chemically initiated free radical polymerization using azobisisobutyronitrile (AIBN). For lead vinylbenzoate copolymers produced with and without sonication, it was determined that sonication results in a more homogeneous product, but with only a slight increase in capacity over the polymers obtained by heating in an oil bath. On the other hand, examination of some uranium containing copolymers has shown larger improvements (greater than tenfold) in capacity due to sonication. The expectation was that as well as improving solubility, the sonication would result in more complete incorporation of the metal ion complex in the copolymer. It is apparent that the effectiveness of sonication varies with the choice of metal ion complex and may or may not be simply a function of miscibility.

When the polymerization is completed, the crosslinked polymer is washed, cryogenically ground to a uniformly fine powder, and extensively eluted with non-polar solvents to remove the unreacted complex. A grinding step is necessary to maximize surface area and allow for access by the various reagents and samples. It requires freezing the polymer in liquid nitrogen and using a ball mill employing a stainless steel capsule. This freezing makes the polymer brittle enough to be ground and prevents distortions of the polymer by the heat of friction. Sieves are used to obtain the uniformly sized particles.

Acetone is used to swell the polymer allowing greater access to the coordinated metal ions. This is necessary since, unlike commercial resins whose large amount of functionalization (ionogenic sites) causes the resin to swell most in polar solvents, the templated resins have a relatively low level of functionalization and are primarily nonionic matrices. Subsequent to the removal of unreacted monomer, a 1 N aqueous acid solution is mixed into the acetone washes with increasing proportions of the aqueous acid phase in each sequential wash to remove the metal ions from the cavities by mass action. A significant fraction of the cations (but not all) can be removed, thus leaving cavities that are cation-specific. Those template ions which cannot be removed even after swelling are probably locked inside the polymer in inaccessible sites. Polymers were prepared with Pb(II) ions and other cations such as H^+ in the form of a blank or another metal ion template to examine inherent ligand selectivity (see below). After repeated loading and washing, recoveries for rebound lead ions have been measured and found to range from 88% (5% crosslinked polymer) to ~100% (2% polymer).

Testing for Cation Selectivity and Capacity. Selectivity of the resins was assessed by employing batch extraction techniques for measuring the uptake of various cations by a polymer templated for a given cation. The exchange equilibria are approximated by measuring the capacity of the resin for the various cations involved in the following general reaction:

$$(1/n)M^{n+} + HR \rightarrow (1/n)MR_n + nH^+ \tag{1}$$

where HR represents the protonated resin and M^{n+} is the metal ion. The steps involved in determination of the capacity are given in Figure 1. The data collected from the steps depicted in Figure 1 were used in the equations 2-4 to calculate the metal ion capacity and selectivity parameters.

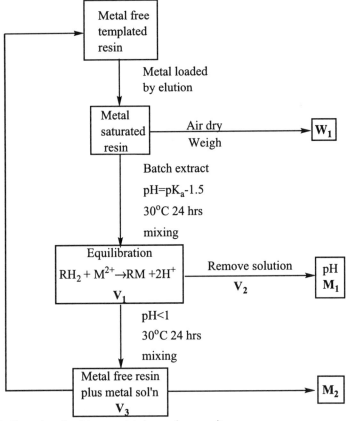

Figure 1. Steps involved in ascertaining resin capacity.

In addition to the cation used to prepare the polymer, other cations with differing charges, sizes, coordination numbers, and/or coordination geometries were used in selectivity quotient measurements to verify specificity. Measurements were also made using polymers prepared with a non-metal cation (H^+ or NH_4^+). A pH meter was employed for $[H^+]_a$ determination, and elemental analysis techniques (ETA-AAS, ICP-OES and ICP-MS) were used to measure $[M^{n+}]_a$. Uranium, in the form of the uranyl ion, was measured by luminescence and ICP-MS. Results for the analyses of the lead-templated VBA resins are given in the Tables III and IV.

$$\text{capacity} = \frac{[M^{2+}]_1 V_2 + [M^{2+}]_2 V_3}{W_1} \tag{2}$$

$$K = \frac{[H^+]^2 \left(\text{capacity} - [M^{2+}]_1 V_1 \Big/ W_1 \right)}{[M^{2+}]_1 \left(2[M^{2+}]_1 V_1 \Big/ W_1 \right)^2} \tag{3}$$

for the reaction:
$$M^{2+} + 2HR \rightarrow MR_2 + 2H^+ \tag{4}$$

Table III shows the effects of crosslinking on the capacities of the lead templated resins (P-1—P-3) and a non-templated control polymer (P-0). Although the control resin contains twice the amount of ligand as the templated resins (each Pb(II) ion is associated with two ligand monomers), the latter demonstrate significantly larger ion capacities. The capacity is maximized at about 2% crosslinking, suggesting that a proper balance between site rigidity and accessibility has been obtained. This balance may be more finely tuned by using half integer values for the percent of DVB.

Table III. The Effect of Crosslinking on Lead Ion Capacity

Polymer	Mole % Complex (mole % ligand)	Mole % DVB	Capacity (µmole/g)
P-0	(4)	1	0.112
P-1	1	1	0.179
P-2	1	2	0.231
P-3	1	4	0.127

A feature of the lead-templated resins is that the increase of the mole fraction of template complex produces a logarithmic increase in the resin capacity when the template content is below 5%. In a like fashion, the non-templated polymers gave a logarithmic increase as the ligand monomer content increased from 3 to 5%, but the capacity plot for these polymers lagged about one logarithmic unit behind the plot for a templated polymer. The increase in capacity of the non-templated polymers is likely due to the increase in the number of accessible binding sites. If this is the case, a 10-fold difference in capacity between templated and non-templated polymers must result from the template process. The data in Table IV shows this trend.

Table IV. The Effect of the Amount of Template Complex on Polymer Capacity

Polymer	Mole % Complex (mole % ligand)	Mole % DVB	Lead Capacity (µmole/g)
P-0	(4)	1	0.112
P-1	1	1	0.179
P-6	2	1	0.373
P-7	3	1	0.752
P-24	4	1	2.128
P-8	5	1	3.065
P-9	7	1	5.741
P-10	10	1	8.035

To verify that the template metal ions don't just behave as "spacers", producing cavities that are later re-occupied, sorption experiments were performed with other metal ions (the likely interferents) on the lead-templated resins. Table V shows the results of such sequential loadings of similar metal ions on the various resins, which presents a convenient way of assessing their selectivities. The selectivity of the non-templated resin (P-0) confirms that we have chosen a good ligand for lead. The enhancement of the selectivity by templating (polymer P-1) indicates that the cavities are specifically selective for lead ions and are not just holes in the polymer for metal ions to occupy. The selectivity factor $\alpha_{Pb,M}$ is greater toward Cd(II) than Cu(II) suggesting that the "hardness", where Cu(II) more closely resembles the Pb(II) ion, is of greater importance than the size, where Cd(II) is closer to the Pb(II) .

Table V. Capacity and Selectivity of Resins in Single Species Sorption Experiment

Polymer	Rebinding Ion	Capacity (µmole/g)	Selectivity $\alpha_{Pb,M}$
P-0	Pb	0.112	
P-0	Cd	0.00112	100
P-0	Cu	0.0233	5
P-1	Pb	0.179	
P-1	Cd	0.00103	174
P-1	Cu	0.0113	16

To ascertain the selectivity of the lead-templated resins, another set of experiments was performed with competitive sorption of two or more metal ions. The results of this study are presented in Table VI. Again, the templated polymer showed better selectivity for lead ion.

Table VI. Capacity and Selectivity of Resins in Competitive Sorption Experiment

Polymer	Capacity (μmole/g)			Selectivity	
	Pb	Cd	Cu	$\alpha_{Pb,Cd}$	$\alpha_{Pb,Cu}$
P-0	0.162	0.00463		35	
P-1	0.119	0.0022		54	
P-6	0.107	0.0012	0.00931	90	12

Further characterization has included the determination of the acidity constants for some of the resins. The titration of a resin allows the determination of the number of ionogenic sites that become accessible to H^+ ion as a result of swelling and cleaning. This proton-based capacity is indicative of the relative amount of lead ion complex that has been incorporated in the copolymerization reaction. A calculation of the expected load of lead vinyl benzoate complex in a polymer containing 5.00 mole % complex (with 1.00 mole % crosslinking) is 416 μmole/gram. A titration of this copolymer yields a value of 20.2 μmole/gram for H^+, or 10.1 μmole/gram for a divalent lead ion. This corresponds to only 2.5 percent of the metal ion complex being incorporated into the copolymer. The methods we intend to pursue to increase this value are discussed below. The equilibrium constants for the resins were determined by the following:

$$K = \frac{[R_2 M][H^+]^2}{[M^{2+}][RH]^2} = \left(\frac{[R_2 M]}{[M^{2+}][R^-]^2}\right) \times \left(\frac{[R^-]^2[H^+]^2}{[RH]^2}\right) = K_s \times K_a^{\ 2} \qquad (5)$$

where K_s is the association constant of metal ligand complex and K_a is the dissociation constant of polymeric acid. The association constants as a function of complex content are exhibited in Table VII.

Table VII. Effect of Template Complex Content on the Association Constant

Polymer	Mole % Complex (mole % ligand)	Mole % DVB	Log K_s
P-0	(4)	1	2.31
P-1	1	1	4.93
P-5	2	1	4.83
P-6	3	1	5.21
P-7	4	1	3.67
P-8	5	1	3.29
P-9	7	1	3.09
P-10	10	1	3.07

Structural Characterization of the Templated Polymers. The ground polymer particles were examined by means of electron microscopy and X-ray spectroscopy. Cryogenic grinding of the polymer resulted in a powder consisting of small roughly spherical particles. These particles were sieved to give samples of uniform size. The theoretical capacity is the exchange capacity expected if the entire amount of template complex in the polymer feed were utilized for exchange. The measured capacity of the resins is considerably smaller than the theoretical, since much of the template complex is buried inside the particle. A semi-quantitative value of the active, or accessible, depth of the surface of the polymer particles can be estimated on the assumption that all of the template complex used in the copolymerization reaction is homogeneously dispersed in the bulk polymer. The calculation is based on the following equations:

$$V = \frac{4}{3}\pi r^3; \quad dV = 4\pi r^2 dr; \quad \frac{dV}{V} = 3\frac{dr}{r} \tag{6}$$

Reasoning behind the above calculation rests on the assumption that for a roughly spherical particle total volume (V) represents its theoretical capacity, of which only a portion (dV) is available for exchange, corresponding to the experimental capacity. Then we solve for the exchange depth (dr), which is a function of the exchange volume. The experimental value for the capacity of the 5 mole % resin is 3.1 μmole/g. The amount of complex loading based on the polymer feed is 418 μmole/g. The volume of a 100 mesh bead (r = 0.0050 in = 0.0127 cm) is calculated to be 8.58×10^{-6} cm^3. The portion of the surface volume equal to the experimental exchange capacity yields an active surface depth of about 3.1×10^{-5} cm or 310 nm. This depth corresponds to a distance of the order of 5 template complex molecules. The model suggests that the polymer particles can be viewed as similar to an interwoven mass of fibers whose frayed ends are accessible for ion exchange.

The above hypothesis was examined using a scanning electron microscope equipped with an X-ray analyzer. The electron micrograph of the sieved resin showed that the sieved particles are of fairly uniform size with a diameter slightly exceeding 200 μm. Analysis of X-rays emitted by the particles showed that at low accelerating voltages no lead appeared to be present. As the acceleration voltage was increased, resulting in a deeper penetration of the electron beam, a clear lead signal was obtained. The micrograph reveals a rough granular structure of the particles that may tend to give the calculations a low estimate for the surface area of the particle. This may suggest that the depth of exchange is somewhat less than the depth calculated.

Analytical Applications of the Metal Ion Templated Polymers

Metal-templated polymers were used in a variety of analytical applications. Lead-selective resins were applied to the analysis of seawater in a head-to-head experiment with the three other ion-selective resins. A fast sensitive method for determination of lead in seawater was developed. Lead and uranyl ion selective membrane electrodes were fabricated using ground polymer particles as sensing elements, and a multitude of selectivity studies were conducted therewith. Finally, a fluorescent lead-templated polymer was used as an optical sensor (optrode) for determination of Pb(II) ion in aqueous solutions.

Templated Polymers as Sequestering Agents. The first application was a comparative study of ion-exchange materials as a means of extraction and preconcentration of Pb(II) ion from complex matrices followed by analysis by ICP-AES or ICP-MS. Pb(II) ion was extracted from seawater by four different ion-exchange resins: our lead-templated VBA resin; a proprietary NASA polyacrylic resin; Chelex-100; and a commercial thiol-based ion exchange resin. Measurements of the Pb(II) concentrations, as well as the concentrations of possible interfering species, were performed first with ICP-AES and later with ICP-MS. Recoveries for Pb(II) were determined to establish the suitability of the various resins as sequestering agents and to evaluate any advantage in selectivity the templated ion exchange material might offer. The pH of the seawater was unmodified, and a flow rate of 1.0 mL/min was maintained for each extraction to ensure uniform testing conditions. The results of this experiment, normalized to Pb(II) ion concentration, are summarized in Table VIII. Table IX presents the results of the seawater lead recovery study.

Table VIII. Metal Ion Concentrations in the Seawater Eluents from Resin Columns

Metal Ion	Ion-Exchange Resin Column			
	Templated	NASA	Chelex-100	Thiol
Li(II)	1.3 ppb	0.723 ppb	65.5 ppb	0.710 ppb
Na(II)	6.0 ppm	> 100 ppm	>100 ppm	>100 ppm
Mg(II)	1.9 ppm	> 100 ppm	>100 ppm	>100 ppm
Ca(II)	1.40 ppm	37.57 ppm	5.52 ppm	20.4 ppm
Cu(II)	1.0 ppm	1.88 ppm	1.31 ppm	4.66 ppm
Zn(II)	940 ppb	1100 ppb	931 ppb	2220 ppb
Cd(II)	1.1 ppb	0.376 ppb	0.414 ppb	0.511 ppb
Hg(II)	0.41 ppb	109.8 ppb	44.83 ppb	0.02 ppb
Pb(II)	1.20 ppm	1.20 ppm	1.20 ppm	1.20 ppm

Table IX. Recovery of Trace Lead from Seawater

Resin	% Recovery
Templated (3%)	95.1 ± 0.04
NASA	78.9 ± 0.02
Chelex-100	95.4 ± 0.06
Thiol	95.0 ± 0.20

Due to the relative purity (absence of the significant amounts of interferents) of the seawater extract obtained from the templated resin, it was tested for lead by dithizone complexation with spectrophotometric detection (measuring absorbance at 522 nm). The results of the photometric analysis agreed with the ICP-AES data. The use of the templated resin in this manner provides an inexpensive assay for lead ion in aqueous solutions with a detection limit of <100 ppb and a linear range from 100 ppb to 2.00 ppm. The use of the resin with larger volumes of analyte would allow for preconcentration and much lower limits of detection.

The limitation on the use of the templated resins fabricated by the methods above is their relatively small ion-exchange capacity. The low capacity is due to the majority of the ionogenic sites being buried within the polymer particles and thus being unavailable for exchange. In part, this handicap is compensated by the selectivity of the resin. Since the resin is so selective for the metal of interest, few of the available sites are occupied by interferents, and so less capacity is required for most applications. Another consequence of the activity of the resin being restricted to the surface of the particle is the rapidity of ion extraction, which allows for higher flow rates for extractions and is particularly attractive for analytical preconcentration.

Templated Polymers in Ion-Selective Electrodes. The membranes for ion-selective electrodes were prepared by mixing 90 mg of PVC powder and 30 mg of lead-imprinted ground polymer particles with 0.2 mL of an ionic mediator, dioctylphenylphosphonate. After mixing the dry ingredients, the mixture was thoroughly moistened with THF and subsequently dissolved in 3.0 mL of THF, followed by stirring for *ca.* 20 minutes with a magnetic stirrer. The resulting solution was poured into a glass ring with an internal diameter of 45 mm on a glass plate and the THF was allowed to evaporate at room temperature for a period of about 2 days. A section of the membrane was attached to a PVC tube with diameter of 13 mm or 6.5 mm using a viscous solution of PVC in THF as an adhesive. The bond was allowed to cure overnight. The electrode was then filled with a solution of 1 mM $Pb(NO_3)_2$ and 1 mM NaCl in demineralized water and conditioned for 24 hours by soaking in a 1 mM $Pb(NO_3)_2$ solution. A typical response curve for the templated polymer electrode for Pb(II) is shown in Figure 2.

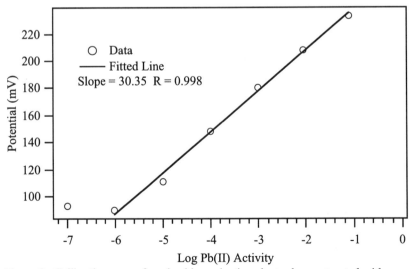

Figure 2. Calibration curve for a lead ion selective electrode constructed with a templated polymer-impregnated PVC membrane.

The potential responses of the electrode to a variety of common interference ions are shown in Figure 3. As expected, the electrode based on the Pb(II) ion imprinted polymer membrane showed strong preference to Pb(II) ion. Membranes based on a 3 mole % equivalent VBA non-templated polymer and on benzoic acid were also prepared and examined. The membrane based on Pb(II) templated polymer was found to be more sensitive than either the membrane based on non-templated polymer or the one prepared with dissolved benzoic acid.

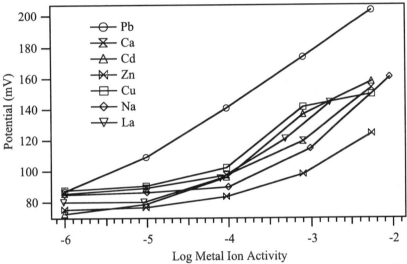

Figure 3. The response of the templated polymer ion selective electrode to lead(II) ion and the most likely interferent ions.

Effect of Crosslinking. The degree of crosslinking has played an important role in the optimization of capacity and selectivity of metal ion imprinted polymers. In the study of Pb(II)-templated ion-exchange resins, the selectivity calculations were based on metal ion capacity (9). In terms of capacity, maximum selectivity was obtained with 1-2% DVB crosslinking. Since the templated metal ion sensors reported by other investigators (1) have required much higher levels of crosslinking to maintain selectivity, it was deemed necessary that the selectivity based on capacity be verified as the optimal method for determination of selectivity. For this study, the responses of lead-selective electrodes prepared with the different amounts of crosslinking agent were evaluated. The electrode prepared with 1% DVB crosslinking showed a better response to lead ion than the electrode prepared with 4% DVB, in regard to both sensitivity (slope) and linear dynamic range. For the electrode with 4% DVB crosslinking, Pb(II) concentrations of under 10^{-3} M resulted in a slope of 25.4 mV per decade; the linear response range was limited to 1×10^{-5} — 1×10^{-2} M of Pb(II). The electrode prepared with 1% DVB exhibited a slope of 33 mV per decade and linear response range of 1 x 10^{-6} — 1 x 10^{-2} M of Pb(II). This result is consistent with the previous study (9) which showed that lower amounts of crosslinking improve the selectivity of templated ion-exchange polymers for metal ions that exhibit directed bonding.

Operational Longevity Studies. A common problem associated with electrodes fabricated by the dissolution of an organic ionophore in a PVC membrane is a loss of active agent due to its finite water solubility. Therefore, a significant advantage of a chemically bound ionophore should be the extended longevity of the electrode. The operational longevity of the templated polymer electrode was investigated in order to verify the above hypothesis. For comparison, an electrode membrane was fabricated by dissolution of benzoic acid in PVC. After using the templated-polymer electrode for one month, the sensitivity, as reflected by the slope of the calibration curve, and the linear range of the response to Pb(II) remained almost unchanged. In contrast, the membrane prepared with benzoic acid lost all response to Pb(II) after only one week's usage. A different electrode based on the templated polymer was tested after being immersed in solution for a period of 3 months. It was noted that the response of that electrode in the region of high concentration exhibited a reduced sensitivity. Thus, a relatively long life-span is one of the definite advantages of the chemically bound ionophore electrode over other membrane ISEs, for which the gradual loss of sensing elements to the test solution or the effects of aging on the asymmetry potential of the membrane cause the decrease in sensitivity and eventual loss of activity.

Selectivity Studies. The most critical characteristic of an ISE is selectivity. Historically, there has been little agreement in the literature regarding optimal methods of determining selectivity. This difficulty arises partly from a lack of systematic study and from the tendency to report selectivity under a single set of conditions. The selectivity coefficients (log $K^{pot}_{Pb,M}$) for the templated lead ion electrode were evaluated by the separate solution method (16), the fixed interference method (16) and the matched potential method (17).

Table X lists the selectivity coefficients of electrodes based on the imprinted polymer evaluated by the separate solution method. The selectivity coefficients were found to exhibit an activity dependence and were observed to decrease with increasing activity. This behavior indicates that the electrode shows a lower sensitivity to interference ions than to the lead ion, a situation which is desired in practice. The selectivity coefficients for Pb(II) over Cd(II), Cu(II), Ca(II), Na(I) (without power term), and Zn(II) are within the activity range of 1 x 10^{-5} — 1 x 10^{-2}. All coefficients are found to be negative, indicating that the electrode exhibits the highest selectivity for Pb(II) ion. The selectivity coefficient for sodium ion is unrealistically large when a power term is included, since the separate solution method is based on the empirical Nicolsky-Eisenman equation:

$$E = E^{o} + \left(\frac{RT}{Z_I F}\right) \ln\left[a_I(IJ) + K_{IJ}^{pot}\left(a_j(IJ)\right)^{Z_1/Z_2}\right] \qquad (7)$$

This equation is only adequate when ions under consideration have the same charge, whereas for the ions of unequal charge the N-E equation is inconsistent and can lead to an incorrect interpretation of the observed EMF electrode response function. Therefore, despite the seemingly high selectivity coefficient, sodium ion may not interfere unless present at a high concentration, as demonstrated by calculations which neglect the power term.

Table X.   Selectivity Coefficients Determined by the Separate Solution
            Method (T=20°C)

Concentration	log $K_{Pb,M}$					
(M)	Cd(II)	Cu(II)	Zn(II)	Ca(II)	Na(I)	Na(I)[a]
1x10^{-2}	-1.0	-1.3	-2.2	-1.2	1.4	-0.9
1x10^{-3}	-0.7	-0.5	-2.0	-1.3	1.6	-1.5
1x10^{-4}	-1.0	-0.7	-1.4	-0.9	2.9	-1.2
1x10^{-5}	-0.5	-0.1	-0.5	-0.1	4.8	-0.2

[a]Without a power term.

Selectivity coefficients were also evaluated by the fixed interference method. In this method, the potential response is measured using solutions with a constant activity of interfering ion a_J(IJ) and varying activity of the primary ion a_I (IJ). The potential values obtained are plotted versus the activity of the primary ion. The intersection of the two extrapolated linear portions of this curve provides a value of a_I(IJ) that is used to calculate K_{IJ} ^{pot} from the following equation:

$$K_{IJ}{}^{pot} = \frac{a_I(I)}{\left[a_J(J)\right]^{Z_I/Z_J}} \tag{8}$$

As it is apparent from Equation 7, the fixed interference method also suffers from inadequacies associated with differences in ionic charge. Figure 4 shows the selectivity coefficients of a lead ISE based on Pb(II) ion-templated polymer compared to the results reported by Srivastava and coworkers for a lead electrode based on 15-crown-5, where log $K_{Pb,M}$ values for all the interfering ions were larger than -1 (18). The electrode based on the templated polymer clearly demonstrates better selectivity for Pb(II). This confirms the ability of ion templating technology to enhance the selectivity, since the templating process may impart geometric, as well as size constraints, to the templated cavity.

The matched potential method was recommended by IUPAC in 1996 for the determination of selectivity coefficients of ion-selective electrodes (19). This method is independent of the N-E equation and based purely on empirical observations. The matched potential method was applied to determination of selectivities at various Pb(II) concentrations and at various potential changes. Results are summarized in Tables XI and XII respectively.

232

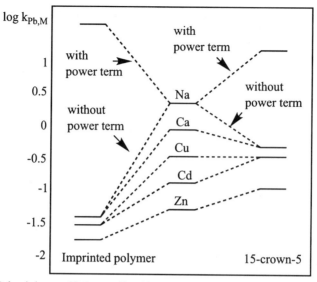

Figure 4. Selectivity coefficients of lead(II) templated ion-selective electrode obtained by fixed interference method compared to the observations by Srivastava *et al* (*18*).

Table XI. Selectivity Coefficients Using the Matched Potential Method

Concentration of Pb(II), M	log $K_{Pb,M}$			
	Ca(II)	Cu(II)	Cd(II)	Na(I)
1×10^{-6}	-0.12 ± 0.01	-0.96 ± 0.01	-0.96 ± 0.01	-1.16 ± 0.01
1×10^{-5}	-1.08 ± 0.01	-1.09 ± 0.01	-1.15 ± 0.01	
1×10^{-4}	-1.11 ± 0.01	-1.21 ± 0.01		

Table XII. Selectivity Coefficients Using the Matched Potential Method and Varying Potential Changes[a]

Metal Ion	log $K_{Pb,M}$ for a potential change of			
	10 mV	20 mV	30 mV	40 mV
Ca(II)	-0.96 ± 0.06	-0.99 ± 0.02	-1.04 ± 0.01	-1.12 ± 0.01
Zn(II)	-1.69 ± 0.04	-2.00 ± 0.02	-1.87 ± 0.01	-1.93 ± 0.01
Na(I)	-1.22 ± 0.05	-1.24 ± 0.02	-1.17 ± 0.01	-1.16 ± 0.01
Cd(II)	-1.15 ± 0.05	-1.08 ± 0.02	-1.09 ± 0.01	-1.13 ± 0.01
Cu(II)	-0.86 ± 0.07	-0.93 ± 0.03	-0.99 ± 0.01	-1.10 ± 0.01

[a]The concentration of Pb(II) was fixed at 1×10^{-5} M and all measurements conducted at T=20°C.

The results obtained by the three different methods are essentially consistent. When compared to the separate solution method or the fixed interference method, the matched potential method has the advantage of no power term inconsistencies for ions of unequal charge, e.g., Na(I). This method, however, is laborious and time-consuming. The potential change resulting from the addition of an interference ion is difficult to match to the reference solution value due to the instability associated with reading an ion analyzer. In the case of a low interference ion concentration (*e.g.*, Na(I) and Zn(II) in this study) and a high primary ion concentration, to produce the same potential change as in the reference solution the amount of interference ion needed is so large that the change in volume can no longer be neglected.

Polymers Templated for Other Metal Ions. The production of polymers templated for other metal ions has begun with uranyl ion, UO_2^{2+}. Figure 5 shows the response of a membrane prepared with a uranyl templated polymer. The result was astounding in that no interference has yet been detected. It was thought that the unique shape of the uranyl ion would make it a prime candidate for the template process, and this expectation has been realized. The influence of the ionic shape and other factors are currently being investigated for uranyl templated polymers.

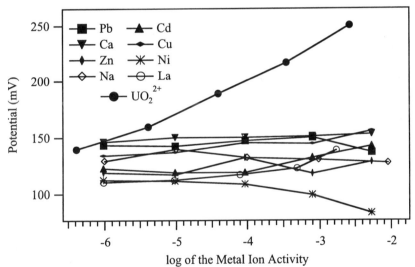

Figure 5. Response of a uranyl ion-templated electrode.

Templated Polymers in Optical Sensing. During the initial screening of coordinating monomers for possible ion selective electrode enhancements, it was observed that methyl 3,5-divinylbenzoate (DVMB) luminesces blue, and that the luminescence becomes yellow-green when a lead complex is formed. This observation suggests phosphorescence due to an external heavy atom effect. Since the lead-DVMB complexes rapidly polymerize in solution, it was decided to investigate the optical properties of the ligand precursor, methyl 3,5-dimethylbenzoate. The lead complex was prepared by the method outlined previously and dissolved in hexane to prevent dissociation. Luminescence spectra of the lead complex and free ligand were obtained with a spectrofluorometer using a 467 excitation wavelength, as shown in Figure 6. Serial dilutions were used to obtain analytical figures of merit for a solution assay of the lead complex in hexane. The calibration curve was linear for four decades of concentration from 70 to 7.0×10^5 ppb of complex with a detection limit of 50 ppb of complex (equivalent to 20 ppb of lead) and a correlation coefficient of 0.997.

These observations, along with similar results obtained for a variety of complexing ligands, strongly suggested that a lead sensing optrode could be made. The 3% lead-DVMB complex (2% DVB) was bound by *in situ* copolymerization to a vinylized 400 μm optical fiber surface. The fiber was then used in the setup schematically presented in Figure 7. The luminescence spectrum of the polymer-coated optical fiber showed the characteristic band of the Pb(II) complex, virtually identical to the one displayed in Figure 6. The Pb(II) ion was then removed from the polymer by first swelling it in a mixture of methanol and water and then soaking in a stirred solution of EDTA; each process took about one hour. The luminescence spectrum of the cleaned polymer no longer exhibited the characteristic lead complex luminescence band.

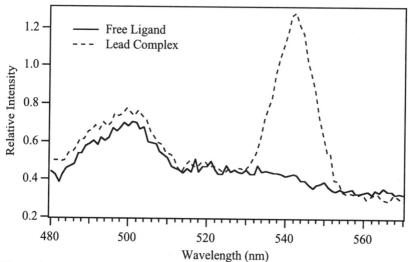

Figure 6. Luminescence spectra of a free ligand (methyl 3,5-dimethylbenzoate) and its lead complex excited at 467 nm.

The standard curve generated by the first prototype optrode is given in Figure 8. The luminescence was excited by the 488 nm line of an argon-ion laser at a power of about 10 mW. The device is being tested for longevity and interference effects and will be applied to field trails in an environmental project aimed at identifying sources of waste water seepage into streams.

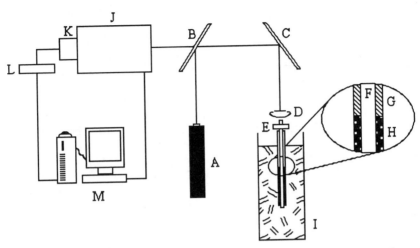

Figure 7. Prototype of a lead ion measuring optrode system. A) Ar ion laser; B) dichroic mirror; C) mirror; D) lens; E) fiber mount; F) optical fiber core; G) fiber cladding; H) templated polymer coating; I) sample; J) monochromator; K) detector; L) A/D converter; M) computer.

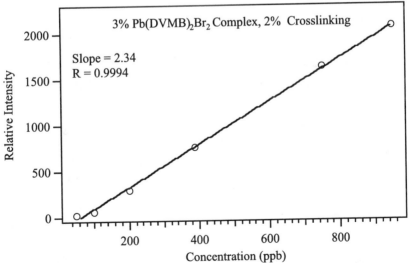

Figure 8. Calibration curve obtained for the first prototype lead optrode.

Conclusions

Ionic templating is an emerging technology for the creation of selective binding sites in synthetic polymers. Ion-selective sensors have the advantages of being simple, rapid and inexpensive. This research explores the utilization of metal ion templated polymers as selective sequestering agents and as components of electrochemical and optical sensors. Results from the characterization of the metal ion templated electrodes for lead and uranyl ions demonstrate a broad linear range of measurement with a near-Nernstian response and a strong preference for the template ion over the other metal ions tested. Selectivity coefficients of the lead ISE were systematically evaluated by the separate solution method, the fixed interference method and the matched potential method under various conditions. The life-time of the electrode is relatively long, which is one of the advantages rendered by templated polymers with the functional groups covalently attached to the polymer matrix. An optical sensor based on a templated polymer has been shown to determine lead ion in the parts per billion range. The information obtained from these studies will be used to produce a variety of new metal ion selective polymers and to fabricate metal sensors.

Acknowledgment

We wish to acknowledge Dr. O. Manuel Uy and Mr. Steven Wajer of the Johns Hopkins Applied Physics Research Lab for the use of the ICP-MS and help in obtaining scanning electron micrographs and X-ray emission spectra.

Literature Cited

1. Rosatzin, T.; Andersson, L. I.; Simon, W.; Mosbach, K. *J. Chem. Soc. Perkin Trans.*, **1991**, *2*, 1261-1265.
2. Gupta, S. N.; Neckers, D. C. *J. Polym. Sci., Polym. Chem. Edn.*, **1982**, *20*, 1609-1615.
3. Wulff, G. *Angew. Chem., Int. Edn.*, **1995**, *34*, 1812-1832.
4. Efendiev, A. A.; Kabanov, V. A. *Pure Appl. Chem.*, **1982**, *54*, 2077-2080.
5. Nishide, H.; Deguchi, J.; Tsuchida, E. *J. Polym. Sci., Polym. Chem. Edn.*, **1977**, *15*, 3023-3029.
6. Harkins, D. A.; Schweitzer, G. K. *Sep. Sci. Tech.*, **1991**, *26*, 345-354.
7. Kuchen, W.; Schram, J. *Angew. Chem., Int. Edn.*, **1988**, *27*, 1695-1697.
8. Chen, H.; Olmstead, M.M.; Albright, R.L.; Devenyj, J.; Fish, R.H. *Angew. Chem. Int. Ed. Engl.*, **1997**, *36*, 642.
9. Zeng, X.; Murray, G. M. *Sep. Sci. Tech.*, **1996**, *31*, 2403-2418.
10. Yatsimirskii, K. B.; Vasiliev,V. P. *Instability Constants of Complex Compounds*, Plenum: New York, NY, 1960.
11. See pertinent references in Beilstein (Series H, EI, EII, EIII, EIV), System Nos. 25, 90, 163, 325, 404, 405, 534, 644, 949, 1524, 1709, 2292.

12. Sillen, L. G.; Martell, A. E. *Stability Constants of Metal Ion Complexes*; The Chemical Society, London, England, 2nd edition, 1964.
13. Broos, R.; Tavernier, D.; Anteunis, M. *J. Chem. Educ.*, **1978**, *55*, 813-814.
14. Shea, K. J.; Stoddard, G. J. *Macromolecules*, **1991**, *24*, 1207-1210.
15. Wulff, G.; Ahmed, A. *Macromol. Chem.*, **1979**, *174*, 2647-2651.
16. Guilbault, G. G.; Durst, R. A.; Frant, M. S.; Freiser, H.; Hansen, E. H.; Light, T. S.; Pungor, E.; Rechnitz, G.; Rice, N. M.; Rohm, T.J.; Simon, W.; Thomas, J. D. R. *Pure Appl. Chem.*, **1976**, *46*, 129-132.
17. Gadzekpo, V. P. Y.; Christian, G. D. *Anal. Chim. Acta,* **1984**, *164*, 279-282.
18. Srivastava, S. K.; Gupta, V. K.; Jain, S. *Analyst*, **1995**, *120*, 495-498.
19. Umezawa, Y.; Umezawa, K.; Sato, H. *Pure Appl. Chem.*, **1995**, *67*, 507-518.

Chapter 16

Metal Ion Templated Polymers

Studies of *N*-(4-Vinylbenzyl)-1,4,7-Triazacyclononane-Metal Ion Complexes and Their Polymerization with Divinylbenzene: The Importance of Thermodynamic and Imprinting Parameters in Metal Ion Selectivity Studies of the Demetalated, Templated Polymers

Richard H. Fish

Lawrence Berkeley National Laboratory, University of California, Berkeley, CA 94720

A Zn^{2+} templated polymer, containing a sandwich arrangement of the well known 1,4,7-triazacyclononane (TACN) ligand, was found to be highly selective toward reintroduced Cu^{2+} in the presence of Fe^{3+}, Co^{2+}, Ni^{2+}, Zn^{2+}, and Mn^{2+}, after the removal of the Zn^{2+} ion template with 6 N HCl; while a similar Hg^{2+} templated TACN polymer was found to be highly selective for Hg^{2+} in the presence of Cd^{2+}, Pb^{2+}, Ag^+, Cu^{2+}, and Fe^{3+}. The following parameters were addressed: structures of the monomers of the N-(4-vinylbenzyl)TACN-Zn^{2+}/Hg^{2+} complexes; polymerization parameters; structure of the polymer TACN-template metal ion complex; and selectivity to reintroduced metal ions as a function of the thermodynamic stability and the metal ion templating parameters for the polymer TACN-metal ion complexes that are formed.

The pioneering studies of Wulff, Mosbach, and Shea, among others (*1-6*), introduced the concept of templated polymers, that initially involves a host-guest relationship between a vinyl monomer host and a bound guest acting as a template. The vinyl monomer host-guest template complex is then polymerized with a crosslinking agent, which then provides spatially regulated sites in the polymer matrix, and hopefully, further allows continual recognition of the removable guest template in subsequent reintroduction experiments (Figure 1).

Among the guest templates that have been studied, in this elegant approach for highly selective polymers of defined architecture, were various metal ions (*7-16*). However, the area of metal ion templated polymers has received much less attention than organic compound templated polymers. For example, the following parameters need attention: [1] the unequivocal structures of the starting vinyl monomer ligand-metal ion complexes; [2] the polymerization parameters such as the effect of the metal ion, ligand structure, and the nature of the crosslinking agent; [3] the structure of the templated metal ion-polymer ligand site; and [4] whether subsequent selectivity, following removal of the initial metal ion template, was a consequence of the metal ion recognition process or, in fact, the thermodynamic stability of the resulting polymeric ligand-metal ion complex.

These above-mentioned aspects, [1-4], were addressed and the most important find was that it is tentatively possible to spatially regulate a sandwich arrangement of TACN ligands by the ionic radius of the metal ion template. Thus, it may be feasible, in the future, to separate metal ions on this basis; however, the thermodynamic stability of the polymer TACN-metal ion complexes of similar ionic radius appears to be dominant.

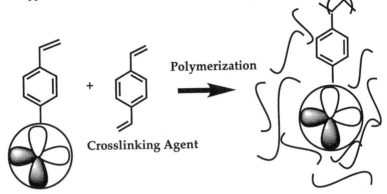

Vinyl Templated Monomer   **Spatially Regulated Templated Polymer**

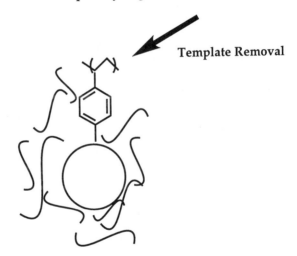

Polymer of Defined Architecture for Template Reintroduction

Figure 1. Metal Ion Templated Polymer Concept.

Results and Discussion

In this chapter, we provide an overview of our initial findings in the relatively new field of metal ion templated polymers and define a novel, highly selective system to Cu^{2+} ions in the presence of Fe^{3+}, Co^{2+}, Ni^{2+}, Zn^{2+}, and Mn^{2+} utilizing the 1,4,7-triazacyclononane (TACN) ligand that has been widely studied with a variety of metal ions (17). As well, a similar Hg^{2+} templated polymer with TACN was found to be highly selective to Hg^{2+} in the presence of Cd^{2+}, Pb^{2+}, Ag^+, Cu^{2+} and Fe^{3+}

ions. We will show the structures of the monomers of N-(4-vinylbenzyl)TACN-template metal ion complexes, polymerization parameters, structure of the polymer TACN-template metal ion complex, and selectivity to reintroduced metal ions as a function of the thermodynamic stability and the metal ion templating parameters for the polymer TACN-metal ion complexes that are formed (18).

The N-(4-vinylbenzyl)TACN ligands, 1 and 2, were readily synthesized in high yields by reacting 1 or 3 moles of 4-vinylbenzyl chloride with TACN in ethanol/H_2O with LiOH or in toluene with K_2CO_3 as the base; respectively.

The Zn^{2+} complexes 3 and 4, from 1 and 2, respectively, were formed by reaction of $Zn(NO_3)_2$ in methanol, while the unequivocal structure of complex 4 was determined by single crystal X-ray analysis. What was pertinent about the structure of complex 4 was the distorted octahedral configuration around the Zn^{2+} metal ion center, and the bonding nitrate and methanol groups that apparently prevents the formation of a possible 2:1 TACN:Zn^{2+} sandwich complex (5). In addition, we assign complex 3 as a similar structure to 4 based on the analytical and spectroscopic data obtained for each complex; in solution, the FAB/MS for 3 and 4 also provide evidence for their 1:1 Zn:TACN structures.

4

The copolymerization of Zn^{2+} complexes **3** and **4** with the crosslinking agent, 80 % divinylbenzene (DVB), using AIBN in MeOH at 78 °C, provided highly crosslinked, macroporous polymers with a ratio of **1** or **2**/DVB of ~ 1:2.5-3.0 and a TACN concentration of 1.33 and 1.15 mmol/g; respectively, designated as polymers **5** and **6**. More importantly, the ratio of TACN:Zn^{2+} in both polymers **5** and **6** was 2:1. *Thus, it was evident that free Zn^{2+} ions were extruded (loss of Zn^{2+}: 5, 43% ; 6, 50%) during the polymerization process to provide a possible $TACN_2Zn^{2+}$ sandwich complex in the polymer matrix.* It is also important to note that preparation of the corresponding Cu^{2+} complexes of **1** and **2**, and subsequent attempts to polymerize the formed Cu^{2+} monomers, under similar conditions described for **3** and **4**, were not successful, presumably because electron transfer from the developing carbon radical to the Cu^{2+} metal ion center that quenched/terminated the propagation step of the free radical chain reaction (7). Moreover, attempts to prepare an untemplated polymer with **1** and divinylbenzene, under similar reaction conditions as described for **3**, to compare to polymer **5** for metal ion selectivity, were unfortunately not successful, since no polymeric material formed during this reaction; only oligomers appeared to be formed (FAB/MS). This latter result confirms the unique role of the templated metal ion in the kinetics of the polymerization process (7).

5

To provide further evidence for the formation of a templated, polymeric sandwich complex, during the polymerization reactions of **3** and **4**, we synthesized and characterized by ^1H NMR, electrospray ionization/MS, and elemental analysis, the monomer, sandwich complex, [(mono-N-(4-vinylbenzyl)-1,4,7-triazacyclononane)$_2$Zn](OTf)$_2$, **7**. Subsequent formation of polymer **5'** (TACN concentration, 1.33 mmol/g) from **7**, under similar polymerization conditions as described for **3** and **4** with the crosslinking agent DVB, *clearly showed minor loss of Zn^{2+} metal ions (< 15%), and the expected 2:1 ratio of TACN:Zn^{2+} in the polymer matrix* (eq 1). This result further substantiates the formation of the 2:1 TACN:Zn^{2+} sandwich complexes in the polymer matrices of **5** and **6**, possibly on thermodynamic grounds (7-16, 18).

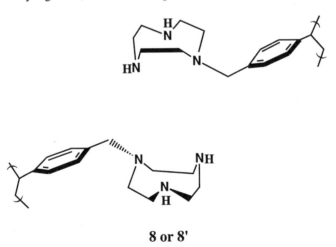

$$\text{Zn}^{2+}\ (\text{OTf}^-)_2 \xrightarrow[\text{DVB, MeOH}]{\text{AIBN, 78 °C}} \textbf{5'}$$

(1)

7

The Zn^{2+} templated polymers, **5, 5'**, and **6**, were reacted with 6N HCl at ambient temperature, which provided the facile (rapid kinetics) and total removal (>99%) of the templated Zn^{2+} ions. Selectivity studies were carried out on the these Zn^{2+} demetalized polymers, designated as **8, 8'**, and **9**, respectively, to answer the question of whether the more important parameter was templating or thermodynamic stability in the ensuing reintroduction step of various metal ions with the spatially regulated, sandwich arrangement of TACN ligands.

8 or 8'

Thus, we competed the original Zn^{2+} ion template with divalent metal ions, Cu^{2+}, Co^{2+}, Ni^{2+}, and Mn^{2+}, at pH 4.5 utilizing polymers **8** and **8'**; polymer **9** results are reported in reference *18*. Figure 2 shows the results and what we find is that the thermodynamic stability of the newly formed metal(TACN)$_2^{2+}$ complexes appears to dominate over the templating factor; i .e., the $\text{Cu}^{2+}/\text{Zn}^{2+}$ selectivity ratio for **8, 8'** was ~157. The overall order of metal ion selectivity for polymers **8, 8'** is: $\text{Mn}^{2+} \le \text{Ni}^{2+} \le \text{Zn}^{2+} < \text{Co}^{2+} <<< \text{Cu}^{2+}$; somewhat, but not exactly, following the Irving-Williams order of stability (*19*).

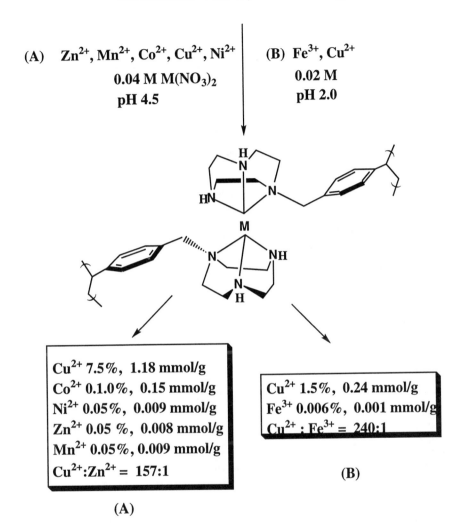

Polymer 8 or 8'

Zn Content: 0.02 %, 0.0031 mmol/g

(A) $Zn^{2+}, Mn^{2+}, Co^{2+}, Cu^{2+}, Ni^{2+}$ (B) Fe^{3+}, Cu^{2+}

0.04 M $M(NO_3)_2$ 0.02 M

pH 4.5   pH 2.0

Cu^{2+} 7.5%, 1.18 mmol/g
Co^{2+} 0.1.0%, 0.15 mmol/g
Ni^{2+} 0.05%, 0.009 mmol/g
Zn^{2+} 0.05 %, 0.008 mmol/g
Mn^{2+} 0.05%, 0.009 mmol/g
$Cu^{2+}:Zn^{2+}$ = 157:1

(A)

Cu^{2+} 1.5%, 0.24 mmol/g
Fe^{3+} 0.006%, 0.001 mmol/g
$Cu^{2+} : Fe^{3+} = 240:1$

(B)

Figure 2. Metal Ion Selectivity Studies with the Demetallated Polymers 8 and 8'

It is interesting to note that if we compete the template Zn^{2+} ion at pH 4.5 with Co^{2+}, Ni^{2+}, and Mn^{2+}, *in the absence of Cu^{2+}*, with polymers **8** and **8'**, the overall order is: $Mn^{2+} < Ni^{2+} < Co^{2+} <<< Zn^{2+}$; in this case, templating appears more important than stability, if the Irving-Williams order has any significance in metal ion templated polymers (*19*). More importantly, the stability constants for the formation of the $M(TACN)_2^{n+}$ complexes shows the following order, $Co^{2+} > Ni^{2+}$

> Zn^{2+}, and further supports the templating effect of Zn^{2+} in the absence of Cu^{2+} (17).

We also competed Cu^{2+} and Fe^{3+} at pH 2.0, in order to define a possible example of a divalent metal ion being selective over a trivalent metal ion. The selectivity data in Figure 2 for polymers **8** and **8'** allows calculations of Cu^{2+}/Fe^{3+} ion selectivity (µmol/µmol/g) and the equilibrium selectivity coefficient (K) ~240 ($K_{Cu^{2+}/Fe^{3+}} = 7 \times 10^5$); respectively, demonstrating surprisingly high selectivity of a divalent to a trivalent metal ion, while polymer **9** had a low ratio of 3 ($K_{Cu^{2+}/Fe^{3+}} = 1.5$). This exciting result might imply that one parameter for templated metal ion selectivity is a consequence of the ionic radius of the original metal ion template (Zn^{2+}, 0.69 Å; Cu^{2+}, 0.72 Å; and Fe^{3+}, 0.53 Å); this order of selectivity, $Cu^{2+} >>> Fe^{3+}$, is diametrically opposite to that found for the bond strengths of known $M(TACN)_2^{2+/3+}$ complexes as studied by X-ray crystallographic and infrared techniques ($Fe^{2+} > Co^{2+} > Fe^{3+} > Cr^{3+} >> Ni^{2+} > Cu^{2+} > Zn^{2+}$); however, more examples have to be studied before this ionic radius concept can be proven unequivocally (20). What is also apparent from the results is that even though polymers **8** and **8'** were prepared from different Zn^{2+} monomer precursors, **3** and **7**, they provide very similar metal ion selectivity orders and metal ion capacities upon reintroduction of various metal ions. The reason, we believe, is that both **8** and **8'** have very similar spatially regulated polymeric sites that are readily able to form stable, metal ion sandwich complexes, while polymer **9** appears to form a more highly crosslinked, hydrophobic polymer site, with dramatically reduced selectivities and capacities for all metal ions studied.

To further determine whether this novel result was a consequence of the ionic radius of the template metal ion used in the original polymerization reaction, we also prepared a Hg^{2+} (0.93 Å) templated sandwich monomer, [(mono-N-(4-vinylbenzyl)-1,4,7-TACN)$_2$Hg](OTf)$_2$ and polymerized it with the crosslinking agent, DVB, Figure 3 (21). After removal of the template Hg^{2+} ion with 6N HCl, a bulk polymer resulted that had very high selectivity to reintroduced Hg^{2+} ions in competition with Cd^{2+}(1.03 Å), 389:1; Pb^{2+} (1.17 Å), 259:1; Ag^+ (1.13 Å), 1111:1; Cu^{2+} (0.72 Å), 165:1; and Fe^{3+} ions (0.53 Å), 24:1, at pH 2.0 (Figure 4).

It tentatively appears that the template Hg^{2+} ion was able to compete effectively with metal ions with similar ionic radii as well as those that were smaller. Thus, we feel that the selectivity for Hg^{2+}, in this case, might also be predicated on the thermodynamic stability of the complex between Hg^{2+} and the polymeric sandwich arrangement of TACN ligands. Other parameters could include the softness of the Hg^{2+} ion and its affinity for the relatively soft nitrogen atom, as well as the geometry of the TACN ligands in the surrounding coordination sphere.

Recently, we published several papers on polymer pendant ligand chemistry that had the pendant ligand anchored to modified 6% crosslinked **PS-DVB** beads (22-27). Moreover, in contrast to this metal ion template approach for selectivity, we found that simply anchoring a TACN ligand on a 6% crosslinked chloromethylated **PS-DVB** bead, **PS-TACN** (0.7 mmol/g, eq 2), showed dramatically less selectivity for all metal ions as well as a lower capacity. This data is shown in the following metal ion studies (Figure 5) with **PS-TACN** being considerably less selective to Cu^{2+} over Zn^{2+} (only 26 %) in comparison to the sandwich arrangement of TACN ligands (Figure 2) and that the metal ion capacities are also greatly reduced; i. e., a ratio of 0.014! These **PS-TACN** results further strengthen the concept of the spatial sandwich arrangement of TACN ligands in a templated polymer matrix being a novel method for metal ion selectivity as well as for the removal of higher capacities of selective metal ions from aqueous solution.

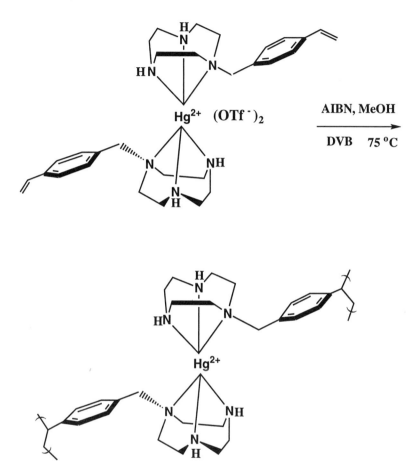

62 % Crosslinked, 8% Hg, TACN, 0.84 mmol/g

TACN:Hg = 2.1

Figure 3. Polymerization of Hg(VBTACN)$_2^{2+}$ with DVB

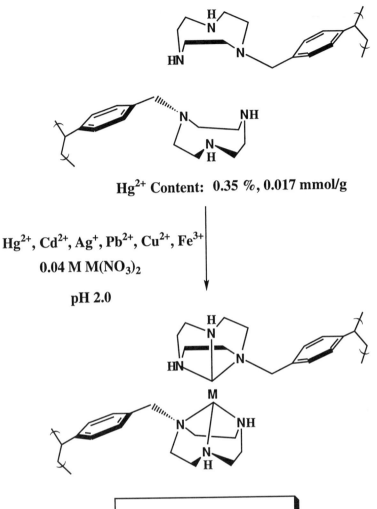

Figure 4. Selectivity Studies with a Demetalated Hg²⁺ Polymer

0.7 mmol TACN/g resin

(2)

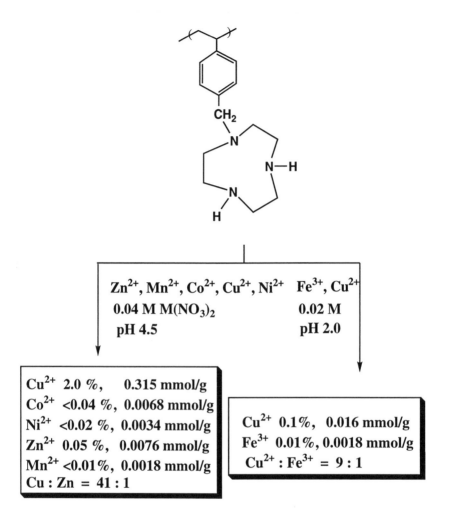

Figure 5. Metal Ion Selectivity Studies with the Polymer Supported TACN ligand, PS-TACN.

In conclusion, we have described significant new findings in the area of metal ion templated polymers that encompass: (1) synthesis of polymeric, template metal ion sandwich complexes from precursor 1:1 mono and tris-N-(4-vinylbenzyl)-1,4,7-TACN-Zn monomer complexes, and from an authentic 2:1 [mono-N-(4-vinylbenzyl)-1,4,7-TACN]$_2$-Zn^{2+} sandwich monomer with DVB; (2) thermodynamic stability to explain the high Cu^{2+} ion selectivity over Zn^{2+}, Ni^{2+}, Co^{2+}, and Mn^{2+} ions reintroduced to the Zn^{2+} demetalized polymer sites; (3) metal ion templating as a possibly important parameter in the reintroduction of the Zn^{2+} ion template and other divalent metal ions (Mn^{2+}, Ni^{2+}, and Co^{2+}, *in the absence of Cu^{2+}*); (4) tentative templating (ionic radius of Zn^{2+} and Cu^{2+} being comparable) for Cu^{2+} selectivity over Fe^{3+} ion, the latter, a metal ion that can readily compete for any polymer pendant ligand site due to its favorable charge to ionic radius ratio; (5) synthesis of a polymeric 2:1 [mono-N-(4-vinylbenzyl)-1,4,7-TACN]$_2$-Hg^{2+} sandwich complex that was found to be highly selective, after demetalation, to reintroduced Hg^{2+} ions in the presence of Cd^{2+}, Pb^{2+}, Ag$^+$, Cu^{2+} and Fe^{3+} ions.

Future studies will focus on the scope of the metal ion templated polymer field with other macrocycles that will hopefully provide high metal ion selectivities to reintroduced metal ions. As well, we need to further define the importance of the templating, thermodynamic, hard/soft heteroatoms/metal ions, geometry of template metal ion-ligands, and other parameters that might help understand metal ion selectivity. Finally, these metal ion templated polymers may have utilization in real world remediation applications and for real-time metal ion sensors in conjunction with fiber optic techniques.

Acknowledgments

The studies at LBNL were generously supported by the Office of Environmental Management, Office of Science and Technology, Office of Technology Systems, Efficient Separations and Processing Cross Cutting Program under the U. S. Department of Energy Contract No. DE-ACO3-76SF00098. RHF thanks Drs. Hong Chen, Robert L. Albright, Jozsef Devenyi, H. Christine Lo, and Marilyn M. Olmstead for discussions and performing the research discussed in this book chapter.

Literature Cited

1. Wulff, G. *Angew. Chem., Int. Ed. Engl.* **1995**, *34*, 1812.
2. Wulff, G. In *Biomimetic Polymers*; Gebelein, C. G. Ed., Plenum Press: New York. 1990, p 1.
3. Wulff, G. in *Polymeric Reagents and Catalysts*: Ford, W. T. Ed., ACS Symposium Series 308, American Chemical Society, Washington. 1986, p 186.
4. Mosbach, K. *Trends Biochem. Sci..* **1994**, *19*, 19 and references therein.
5. Shea, K.*Trends Polym. Sci.* **1994**, *2*, 166 and references therein.
6. Whitcombe, M. J.; Rodriguez, M. E.; Villar, P. ; Vulfson, E. N. *J. Am. Chem. Soc.* **1995**, *117*, 7105.
7. Pomgailo, A. D.; Savost'yanov, V. S. *Synthesis and Polymerization of Metal-Containing Monomers* , CRC Press: Boca Raton. 1994, p 134 and references therein.
8. Dahl, P. K.; Arnold, F. H. *Macromolecules* **1992**, *25*, 7051.
9. Rosatzin, T.; Andersson, L. I.; Simon, W.; Mosbach, K.*J. Chem. Soc. Perkin Trans 2*, **1991**, 1261.
10. Gupta, S. N.; Neckers, D. *J. Poly. Sci. Polym. Chem. Ed.* **1982**, *20*, 1609.
11. Kuchen, W.; Schram, J. *Angew. Chem., Int. Ed. Engl.* **1988**, *27*, 1695.
12. Nishide, H.; Deguchi, J.; Tsuchida, E. *Chem. Lett.* **1976**, 169.
13. Nishide, H.; Tsuchida, E. *Makromol. Chem.* **1976**, *177*, 2295.
14. Kato, M.; Nishide, H. ; Tsuchida, E.; Sasaki, T. *J. Polym. Sci. Polym. Chem. Ed.* **1981**, *19*, 1803.
15. Biswas, M.; Mukherjee, A. in *Advances in Polymer Science*, Springer Verlag Berlin Heidelberg, 1994, vol 115, p 89 and references therein.
16. Zeng, X.; Murry, G. M. *Sep. Sci. Technol.* **1996**, *31*, 2403.
17. Chaudhuri, P.; Wieghardt, K. L. *Prog. Inorg. Chem.* **1987**, *35*, 329 and references therein.
18. Chen, H.; Olmstead, M. M.; Albright, R. L.; Devenyi, J.; Fish, R. H. *Angew. Chem. Int. Ed. Engl.* **1997**, *36*, 642.
19. Irving, H.; Williams, R. J. P. *J. Chem. Soc.* **1953**, 3192.

20. Boeyens, J. C. A.; Forbes, A. G. S.; Hancock, R. D.; Wieghardt, K. L. *Inorg. Chem.* **1985**, *24* , 2926.
21. Lo, H. C.; Fish, R. H., in preparation for publication.
22. Huang, S-P.; Li, W.; Franz, K. J.; Albright, R. L.; Fish, R. H.*Inorg. Chem.* **1995**, *34*, 2813.
23. Huang, S-P.; Franz, K. J.; Olmstead, M. M.; Fish, R. H. *ibid*, 2820.
24. Li, W.; Coughlin, M.; Albright, R. L.; Fish, R. H. *Reactive Polymers*, **1995**, *28*, 89.
25. Li, W.; Olmstead, M. M.; Miggins, D.; Fish, R. H. *Inorg. Chem.* **1996**, *35*, 51.
26. Huang, S.-P.; Franz, K. J.; Arnold, E. H.; Devenyi, J.; Fish, R. H. *Polyhedron*, **1996**,*15*, 4241.
27. Fish, R. H. U.S. Patent 5,662,996, issued April 1997.

Chapter 17

Surface Imprinting: Preparation of Metal Ion-Imprinted Resins by Use of Complexation at the Aqueous-Organic Interface

Kazuhiko Tsukagoshi[1], Kai Yu Yu[2,4], Yoshihisa Ozaki[2], Tohru Miyajima[3], Mizuo Maeda[2], and Makoto Takagi[2]

[1]Department of Chemical Engineering and Material Science, Faculty of Engineering, Doshisha University, Tanabe, Kyoto 610-03, Japan
[2]Department of Chemical Science and Technology, Faculty of Engineering, Kyushu University, Hakozaki, Fukuoka 812-81, Japan
[3]Department of Chemistry, Faculty of Science, Kyushu University, Hakozaki, Fukuoka 812-81, Japan

Based upon the concept of 'Surface Imprinting' proposed by the authors in 1992, metal ion-imprinted microspheres have been prepared by seeded emulsion polymerization of divinylbenzene, styrene, butyl acrylate, and methacrylic acid. The imprinted structure was introduced on the carboxylated microsphere by Surface Imprinting in which the carboxyl groups of mobile, linear polymers are reorganized through complexation with metal ions on the surface and then fixed by crosslinking polymerization in their organized orientation. The 'print' metal ion species was then removed by washing with hydrochloric acid to give Cu(II)-, Ni(II), and Co(II)-imprinted microspheres as spherical submicron particles with average diameters of 0.55-0.60 μm. Without further treatment, such as grinding or sieving, the imprinted microspheres were used as metal ion-selective adsorbents. Their adsorption behavior towards Cu(II), Ni(II), and Co(II) was examined. It was found that the imprinted microspheres readsorbed the 'print' metal ion species more efficiently than did analogous unimprinted microspheres. Some spectroscopic studies were conducted for Cu(II)-loaded microspheres. The data indicate that the origin of the imprinting effect on a Cu(II)-imprinted microsphere is interaction between Cu(II) and carboxylate groups at the aqueous-organic interface, which supports the concept of Surface Imprinting.

The molecular imprinting technique has attracted much attention as a method for preparint novel polymeric materials which may be utilized for the facile, rapid, inexpensive, and large-scale separation and concentration of chemical species. This technique has been demonstrated to be effective for metal ions (1,2), saccharides (3,4), amino acid derivatives (5,6), organic compounds of low molecular weight (7,8), and so on. These preparations are mainly based on solution polymerization or bulk polymerization techniques which distribute the imprinted sites throughout the

[4]Permanent address: Department of Applied Chemistry, Taiyuan University of Technology, 11 Yingze Street, Taiyuan Shanxi 030024, China.

bulk resin. These processes suffer from some fundamental difficulties: 1) Additional processing of the resin by grinding and sieving is necessary to obtain small particles; 2) Partial destruction of the imprinted sites is inevitable in this grinding process; 3) Even after grinding a portion of the imprinted sites will be buried within the resin and unavailable for exchange; 4) It is difficult to accommodate biologically important print species which are water-soluble organic molecules.

In 1992, we proposed the new imprinting technique of 'Surface Imprinting' to solve these problems (9-12). In this technique, the aqueous-organic interface in an emulsified resin suspension (latex) is used as the recognition site for the print species during the imprinting polymerization. The concept of this technique is shown in Scheme 1. First, a seed microsphere emulsion is prepared. The microspheres consist of linear-chain polymers (in I) with functional groups which can interact with the guest (imprinting) species. Divinyl-type monomers are then added to the emulsion which swell the seed microspheres and provide substantial mobility for the polymer chains. When the guest species G is introduced into the emulsion, the functional groups interact with them at the aqueous-organic interface to form complexes (in II). The structures thus obtained are then immobilized by crosslinking polymerization of the divinyl-type monomers (in III). The guest (print) chemical species is then removed to leave imprinted sites on the surface of the microspheres.

In this chapter, we describe the preparation by Surface Imprinting of Cu(II)-, Ni(II)-, and Co(II) imprinted microspheres. These microspheres are uniform spheres which can be used directly for metal ion sorption without grinding and/or sieving. It is anticipated that imprinted microspheres will find applications in chemical analysis, such as those reported for other imprinted polymeric materials as selective packings for HPLC (13-16).

Experimental Section.

Reagents and Apparatus.   Divinylbenzene (DVB) was a gift from Sankyo Chemical Industries, Ltd. Styrene (S), methacrylic acid (MAA), and butyl acrylate (BA) were purchased from Tokyo Kasei Kogyo Company, Ltd. Pure water for the metal ion adsorption tests was obtained with a MILLI-Q Water Purification System (Nippon Millipore Ltd.) and ordinary deionized water was used for other experiments. Metal ion standard solution for atomic absorption analysis were obtained from Wako Pure Chemical Industries, Ltd. Other chemicals were reagent grade and were used as obtained from commercial sources.

Particle grading analysis and FT-IR spectral measurements were performed with Nikkiso Microtrac CM and Nicolet 510M instruments, respectively. The scanning electron microscopy (SEM) was conducted with a Hitachi S-900 FE-SEM instrument. EPR spectra were obtained with a Jeol JEX-FE1XG instrument. Metal ion concentrations were determined with a Seiko Instruments SAS 760 atomic absorption spectrophotometer.

Preparation of Metal Ion-Imprinted Microspheres.   S (20.55 g), BA (2.71 g), MAA (1.64 g), and water (75 mL) in a reaction flask were polymerized under nitrogen at 70 °C and pH 2.2 (adjusted with hydrochloric acid) for 7 hours with potassium peroxodisulfate (125 mg) as initiator for the first-step polymerization. The resultant seed emulsion was cooled to room temperature and DVB (29.86 g), BA (2.90 g) and water (95 g) were added. Of the emulsion, 100 g was kept at 2 °C and pH 9.5 (adjusted with potassium hydroxide) for 24 hours. The pH of the emulsion was adjusted to 5.6 with hydrochloric acid. A portion (8.0 mL) of the emulsion was removed and added to a metal ion solution (40 mL of 0.10 M Cu(II), Ni(II), or Cu(II) chloride). This emulsion was polymerized by use of ^{60}Co γ-rays (0.2 Mrad) for 50 hours at room temperature for the second-step polymerization. Polymerization at low temperature is expected to produce a good imprinting effect (17,18). The

linear-chain polymer microsphere

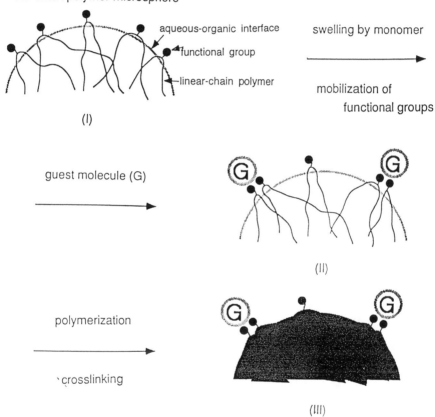

Scheme 1. The Concept of Surface Imprinting.

microspheres were separated from the medium by centrifugation (10,000 rpm for 1 hour) and washed with 150 mL of 0.10 M hydrochloric acid to exchange the print metal ions with protons. The resultant imprinted microspheres were dried *in vacuo*.

For comparison, unimprinted microspheres were synthesized in similar fashion, but in the absence of metal ions.

Measurement of Metal Ion Adsorption. The equilibrium adsorption of metal ions onto the microspheres was determined in the following way. The microsphere sample (0.050 g) was placed in a 1.5 mL-plastic tube and 1.0 mL of a 5.0×10^{-4} M solution of Cu(II), Ni(II), or Co(II) chloride was added. The pH was adjusted to 5.6 with hydrochloric acid or potassium hydroxide. The mixture was shaken at 20 °C for a specified period of time. The equilibrated mixture was centrifuged (10,000 rpm for 1 hour) and the metal ion concentration in the supernatant was determined by atomic absorption spectroscopy. For each combination of metal ion and microsphere species, 3-5 adsorption samples were utilized. The adsorption percentage [{(initial metal ion concentration)-(final metal ion concentration)}/[(initial metal ion concentration) x 100 %] was calculated.

Preparation of Cu(II)-Loaded Microspheres for the FT-IR and EPR Spectroscopic Studies. To probe the origin of the imprinting effect by FT-IR and EPR spectroscopy, three types of Cu(II) loaded microspheres (Samples 1-3) were prepared. The unimprinted (precursor to Sample 1) and Cu(II)-imprinted (precursor to Sample 2) microspheres (0.15 g) which were produced by the second-step polymerization and treatment with acid (*vide supra*) were added to 3.0 mL of a 5.0×10^{-4} M solution of Cu(II) chloride. The pH of the solution was adjusted to 5.6 with hydrochloric acid or sodium hydroxide and the mixture was shaken at 20 °C for 18 hours. The equilibrated mixture was centrifuged (10,000 rpm for 1 hour). The Cu(II)-loaded microspheres were collected and dried *in vacuo*. The amount of Cu(II) adsorbed on the microspheres was assessed by determining the concentration of residual metal ions in the supernatant. The Cu(II)-loaded, unimprinted and Cu(II)-loaded, Cu(II)-imprinted microspheres are identified as Samples 1 and 2, respectively. Sample 3 was the Cu(II)-imprinted microsphere product from the second-step polymerization (*vide supra*) which was isolated by simple centrifugation of the reaction mixture and drying *in vacuo*.

Results and Discussion.

Historical Survey of Conventional Metal Ion-Imprinted Resins. The molecular imprinting technique has been demonstrated for be effective for metal ions as well as for organic molecules. The Nishide and Kabanov groups reported the first examples of metal ion-imprinted resins in 1976 (*1,19*). In 1977, Nishide *et al.* reported imprinted resins which could readsorb the print metal ion selectively from a weakly acidic solution (*20*). This feature is in sharp contrast with conventional metal ion-chelating resins, which are usually designed for use in more acidic solutions. Since their successful demonstration of the concept, a variety of metal ion species have been examined as imprinting guest species including Cu(II), Zn(II), Ni(II), Co(II), Cd(II), Fe(II), Hg(II), and Ca(II) (*21-27*).

The metal ion-imprinted resins reported to date may be classified as one of two types depending upon their mode of preparation: 1) Method 1 - Linear chain polymers with metal-binding groups which are crosslinked with a bifunctional reagent in the presence of metal ions; and 2) Method 2 - Metal ion-complexing monomers are prepared and these complex are polymerized with matrix-forming monomers. As an example of Method 1, Nishide *et al.* crosslinked poly(4-vinylpyridine) and 1,4-dibromobutane in the presence of metal ions (*1,20*). In a second example, Ohga *et al.*

crosslinked chitosan with epichlorohydrin to give an imprinted resin (21) which was utilized to separate metal ions by column chromatography.

The first example of Method 2 was also reported by Nishide *et al.* who polymerized a metal complex of 1-vinylimidazole with 1-vinyl-2-pyrrolidone and divinylbenzene (DVB) (22). Subsequently Gupta *et al.* polymerized DVB with a metal complex of 4-vinyl-4'-methyl-2,2'-bipyridine in methanol (23). Also Rosatzin *et al.* polymerized N,N'-dimethyl-N,N'-bis(4-vinylphenyl)-3-oxapentane diamide (a metal ionophore), DVB, and styrene in chloroform in the presence of metal salts (24). In these resin preparations, polymerization was conducted without isolation of the metal ion-monomer complexes.

In a more sophisticated way of carrying out Method 2, specific metal complexes were prepared from ligands with vinyl groups isolated, and polymerized with matrix-forming monomers. For example, Kuchen *et al.* synthesized a metal ion-acrylic acid complex and then polymerized it with ethylene glycol dimethacrylate in a benzene-methanol solution (25). Fujii *et al.* prepared a metal ion complex of 2-acetyl-5-(*p*-vinylbenzyoxy)phenol and polymerized it with acrylamide and ethylene glycol dimethacrylate (26). Isobe *et al.* used Amberlite XAD-2 as a matrix resin and a 4-vinylpyridine-metal ion complex was introduced on the resin through polymerization (27).

Although all of these imprinting techniques offer successful ways to prepare metal ion-selective adsorbents, they have some common fundamental problems as discussed in the Introduction. To circumvent such difficulties, we have proposed a third method of imprinting (Surface Imprinting) which is conducted on the surface of microspheres.

Surface Imprinting of Microspheres. This concept of surface imprinting is depicted in Scheme 2. In the first-step polymerization (I), a monomer with a chelating group and one or more additional monomers are chemically polymerized to form a seed microsphere emulsion. These microspheres (II) contain linear polymer chains which carry the chelating group. A divinyl-type monomer is then added to the emulsion which swells the microspheres (III) and provides substantial mobility to the linear polymer chains. When the print metal ions are introduced, the chelating groups interact with them at the aqueous-organic interface to form complexes (IV). These structures are then immobilized by radiation-induced crosslinking polymerization of the divinyl-type monomer (V), *i.e.* the second-step polymerization. The print metal ions are then removed from the microsphere surface by treatment of with acid to give microspheres with a metal ion-templated surface.

Synthesis of Metal Ion-Imprinted Microspheres. Cu(II), Ni(II), and Co(II)-imprinted microspheres were prepared in 85-90 % yields by the method described in the Experimental Section. For comparison unimprinted microspheres were synthesized by the same method but in the absence of a metal ion template.

Physicochemical Characterization of the Microspheres. Both the chemical compositions and particle size distributions of the microspheres were determined. The elemental analyses were in good agreement with the chemical compositions of the feed mixtures. Figure 1 shows typical FE-SEM view of the Cu(II)-imprinted microspheres. Similar particle forms and distribution patterns were observed for the other microspheres. The submicron particles are nearly spherical and had average diameters of 0.55-0.60 μm. The SEM results were consistent with those of a particle grading analysis.

Metal Ion Adsorption by the Microspheres. Abilities of the unimprinted and Cu(II)-, Ni(II)-, and Co(II)-imprinted microspheres to adsorb metal ions from aqueous solutions were determined at pH 5.6. The percentage adsorption results are

256

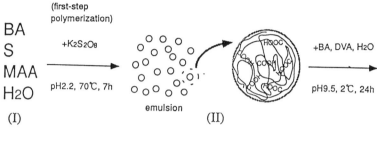

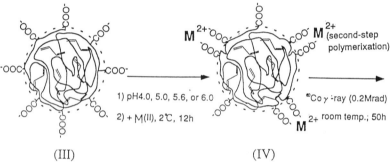

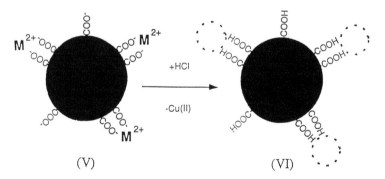

Scheme 2. The Synthesis of Metal Ion-Imprinted Microspheres.

given in Table I. (Under the conditions specified in Table I, 100 % adsorption corresponds to 1.0 x 10^{-2} mmol of metal ion/g of dry weight resin.)

Table I.   Percentage Adsorption of Single Metal Ion Species onto Unimprinted and Metal Ion-Imprinted Microspheres[a,b]

	Adsorption percentage, %		
Imprinted ion	Cu(II)	Ni(II)	Co(II)
None	24	0	0
Cu(II)	40	7	0
Ni(II)	14	17	0
Co(II)	5	0	0

[a]Conditions: pH, 5.6; 0.050 g of microsphere; 1.0 mL of 5.0 x 10^{-4} M metal ion solution; shaking time, 18 hours. [b]For 3-5 samples, the standard deviation was ±2 %.

Considering first the adsorption of Cu(II), this metal ion is most effectively adsorbed by the Cu(II)-imprinted microsphere, followed in order by the unimprinted, Ni(II)-imprinted, and Co(II)-imprinted microspheres. This sequence clearly verifies the effect of Cu(II) imprinting. Thus the carboxyl groups and the polymer chains which carry them are placed on the surface of the Cu(II)-imprinted microsphere in such a way that they match the square planar coordination geometry of Cu(II). However, the diminished affinity of the Ni(II)- and Co(II)-imprinted microspheres compared with the unimprinted microspheres also warrants discussion. For the unimprinted microspheres, the carboxyl groups are distributed on the surface in energetically optimal placements which are determined by such solution conditions as the pH and the counter ion (K^+) concentration. Although this distribution of carboxyl groups happens to favor the accommodation of Cu(II) to a considerable extent (24 % adsorption compared to 40 % for the Cu(II)-imprinted microspheres), the Ni(II)- and Co(II)-imprinting conditions decrease this distribution of the carboxyl groups. Thus , Cu(II) adsorption by the Ni(II) and Co(II)-imprinted microspheres decreases to 14 % and 5 %, respectively. This reveals a negative imprinting effect for a non-print metal ion species.

For Ni(II), only the Cu(II)- and Ni(II)-imprinted microspheres exhibited a binding affinity under these experimental conditions with the latter a giving higher percentage adsorption (17 %) than the former (7 %). Again, this clearly shows the operation of an imprinting effect.

It is important to note that whereas the unimprinted microspheres did not show any adsorption affinity for Ni(II), the Ni(II)-imprinted microspheres exhibit considerable affinity for readsorption of the print metal ion. Obviously the Ni(II)-imprinting process alters the nature of the carboxyl group dispositions on the microsphere surface to accommodate otherwise unadsorbed metal ion species.

None of the microsphere species showed Co(II) binding under these experimental conditions. This is probably due to the fact that the complexation tendency of Co(II) is rather low, as seen in the Irving-Williams series (28). However under other experimental conditions it has been reported that Co(II)-imprinted microspheres adsorbed Co(II) more effectively than did unimprinted microspheres (11).

The Ni(II) -imprinted microspheres exhibit a Ni(II)-binding affinity which exceeds that for Cu(II). This is contrary to expectations based upon the Irving-Williams series ordering (28). To confirm this rather surprising result, a second preparation of the Ni(II)-imprinted microspheres was performed. It was found that the second preparation gave very similar results to those reported in Table I. Thus for the second preparation of the Ni(II)-imprinted resin, the adsorption percentages for Ni(II) and Cu(II) were 18 and 13 %, respectively.

It should be emphasized that these results all arise from the differences in spatial orientation of the carboxyl groups on the surface which is fixed by crosslinking with DVB.

Figure 2 shows the pH dependence of metal ion adsorption onto the Cu(II)- and Ni(II)-imprinted microspheres and the unimprinted microspheres. The pH was adjusted in the region 1.0-5.6 by addition of hydrochloric acid or potassium hydroxide. The imprinted microspheres preferentially readsorbed the print metal ion species over the entire pH range of metal ion adsorption. The percentage adsorption was enhanced with increasing pH for all three types of microspheres. This demonstrates that the carboxyl groups on the microsphere surface participate in metal-ion binding in their ionized carboxylate forms. Although the data given in Table I and Figure 2 were obtained after adsorption equilibration for 18 hours, most of the adsorption was completed within 1 hour.

FT-IR and EPR Spectroscopic Studies of Cu(II)-Loaded Microspheres. These investigations were conducted with microsphere Samples 1-3. As described in the Experimental Section, Sample 1 is unimprinted microspheres which have adsorbed Cu(II). Sample 2 is Cu(II)-imprinted microspheres with readsorbed Cu(II). Sample 3 is the Cu(II)-imprinted polymer which was recovered after the second-step polymerization before treatment with acid to remove the print metal ions. First the amount of Cu(II) carried by the microsphere samples was assessed by measuring the concentration of free Cu(II) in the aqueous solution at adsorption equilibrium. The calculated amounts of Cu(II) were 1.5×10^{-6}, 5.7×10^{-6}, and 1.4×10^{-4} mol g^{-1}.

Obuko et al. (29) reported that the amounts of carboxyl groups on the surface of the microsphere and inside of the microsphere were 2.2×10^{-4} mol g^{-1} and 1.5×10^{-4} mol g^{-1}, respectively, for the surface-carboxylated microspheres which they developed. Since we used their recipe for the preparation of our microspheres, the amounts of carboxyl groups in the two locations are expected to be similar to the values determined by Obuko et al.

We now make the assumption that the total amount of surface carboxyl groups for our microspheres is 2.2×10^{-4} mol g^{-1} and that each Cu(II) is bound by two carboxylate groups. Then, the mole fraction of carboxyl groups bound to Cu(II) of the carboxyl groups on the surface is calculated to be: Sample 1, 1/70; Sample 2, 1/20; and Sample 3, 1/1. It is emphasized that all of the surface carboxyl groups for Sample 3 appear to be complexed with Cu(II).

The FT-IR spectrum of Sample 3 is shown in Figure 3, as an example. When the absorption band intensity at 1730 cm^{-1} for the ester group from the butyl acrylate monomer was taken as a standard with a normalized intensity of 1.0, the absorption band intensities for dimeric carboxyl groups at 1705 cm^{-1} and the carboxylate anion at 1601 cm^{-1}, respectively, were found to be: Sample 1, 0.41 and 0.52; Sample 2, 0.41 and 0.54; and Sample 3, 0.31 and 0.62. The values noted for Samples 1 and 2 were very similar to the values of 0.41 and 0.52 which were calculated for unimprinted microspheres which had not been loaded with Cu(II). For Samples 1 and 2, the fraction of the Cu(II)-bound carboxylate groups to the total carboxyl groups (i.e. 2.2×10^{-4} mol g^{-1} + 1.5×10^{-4} mol g^{-1}) was calculated to be about 1/120 and 1/30, respectively. These low values should be the reason for no significant change in absorption intensity at 1705 and 1601 cm^{-1} between Samples 1 and 2. On the other hand, almost all of the surface carboxyl groups in Sample 3 are proposed to be used for Cu(II) adsorption. Thus the band intensity for the dimeric carboxyl groups decreased and that for the carboxylate groups increased. This demonstrates that carboxyl groups on the microsphere surface participate in Cu(II) binding in their ionized carboxylate forms. That the band intensity for the dimeric carboxyl groups did not decrease beyond a certain limit is reasonable if some fraction of the carboxyl groups are located deep inside the microsphere where they cannot interact with Cu(II) in aqueous solution. It should also be noted that the 1601 cm^{-1} adsorption did not

1.0 μ m

Figure 1. Field Emission Scanning Electron Micrograph for Cu(II)-Loaded Microspheres (Sample 3).

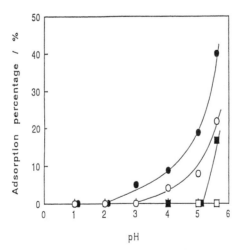

Figure 2. Effect of pH on the Metal Ion Adsorption by Unimprinted and Metal Ion-Imprinted Microspheres.

● = Adsorption of Cu(II) by Cu(II)-imprinted microspheres

O = Adsorption of Cu(II) by unimprinted microspheres

■ = Adsorption of Ni(II) by Ni(II)-imprinted microspheres

□ = Adsorption of Ni(II) by unimprinted microspheres

Conditions: 0.050 g of microspheres; 1.0 mL of 5.0×10^{-4} M metal ion solution; and shaking for 12-18 hours.

decrease below 0.51 for any of the microsphere preparations tested in the present study. This is ascribed to overlapping of the band with the absorption due to a skeletal vibration of the benzene ring.

EPR spectra were obtained for Samples 1-3. The spectrum for Sample 3 is shown in Figure 4. Since these spectra suffered from considerable noise due to the small amount of Cu(II), $g^{//}$ and $A^{//}$ values could not be determined precisely. However all three spectra indicated the peculiar pattern (*30,31*) for square-planar Cu(II) complexes with four coordinating oxygen atoms ($g^1 = g^{//}$, g^2, $g^3 = g^{\perp}$, axis symmetrical spectra; relatively large values of $g^{//}$ and $A^{//}$; and four splitting patterns of hyperfine splitting). The Cu(II) spectral intensities for Samples 1-3 were in the order: Sample 3 > Sample 2 > Sample 1. Although EPR spectroscopy is, in general only semi-quantitative in nature, the ordering of spectral intensities is consistent with the amount of Cu(II) on the microspheres as determined in the Cu(II) adsorption study. The imprinting effect or an enhanced Cu(II) adsorption behavior in the metal ion adsorption test is in a sense indirect because it only measures the decrease in Cu(II) concentration from the adsorption solution. On the other hand, the EPR spectra indicate directly such the enhanced Cu(II) adsorption, as well as the chemical structure of the adsorbed Cu(II) species in the Cu(II)-imprinted microspheres.

Complex Formation Equilibria and the Imprinting Effect at the Interfacial Domain of Microspheres. The IR and EPR spectral results also confirm the interaction of Cu(II) with the carboxylate groups. In the previous section, it was suggested that the Cu(II) is bound to the microsphere with a Cu(II):carboxyl group molar ratio of approximately 1:2. In an, earlier, related study of Cu(II)-imprinting (*32*), formation of a 1:2 Cu(II) carboxylated complex was indicated to be a key step for the introduction of the imprinted structure on the microsphere surface.

However, the supposition that such 1:2 complexes are formed in the imprinting reaction mixture is seemingly inconsistent with the concentrations of Cu(II) in solution and carboxyl groups on the microsphere surface (8.3 x 10^{-3} M and 7.5 x 10^{-3} M, respectively, if the polymerization mixture is assumed to have a homogeneous distribution of species). Thus formation of 1:1 complexes rather than 1:2 complexes would be anticipated. Therefore, one has to focus on some restricted domain at the aqueous solution-microsphere interface, where all of the complexation reactions are taking place. In this domain, there will be a highly concentrated aqueous solution of carboxyl groups (from the polymer matrix) and a certain concentration of free Cu(II) (the aquo-Cu(II) complex) which is similar to the concentration in the bulk aqueous solution.

To provide more information about this system, a model calculation was performed for Cu(II) complexation by carboxylate anion (RCO_2^-) in aqueous solution using the complex stability constant data (*33*) for the Cu(II)-acetate system. In Figure 5 are plotted the percentages of different Cu(II)-containing species as a function of the carboxylate concentration. We employed the data from this model calculation in consideration of the reactions taking place in the interfacial domain during the second-step polymerization.

After the second-step polymerization and centrifugation, the free Cu(II) concentration in the supernatant solution was determined. Compared with the initial Cu(II) concentration for the polymerization, 19 % remained in the supernatant as free Cu(II). Since Cu(II) is obviously heavily concentrated in the interfacial domain, the level of free Cu(II) in the total Cu(II) concentration in this region should be much lower than 19 %. The dashed lines in Figure 5 shows that with 19 % of free Cu(II), the remainder of the Cu(II) species consist of 53 % of $[Cu(RCO_2)]^+$, 26 % of $[Cu(RCO_2)_2]$, and 2 % of $[Cu(RCO_2)_3]^-$. Since 19 % is too high of an estimate for the free Cu(II) concentration in the interfacial region, the fraction of Cu(II)-carboxylate species in the interfacial domain should be even higher than these levels. The dashed lines in Figure 5 also show that with 19 % of free Cu(II), the concentration of

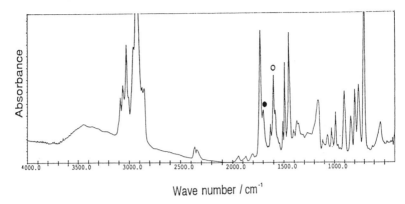

Figure 3. FT-IR Spectrum for Cu(II)-Loaded Microspheres (Sample 3).
● = Absorption band for the dimeric carboxyl group
O = Absorption band for the carboxylate group

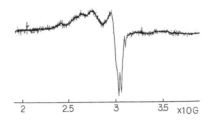

Figure 4. EPR spectrum for Cu(II)-Loaded Microspheres (Sample 3). Conditions: frequency, 9227.0 MHz; magnetic field, 3100 ± 1000 G; room temperature; sweep time, 8 minutes; modulation, 100 Hz - 63 G; amplitude, 1.25 x 10^{-3}; and response, 0.03.

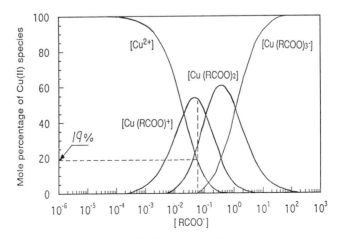

Figure 5. Relationship between $[RCO_2^-]$ at equilibrium and the percentages of Cu(II)-containing species.

carboxylate groups is 6 x 10^{-2} M. This implies a high concentration of carboxylate anions in the interfacial domain. Thus consideration of the model supports the proposal that a substantial fraction of 1:2 complex is formed at the interfacial region during the second-step polymerization (*i.e.* the imprint structure-forming step).

Formation of the Cu(II)-imprinted structure actually takes place through interaction of Cu(II) with the carboxylate groups on the linear polymer chains with subsequent chain immobilization by crosslinking. Describing the imprinting process using the Cu(II)-acetate complex in aqueous solution as a model is an obvious over-simplification. However, it does provide clues to understanding the chemical behavior of imprinted microspheres and the operational parameters, such as pH and the concentration of the imprinting metal ion, for their preparation.

Acknowledgments

The authors wish to express their appreciation to Professor M. Okubo of Kobe University for his valuable discussions and to Professor Hideaki Maeda of Kyushu University for performing the FE-SEM examinations of the microspheres. We would also like to thank Professor F. Nakashio's group at Kyushu University, the Center for Advanced Instrumental Analysis at Kyushu University, and the ^{60}Co Irradiation Laboratory at Kyushu University for their assistance in conducting the experiments. This work was partially supported by The Asahi Glass Foundation. Financial support by a Grant-in-aid for Scientific Research from the Ministry of Education, Science and Culture of Japan is also acknowledged.

Literature Cited.

1. Nishide, H.; Tsuchida, E. *Makromol. Chem.* **1976**, *117*, 2295.
2. Kabanov, V. A.; Efendiev, A. A.; Orujev, D. D. *J. Appl. Polym. Sci.* **1979**, *24*, 259.
3. Wulff, G.; *J. Makromol. Chem.* **1991**, *192*, 1329.
4. Wulff, G.; Schauhoff, S. *J. Org. Chem.* **1991**, *56*, 395.
5. Andersson, L.; Oshannessy, D. J.; Mosbach, K. *J. Chromatogr.* **1990**, *513*, 167.
6. Andersson, L.; Mosbach, K. *J. Chromatogr.* **1990**, *516*, 313.
7. Shea, K. J.; Sasaki, D. Y. *J. Am. Chem. Soc.* **1989**, *111*, 3442.
8. Shea, K. J.; Sasaki, D. Y. *J. Am. Chem. Soc.* **1991**, *113*, 4109.
9. Yu, K. Y.; Tsukagoshi, K.; Maeda, M.; Takagi, M. *Anal. Sci.* **1992**, *8*, 701.
10. Tsukagoshi, K.; Yu, K. Y.; Maeda, M.; Takagi, M. *Bull. Chem. Soc. Jpn.* **1993**, *66*, 114.
11. Tsukagoshi, K.; Yu, K. Y.; Maeda, M.; Takagi, M. *Kobunshi Ronbunshu* **1993**, *50*, 455.
12. Tsukagoshi, K.; Yu, K. Y.; Maeda, M.; Takagi, M.; Miyajima, T. *Bull. Chem. Soc. Jpn.* **1995**, *68*, 3095.
13. Hosoya, K.; Yoshizako, K.; Tanaka, N.; Kimata, K.; Araki, T.; Haginaka, J. *Chem. Lett.* **1994**, 1437.
14. Sellergren, B. *J. Chromatogr. A* **1994**, *673*, 133.
15. Plunkett, S. D.; Arnold, F. H. *J. Chromatogr. A* **1995**, *708*, 19.
16. Mayes, A. G.; Mosbach, K.; *Anal. Chem.* **1996**, *68*, 3769.
17. Wulff, G.; Haarer, J. *Makromol. Chem.* **1991**, *192*, 1329.
18. O'Shannessy, D. J.; Ekberg, B.; Mosbach, K. *Anal. Biochem.* **1989**, *177*, 144.
19. Kabanov, V. A.; Efendiov, A. A.; Orujev, J. J. *Bull. Inv.* **1976**, *6*, 58.
20. Nishide, H.; Deguchi, J.; Tsuchida, E. *J. Polym. Sci., Poly. Chem. Ed.* **1977**, *15*, 3023.
21. Ohga, K.; Kurauchi, Y.; Yanase, H. *Bull. Chem. Soc. Jpn.* **1987**, *60*, 444.
22. Kato, M.; Nishide, H.; Tsuchida, E. *J. Polym. Sci., Polym. Chem. Ed.* **1981**, *19*, 1803.

23. Gupta, S. N.; Neckers, D. C. *J. Polym. Sci., Polym. Chem. Ed.,* **1982**, *20,* 1609.
24. Rosatzin, T.; Andersson, L. I.; Simon, W.; Mosbach, K. *J. Chem. Soc., Perkin Trans. 2* **1991**, 1261.
25. Kuchen, W.; Schram, J. *Angew. Chem,. Int. Ed. Engl.* **1988**, *27,* 1695.
26. Fujii, Y.; Maie, H.; Kumagai, S.; Sugai, T. *Chem. Lett.* **1992**, 995.
27. Nakashima, A.; Isobe, T. *Memoirs of the Faculty of Science, Kyushu University* **1987**, *16*, 33.
28. Irving, H.; Williams, R. J. P. *J. Chem. Soc.* **1953**, 3912.
29. Okubo, M.; Kanaida, K.; Matsumoto, T. *J. Appl. Poly. Sci.* **1987**, *33*, 1511.
30. Wanatabe, T. *Bunseki* **1977**, *11*, 742.
31. Yokoi, H.; Kawata, S.; Iwaizumi, M. *J. Am. Chem. Soc.* **1986**, *108*, 3361.
32. Kido, H.; Sonoda, H.; Tsukagoshi, K.; Maeda, M.; Takagi, M.; Maki, H.; Miyajima, T. *Kobunshi Ronbunshu* **1993**, *50*, 403.
33. Forsberg, J. H.; Moller, T. *J. Am. Chem. Soc.* **1968**, *90*, 1932

Chapter 18

Selective Adsorption of Metal Ions to Surface-Templated Resins Prepared by Emulsion Polymerization Using a Functional Surfactant

Yoshifumi Koide[1], Hideto Shosenji[1], Mizuo Maeda[2], and Makoto Takagi[2]

[1]Department of Applied Chemistry, Faculty of Engineering, Kumamoto University, Kurokami, Kumamoto 860, Japan
[2]Department of Applied Chemistry, Faculty of Engineering, Kyushu University, Hakozaki, Fukuoka 860, Japan

Monomer-type surfactants which can function as a ligand, 10-(p-vinylphenyl)decanoic acid (**Rac**) and 2-(p-vinylbenzylamino)alkanoic acids (**R$_n$NAc**), have been used as emulsifiers for the preparation of surface-templated resins. The surfactants adsorb at the oil-water interface and emulsify divinylbenzene-styrene monomers. Emulsion polymerization using a potassium persulfate initiator or by irradiation with γ-rays gave fine particles which were 200—300 nm in diameter. The metal-imprinted resins prepared with **Rac** were 1.8 times more effective than the unimprinted resins for adsorption of Cu(II) and Zn(II)-imprinted resins showed highly effective adsorption of Zn(II). Such surface-template effects were also seen for metal-imprinted resins prepared with **R$_n$NAc**, but the effect was sensitive to the alkyl chain length. The **R$_{18}$NAc** resin was the most effective. The Cu(II)/Zn(II) ratio in competitive sorption was 3.7 for the Cu(II)-imprinted resins prepared with **Rac** and 4.2 with **R$_{18}$NAc**

Molecular imprinting methods (*1-3*) have been used for selective recovery of specific chemical species, such as organic compounds (*4-5*) and metal ions (*6-8*). However, the techniques reported so far lack generality, especially for water-soluble guests. Since host monomers bound to specific guest molecules are chemically solidified with matrix monomers by crosslinking polymerization, the host-guest complexes must be soluble in the matrix monomers. When the target molecules are polar, the complexes will usually have low solubility in the matrix monomers. Moreover, the resulting molecular-imprinted resins must be ground and sieved to produce finely divided resins, since most of the imprinted sites are formed in the inner portions of the bulk resins. For the same reason, the resins may require long times for the separation of guest species. Some surface-templated resins have been developed to reduce such problems (*9-15*). A technique which utilizes the orientation of a surfactant is an efficient method for the preparation of surface-templated resins (*16*). A functional surfactant (emulsifier) that is capable of binding metal ions and functions as a vinyl monomer must orient at the interface between the matrix monomers and water, and emulsify these solutions. The surfactant will be polymerized with the matrix monomers in the form of a metal complex. Thus, a functional surfactant should make the surface imprinting easy and reliable. However, imprinting sites on the resins will be influenced by the hydrophilic lipophilic balance (HLB) of the surfactant and the surface-template effect, *i.e.* the metal ion selectivity should be affected by the location of the templated sites.

In this chapter, monomer-type surfactants which can function ligands, 10-(*p*-vinylphenyl)decanoic acid (**Rac**), 2-(*p*-vinylbenzylamino)alkanoic acids (**R$_n$NAc**), and N-alkyl 2-(*p*-vinylbenzylamino)dodecanoic acid derivatives (**RR$_n$NAc**), have been used as emulsifiers for the preparation of surface-templated resins. Effects of interfacial phenomena on the imprinting effect are probed.

Functional Surfactants for the Preparation of Surface-Templated Resins

A monomer-type surfactant called New Frontier A-229ER, which is a phosphonic ester of acrylate oligomer (Dai-Ichi Kogyo Seiyaku Co., Ltd.), has been used as an emulsifier in emulsion polymerization by Yamasaki *et al.* (*17*). However, surface-imprinting was not attempted. In this study, the monomer-type surfactants (**Rac**, **R$_n$NAc**, and **RR$_n$NAc**) have been synthesized and used as functional surfactants for the preparation of surface-templated resins.

Functional Surfactants. The monomer-type surfactant **Rac** with mp 55-57 °C was obtained in 16 % yield by a procedure (Figure 1) analogous to that reported by Freedman *et al.* (*18-21*) for the synthesis of branched 10-(*p*-vinylphenyl)undecanoic acids. IR (KBr): 3200 (OH); 1690 (C=O) cm^{-1}. ^{1}H-NMR (CCl$_4$): δ = 1.25 (14H, -CH$_2$-), 2.15 (2H, -CH$_2$Ac), 2.43 (2H, -CH$_2$-Ar), 4.94-5.11 (1H, *trans* CH=CAr), 5.49-5.73 (1H, *cis* CH=CAr), 6.49-6.73 (1H, C=CHAr), 7.15 (4H, ArH), 10.38 (COOH). Calculated for C$_{18}$H$_{26}$O$_2$: C, 78.80; H, 9.55. Found: C, 78.10; H, 9.00.

R$_n$NAc and **RR$_n$NAc** were obtained by a procedure analogous to that employed by Freeman *et al.* (*22-24*) for the preparation of *N*-(vinylbenzylimino)diacetic acid. The precursor 2-bromoalkanoic acids were synthesized from the corresponding alkanoic acids and converted into 2-aminoalkanoic acids or 2-alkylaminoalkanoic acids by the reaction with ammonia water or an alkylamine, respectively. By reaction with *p*-chloromethylstyrene, **R$_n$NAc** and **RR$_n$NAc** were produced in 6-30 % yields (Figure 2). **R$_8$NAc**: mp 206-207 °C. IR (KBr): 3400 (NH), 2970 (OH), 1590 (C=O) cm^{-1}. ^{1}H-NMR (CD$_3$OD, NaOD): δ = 0.94 (3H, -CH$_3$, 1.06-1.76 (10H, -CH$_2$-), 3.24 (1H, >CHN-), 3.52-3.85 (2H, -NCH$_2$-), 5.18 (1H, *trans* CH=CAr), 5.73 (1H, *cis* CH=CAr), 6.71 (1H, C=CHAr), 7.04-7.67 (4H, ArH). Calculated for C$_{17}$H$_{25}$NO$_2$: C, 74.18; H, 9.09; N, 5.09. Found: C, 74.16; H, 8.92; N, 4.62. **R$_{18}$NAc**: mp 172-175 °C. IR (KBr): 3400 (NH), 2970 (OH), 1590 (C=O) cm^{-1}. ^{1}H-NMR (CD$_3$OD, NaOD): δ = 0.92 (3H, -CH$_3$), 1.11-1.82 (10H, -CH$_2$-), 3.22 (1H, >CHN-), 3.54-3.93 (2H, -NCH$_2$-), 5.24 (1H, *trans* CH=CAr), 5.73 (1H, *cis* CH=CAr), 6.72 (1H, C=CHAr), 7.04-7.72 (4H, ArH). Calculated for C$_{27}$H$_{45}$NO$_2$: C, 78.07; H, 10.84; N, 3.37. Found: C, 77.76; H, 10.65; N, 3.23. **RR$_6$NAc**: viscous oil. IR (KBr): 3000 (OH), 1590 (C=O) cm^{-1}. ^{1}H NMR (CD$_3$OD): δ = 0.87 (6H, -CH$_3$), 1.12-1.84 (26H, -CH$_2$-), 2.99, -CH$_2$N<), 3.64 (1H, >CHN-), 4.26 (2H, -NCH$_2$Ar), 5.28 (1H, *trans* CH=CAr), 5.84 (1H, *cis* CH=CAr), 6.73 (1H, C=CHAr), 7.25-7.71 (4H, ArH). Calculated for C$_{27}$H$_{45}$NO$_2$•1/2H$_2$O: C, 74.62; H, 10.82; N, 3.29. Found: C, 76.95; H, 10.97; N, 3.17. **RR$_{12}$NAc**: viscous oil. IR (KBr): 3000 (OH), 1590 (C=O) cm^{-1}. ^{1}H-NMR (CH$_3$OD): δ = 0.91 (6H, -CH$_3$), 1.12-2.02 (38H, -CH$_2$-), 2.97 (2H, -CH$_2$N<), 3.56 (1H, >CHN-), 4.25 (2H, -NCH$_2$Ar), 5.28 (1H, *trans* CH=CAr), 5.84 (1H, *cis* CH=CAr), 6.73 (1H,

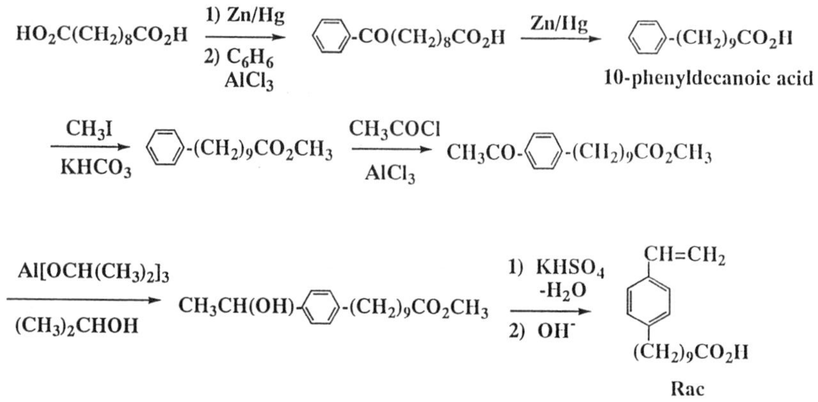

Figure 1. Preparation of **Rac**.

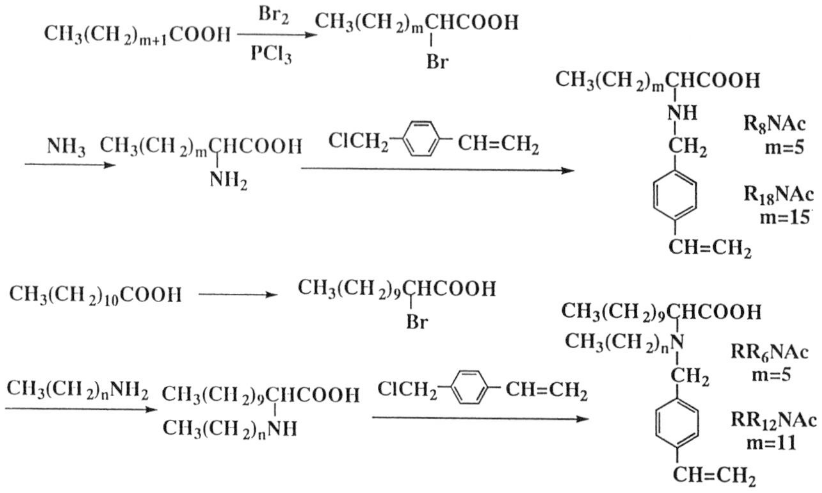

Figure 2. Preparations of **RnNAc** and **RR$_n$NAc**.

C=CHAr), 7.24-7.71 (4H, ArH). Calculated for $C_{33}H_{57}NO_2 \cdot 1/3H_2O$: C, 78.41; H, 11.41; N, 2.77. Found: C, 78.61; H, 11.48; N, 2.64.

Surface Activities of the Functional Surfactants. As measured by a Wilhelmy surface tension balance, an alkaline solution of **Rac** had a surface tension of γcmc = 41 mN m^{-1} and cmc = 1.25 x 10^{-3} M at pH 12. The cross-sectional area of the molecule, as calculated from the Gibbs adsorption equilibrium, is 0.57 nm^2. Thus the cmc of **Rac** is lower than those of $C_{11}H_{23}COONa$ (above 1 x 10^{-2} mol dm^{-3}), $C_{13}H_{27}COOK$ (6 x 10^{-3} M)(25), and $C_{15}H_{31}COONa$ (3.2 x 10^{-3} M at 52 °C)(25). Based on the cmc values, the hydrophobicity of **Rac** should nearly correspond to that of $C_{15}H_{31}COONa$. Hydrophobicity of a phenyl group is known to be equivalent to about three and one-half methylene groups (26). Therefore, the sum of the carbon numbers for **Rac** is estimated to be $C_{12+3.5}$ ($C_{15}H_{31}COONa$ or $C_{16}H_{33}COONa$). The monomer-type surfactant **Rac** is easily polymerized by heating, and an alkaline solution of the resultant polymer (polysoap) shows a lowering of the curve for the surface tension without a break point for the cmc (γ = 47 mN·m^{-1} at 2.5 x 10^{-3} M). A mixture of toluene and water (1:1) containing 0.2 mol % of **Rac** (as the sodium salt) was emulsified by vigorous stirring. The volume (height) of the resulting emulsion at time intervals of five minutes, compared with those for 0.2 mol % of sodium dodecyl sulfate (DS) and some sodium alkanoates was: hexadecanoate > tetradecanoate > **Rac** > DS > dodecanoate >> octadecanoate. Therefore, the chain length of **Rac** seems to correspond to the alkyl chain of 13 carbon number for a sodium alkanoate. This is shorter than the carbon number of 15.5-16 estimated from the cmc values and the calculation. The smaller oleophilicity (larger hydrophilicity) on emulsification is ascribed to the large partitioning of the vinylbenzene moiety between the two liquid phases, since styrene is sparingly soluble in water.

R_nNAc and RR_nNAc (5.0 x 10^{-2} M in toluene) were also emulsified by stirring in a toluene-water mixture (1:1). The order of emulsifying ability for the mixture of toluene-water at pH 5.5 is $R_{18}NAc > R_8NAc > RR_6NAc = RR_{12}NAc$, and that for toluene-Cu(II) solutions (1.0 x 10^{-2} M at pH 5.5) is $R_8NAc > R_{18}NAc > RR_6NAc = RR_{12}NAc$. Emulsification is correlated with the interfacial tension. The interfacial tensions of the toluene-water or toluene-Cu(II) solutions (1.0 x 10^{-2} M at pH 5.5), measured with a Du Nouy type surface tension meter, are shown in Table I. $R_{18}NAc$ exhibits the lowest interfacial tension (γcmc) in water and the lowering by R_8NAc has the lowest interfacial tension in the aqueous Cu(II) solution.

Table I. Lowering of Interfacial Tension by R_nNAc and RR_nNAc

Surfactant	toluene/aqueous phase	γcmc (mN·m^{-1})	cmc (mM)
R_8NAc	water	4.3	0.22
	Cu(II) solution	3.7	0.14
$R_{18}NAc$	water	3.5	0.18
	Cu(II) solution	4.0	0.32
RR_6NAc	water	5.0	0.32
	Cu(II) solution	5.0	0.50
$RR_{12}NAc$	water	5.0	0.35
	Cu(II) solution	5.0	0.63

Preparation of Surface-Templated Resins

The monomer-type surfactants (**Rac**, R_nNAc, and RR_nNAc) orient at the interface between the matrix monomers and water, and emulsify these solutions. Metal-imprinted resins were obtained by both suspension and emulsion polymerization using **Rac**, **R$_n$NAc**, or **RR$_n$NAc**.

Metal Ion-Imprinted Resins. The preparative method was as follows. The monomer-type surfactant **Rac** was dissolved in 80 mL of aqueous Cu(II) solution (pH 6.0) containing 1 wt % of poly(vinyl alcohol) (n= *ca.* 1500, Wako Pure Chemicals). Then, the Cu(II) solution was added to a divinylbenzene:styrene (DVB-ST) (10:1) solution with stirring at 300-400 rpm. The emulsion was heated to 75 °C under nitrogen atmosphere and potassium persulfate was added as the initiator. The monomers in the emulsion gradually polymerized and the polymer (resin) was left in the mixture for 6 hours. After decantation, the white solid resin was separated by filtration with a sintered-glass filter (No. 5). The resin was washed with hot water (90 °C) and then treated with 0.1 M nitric acid for 1 hour to remove the metal ions from the resin (Figure 3). The resin was washed with cold water and dried under vacuum.

The monomer-type surfactants **R$_n$NAc** and **RR$_n$NAc** were dissolved in the DVB-ST solution and added to the Cu(II) solution (pH 5.5) and the mixture was polymerized with potassium persulfate initiator at 80 °C. Removal of the imprinting Cu from the Cu(II)-imprinted resins prepared with **R$_n$NAc** or **RR$_n$NAc** was accomplished by treatment with 5 M HCl.

Polymerization at room temperature was accomplished by irradiating the emulsion with 1.25 x 10^4 C Kg^{-1} of ^{60}Co γ-rays.

Surface-Templated Resins

Orientation of the surfactants at the interface is affected by the hydrophilic lipophilic balance (HLB) and interfacial phenomena should be influenced. The surface-template effect of the resulting resins, *i.e.* the metal ion sorption selectivity, should be highly affected by the orientation and the location of the template sites.

The resins were evaluated in batch metal ion sorption experiments. Resins prepared with **Rac**, **R$_n$NAc**, or **RR$_n$NAc** were added to a Cu(II) solution (pH 5.0 and 6.0) and the suspension was stirred for 2 hours. The resin in the solution was filtered and treated with 0.1 M HCl for desorption of the metal ions. After washing with water, the resins could be reused for metal ion sorption.

Preparation by Suspension Polymerization. Monomer-type surfactants in the matrix monomers (DVB+ST) adsorb at the monomers-Cu(II) solution interface and form the Cu(II) complex. The polymer (resin) formed from these monomers by suspension polymerization will have the same cavities at the surface as the structures of the Cu(II) complexes. Conditions of the resin preparation with **Rac** are summarized in Table II.

Excess DVB (10 times the ST) was copolymerized with ST and **Rac** to prepare resins of tight structure which should prevent the carboxyl groups from moving. Suspension polymerization gave fine resin beads of about 0.08 mm (0.04-0.16 mm) diameter in 31-66 % yields, while polymerization without **Rac** give coarser resin beads with diameters of approximately 0.4 mm in 70-80 % yields. The monomer-type surfactant **Rac** produced finely divided particles because **Rac** lowers the interfacial tension.

After the incorporated Cu(II) was removed from the surface-template resins, the readsorption of Cu(II) on the resins at pH 5-6 reaches equilibrium in 2 hours. The amounts of adsorbed metal ions (mmol/g-resin) increase with enhancement of the molar ratio of **Rac**, as shown in Figure 4. The amounts of **Rac** at the interface should increase with the addition of **Rac** at the time of polymerization. However, the relative amount of **Rac** at the interface is larger than those in the two liquids when **Rac** is polymerized at the lower concentration, and smaller when at a higher concentration because of dissolution of **Rac** in the two liquids in the form of an

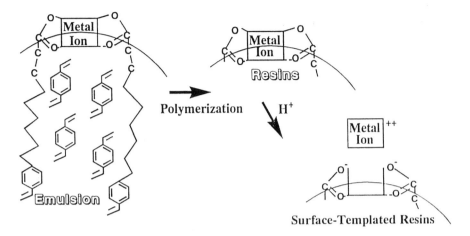

Figure 3. Preparation of Metal Ion-Imprinted Resins.

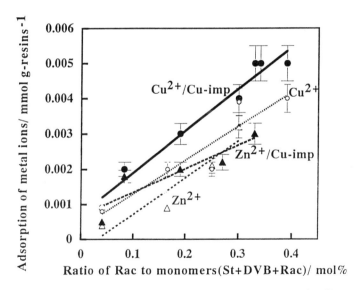

Figure 4. Adsorption of Cu(II) and Zn(iI) on the Cu-Imprinted Resins Prepared by Suspension Polymerization. Initial concentrations of Cu(II) and Zn(II) = 5 x 10^{-5} M, 50 mL, pH 6.0. Resin = 1.0 g. Cu^{2+}/Cu-imp = adsorption of Cu(II) on Cu-imprinted resin; Cu^{2+} = adsorption on unimprinted resin, Zn^{2+}/Cu-imp = adsorption of Zn(II) on Cu-imprinted resin; and Zn^{2+} = adsorption on unimprinted resins.

Table II. Preparation of Cu(II)-Imprinted Resins by Suspension Polymerization[a]

Rac/(DVB+ST), mol %	Rac, mol	Cu(II)/Rac	Yield grams	%
0.04	7.3×10^{-6}	0.5	0.92	41
0.08	1.7×10^{-5}	2	1.68	60
0.19	3.5×10^{-5}	2	1.56	65
0.30	1.8×10^{-5}	1	0.35	44
0.33	3.6×10^{-5}	2	0.68	48
0.34	1.9×10^{-5}	1	0.22	31
0.39	1.8×10^{-5}	1	0.19	31
0.04	7.3×10^{-6}	0	1.47	66
0.17	1.5×10^{-5}	0	0.49	44
0.25	1.9×10^{-5}	0	0.47	45
0.30	1.7×10^{-5}	0	0.23	31
0.39	1.8×10^{-5}	0	0.15	25

[a]Conditions of suspension polymerization: DVB/ST = 10 (mol/mol); initiator, benzoyl peroxide (BPO), 1 wt %; PVA, 2.5 wt %; Cu(II) solution, 80 mL, pH 6.0; stirring of the mixture at 300-400 rpm at 85 °C for 6 hours.

aggregate, micelle, etc. The influence of additional **Rac** decreases at high concentration. Thus, the adsorbed Cu(II) on the resins is 0.021 mmol·g^{-1}·mol %$^{-1}$ at 0.04 mol % of **Rac**, 0.019 mmol·g^{-1}·mol %$^{-1}$ at 0.08 mol % of **Rac**, 0.014 mmol·g^{-1}·mol %$^{-1}$ at 0.19-0.30 mol % of **Rac**, and 0.013 mmol·g^{-1}·mol %$^{-1}$ at 0.39 mol % of **Rac**. Figure 4 also shows that the amount of Cu(II) adsorbed on the Cu(II)-imprinted resin is larger than that on the unimprinted resins. Moreover, the adsorption of Zn(II) on the Cu(II)-imprinted resins is a little larger than that on the unimprinted resins in the range of 0.04-0.25 mol % of **Rac**. Fixation of **Rac** at the interface is effective during the polymerization in the presence of Cu(II) because of the orientation of the carboxyl groups toward the Cu(II) solution, and much of the Zn(II) is also adsorbed on the resin at the lower concentration by cation exchange. The size and the coordination number of the imprinted metal ions should change the conformation of the metal complex at the interface. Therefore, a large adsorption of Cu(II) is brought about by template effects at the surface. However, the resin particles are too large for use in metal ion separations and the amounts of Cu(II) adsorbed per gram of resin are not high. Moreover, lower resin yields may weaken the template effect, since the conformation during the polymerization is variable.

Prepared by Emulsion Polymerization. Emulsion polymerization is expected to produce finer resin particles. **Rac** functions as an emulsifier and gives stable emulsions upon agitation at 300-400 rpm. Three monomers (DVB+ST+**Rac**) in metal-ion solutions were copolymerized with initiation by the water-soluble initiator potassium persulfate or by irradiation with γ-rays. Conditions for the resin preparations are shown in Table III.

The resins prepared with 0.04-0.33 mol % of **Rac** were hard, but those with 1.0 mol % of **Rac** were rubber-like polymers. Polymerization with 0.25 mol % of **Rac** gave fine particles and the size of resin beads was 200-300 nm in diameter, which is similar to that usually obtained by emulsion polymerization. However, the Cu(II)-imprinted resins were obtained in the lowest yields. The polymerization yields were about 3 % for the Cu(II)-imprinted resin, 37 % for the Zn(II)-imprinted resin, 15 % for the Ni(II)-imprinted resin, and 32 % for the unimprinted resin. It is proposed that

Table III. Preparation of Metal Ion-Imprinted Resins by Emulsion Polymerization with Rac[a]

Rac (DVB+St) mol %	Rac, mol	Metal imprinting ion	M(II) Rac	Initiator $K_2S_2O_8$, mg	γ-ray, C Kg^{-1}
0.25	2.0×10^{-5}	Cu(II)	1	20	-
0.25	2.0×10^{-5}	Zn(II)	1	20	-
0.25	2.0×10^{-5}	Ni(II)	1	20	-
0.25	2.0×10^{-5}	none	0	20	-
0.33	3.6×10^{-5}	Cu(II)	1	31	-
0.33	2.7×10^{-5}	Zn(II)	1	21	-
0.33	1.8×10^{-5}	Ni(II)	1	14	-
0.33	3.6×10^{-5}	none	0	31	-
1.0	4.0×10^{-5}	Cu(II)	1	10	-
1.0	4.0×10^{-5}	Zn(II)	1	10	-
1.0	4.0×10^{-5}	Ni(II)	1	10	-
1.0	4.0×10^{-5}	none	0	10	-
0.04	3.7×10^{-6}	Cu(II)	1	-	1.25×10^4
0.04	3.7×10^{-6}	Ni(II)	1	-	1.25×10^4
0.04	3.7×10^{-6}	none	0	-	1.25×10^4

[a]Conditions of emulsion polymerization: DVB/ST = 10 (mol/mol); metal solution = 80 mL, pH 6.0; initiator = $K_2S_2O_8$ at 75 °C or γ-rays at 15-20 °C.

radicals of the initiator are inactivated by the Cu(II) because of the high reduction potential

$$\bullet OSO_3^- + Cu^{2+} \longrightarrow SO_4^{2-} + Cu^+$$

Metal-imprinted resins of **RnNAc** and **RR$_n$NAc** were prepared in the presence of Cu(II) or Zn(II) solution. The resins prepared by emulsion polymerization using **R$_n$NAc** and **RR$_n$NAc** are shown in Table IV.

The resin beads after the desorption of Cu(II) with 5 M HCl were spherical and the size was 0.2-0.8 μm in diameter, which is usual for emulsion polymerization. The Cu(II) would form a complex with a composition of Cu(II)(surfactant)$_2$ in a yield of 70-100 % when the Cu(II) is one half the amount of surfactant and a complex of Cu(II)(surfactant) in 45-54% yield when the ratio of Cu(II) to surfactant is two. A blue-green Cu(II)-imprinted resin means that the complexes are formed in the interior and at the surface of the resin. Complexation at the surface is in following order: **R$_8$NAc > R$_{18}$NAc > RR$_6$NAc > RR$_{12}$NAc**. Since the lipophilic surfactants would be in the bulk monomers, this should reduce complexation at the surface of the resin.

Adsorption of Metal Ions. The Cu(II) adsorbed on 1.0 gram of the resins from emulsion polymerization is 50-100 times as much as that on resins prepared by suspension polymerization (see Figure 4 and Table V). The metal ions are combined with the carboxyl groups on the wider surface of the finer resins. Moreover, the amounts of Cu(II) adsorbed at pH 6.0 are much higher than those at pH 5.0. The monomer-type surfactant **Rac** must dissociate at pH 4-5 and should complex more easily with Cu(II) at pH 6.0, since the acid dissociation of hexanoic acid is 4.63 at 25 °C ($I = 0.1$) (27). The adsorption sites on the unimprinted resins are still unsaturated with the metal ions under the conditions described in Table V because of the increased amount of Cu(II)-sorption at the high concentration. The adsorption of Cu(II) on the Cu(II)-imprinted resins is 1.2-4.5 times (mean = 1.88) higher than that on the unimprinted resins. The Zn(II)- and the Ni(II)-imprinted resins also show high

Table IV. Preparation of Metal-Imprinted Resins by Emulsion Polymerization with R_nNAc or RR_nNAc

Surfactant	Imprinting ion	Yield (%)	Color of resin	Amount of M(II) x 10^5,mol·g^{-1} Inner	Amount of M(II) x 10^5,mol·g^{-1} Surface	Ratio of surface, %
\multicolumn{7}{c}{Surfactant : Metal Ion = 1:0.5}						
R_8NAc	Cu (II)	48	Yellow-Green	4.0	5.7	59
	Zn (II)	52	White	2.0	1.0	29
	None	63	White	----	----	---
$R_{18}NAc$	Cu (II)	48	Blue-Green	7.5	3.3	31
	Zn (II)	68	White	2.9	0.4	12
	None	63	White	----	----	---
RR_6NAc	Cu (II)	42	Green	9.6	4.4	29
	Zn (II)	61	White	4.5	0.5	10
	None	63	White	----	----	---
$RR_{12}NAc$	Cu (II)	58	Green	7.8	3.0	28
	Zn (II)	68	White	4.6	0.5	8
	None	65	White	----	----	---
\multicolumn{7}{c}{Surfactant : Metal Ion = 1:2}						
R_8NAc	Cu (II)	52	Yellow-Green	5.8	20.7	78
	Zn (II)	76	White	3.0	2.0	40
$R_{18}NAc$	Cu (II)	30	Blue-Green	10.1	27.5	73
	Zn (II)	61	White	2.8	1.0	25
RR_6NAc	Cu (II)	28	Yellow-Blue	12.6	30.4	71
	Zn (II)	70	White	4.3	1.2	21
$RR_{12}NAc$	Cu (II)	62	Gray-Green	8.0	13.5	63
	Zn (II)	65	White	4.3	0.7	14

adsorption for Zn(II) and Ni(II), respectively. In addition to the high adsorption, the Zn(II)-imprinted resins show more effective adsorption for Zn(II) than of Cu(II) and Ni(II). Thus 1.0 gram of Zn(II)-imprinted resin prepared with 0.25 mol % of **Rac** adsorbs 0.062 mmol of Cu(II), 0.138 mmol of Zn(II), and 0.095 mmol of Ni(II) at pH 5. Such template effects are also seen for the Cu(II)- and Ni(II)-imprinted resins. The highly effective adsorption of the imprinted metal ions clearly indicates the surface-template effect based on the orientation of **Rac**, and the surface-template effects exceed those of the interior-templated resins (1.3-1.5 times) reported by Nishide *et al.* (6).

Orientation of **Rac** during polymerization by potassium persulfate may be disordered by the high temperature for chemical initiation (75 °C). Resins prepared at room temperature to minimize disorder of the oriented **Rac** were obtained in greater than 90 % yield by irradiation with 1.25 x 10^4 C Kg^{-1} γ-rays. The resins also have similar surface-template effects (see the lower portion of Table V). However, the particles are slightly larger (below 500 nm in diameter) because of the lack of the agitation. Therefore, the metal ion-sorption efficiencies are not as high as those for resins prepared by initiation with potassium persulfate.

The resins prepared with **RnNAc** and **RR_nNAc** also adsorb Cu(II), as shown in Figure 5. The Cu(II)-imprinted resins sorb more Cu(II) than the Zn(II)-imprinted

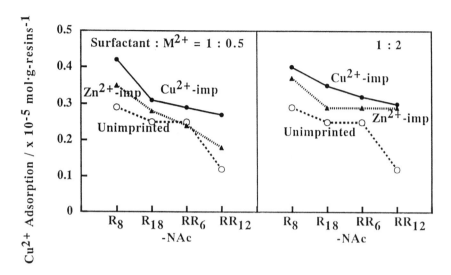

Figure 5. Adsorption of Cu(II). Initial concentration of Cu(II) = 5 x 10^{-4} M, 5 mL, pH 6.0, 25 °C. Resin = 0.10 g.

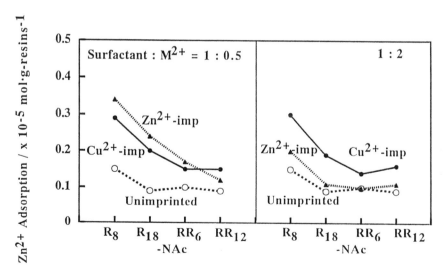

Figure 6. Adsorption of Zn(II). Initial concentration of Zn(II) = 5 x 10^{-4} M, 5 mL, pH 6.0, 25 °C. Resin = 0.10 g.

Table V. Metal Ion Sorption on Metal Ion-Imprinted Resins Prepared by Emulsion Polymerization

Resin	Rac[a], mol%	pH	Amount of metal ion sorbed per gram of resin, mmol/g		
			Cu(II)	Zn(II)	Ni(II)
Suspension polymerization[b]					
Cu(II)-imp	0.33	6.0	0.005±0.001	-	-
Zn(II)-imp	0.33	6.0	-	0.003±0.001	-
Emulsion polymerization (potassium persulfate initiator at 75 °C)					
Cu(II)-imp	0.25	5.0	0.105±0.011(1.6)[c]	-	-
Cu(II)-imp	0.33	5.0	0.318±0.032(4.5)	-	-
Cu(II)-imp	0.25	6.0	0.234±0.024(1.2)	-	-
Zn(II)-imp	0.25	5.0	0.062±0.006	0.138±0.014(1.8)	0.095±0.010
Zn(II)-imp	0.33	5.0	0.074±0.008	0.192±0.019(1.4)	0.108±0.011
Zn(II)-imp	0.25	6.0	0.134±0.014	0.165±0.017(1.8)	0.019±0.002
Ni(II)-imp	0.25	5.0	-	0.052±0.005	0.127±0.013(1.8)
Ni(II)-imp	0.33	5.0	0.065±0.007	0.120±0.012	0.150±0.015(1.5)
Ni(II)-imp	0.25	6.0	0.210±0.021	0.060±0.006	0.167±0.017(1.4)
Unimprinted	0.25	5.0	0.066±0.007	0.075±0.008	0.070±0.007
Unimprinted	0.33	5.0	0.069±0.007	0.139±0.014	0.103±0.010
Unimprinted	0.25	6.0	0.203±0.020	0.093±0.009	0.121±0.012
Emulsion polymerization (γ-ray irradiation[d] at room temperature)					
Cu^{2+}-imp	0.04	6.0	0.139±0.014(1.5)	0.127±0.013	0.043±0.005
Ni^{2+}-imp	0.04	6.0	0.073±0.008	0.053±0.006	0.121±0.012(1.6)
Unimprinted	0.04	6.0	0.092±0.009	0.066±0.007	0.076±0.008

[a]Ratio of **Rac** to the total other monomers (DVB+ST). [b]With 1.0 g of resin for the suspension polymers and 10 mg of resin for the emulsion polymers, metal ion solution = 5.0 x 10^{-5} M, 50 mL. [c]Ratio of adsorbed metal ion on metal-imprinted resins to that on an unimprinted resin. [d]1.25 x10^4 C Kg^{-1} of γ-rays.

resins or than the unimprinted resins. Therefore, the Cu(II)-imprinted resins should have four-coordinate square planar binding sites at the surface of the resins, and adsorb Cu(II) effectively. The sorption order is consistent with that of the complexation of the surfactants at the surface: $R_8NAc > R_{18}NAc > RR_6NAc > RR_{12}NAc$. The sorption of Zn(II) by the metal-imprinted resins prepared under the condition of surfactant:metal ion = 1:0.5 is similar to that of Cu(II) except for the resins with **RR$_{12}$NAc**, as shown in Figure 6. The Zn(II)-imprinted resins, which are considered to have a four-coordinate tetrahedral binding sites, adsorb Zn(II). However, the resins prepared with surfactant:metal ion = 1:2 sorb less Zn(II). The template of a 1:1 complex should be unsaturated with the ligand of the surfactant. Therefore, the lack of coordination to the Zn(II) would not provide a four-coordinate tetrahedral template structure at the surface.

Selective Recovery from a Metal Ion Mixture. The metal-imprinted resins adsorb the target metal ions best, so selective adsorption can be expected. The metal ion sorption from a mixture of Cu(II) and Zn(II) is shown in Table VI.

The selectivities of the resins prepared by suspension polymerization are similar to those of the resins prepared by emulsion polymerization, even though the sorption capacities of the latter are lower. Cu(II) is adsorbed selectively on the Cu(II)-imprinted resins to a much greater extent than on the unimprinted resins. The greatest adsorption capacity and the most selective adsorption for Cu(II) are obtained with the Cu(II)-imprinted resins prepared by emulsion polymerization at 0.25 mol % of **Rac**.

Table VI. Selective Adsorption from a Cu(II)-Zn(II) Mixture

Resin	Rac[a], mol%	pH	Sorption of metal ions per gram of resin[b], mmol/g Cu(II)	Zn(II)	Cu(II)/Zn(II)
Suspension polymerization[b]					
Cu(II)-imprinted	0.25	5.0	0.052±0.008	0.015±0.002	3.5
Cu(II)-imprinted	0.33	5.0	0.068±0.010	0.029±0.004	2.4
Unimprinted	0.25	5.0	0.002±0.001	0.002±0.001	1.0
Emulsion polymerization					
Cu(II)-imprinted**	0.04	5.0	0.090±0.014	0.030±0.005	3.0
Cu(II)-imprinted	0.25	6.0	0.609±0.091	0.165±0.025	3.7
Cu(II)-imprinted	1.0	6.0	0.227±0.034	0.126±0.019	1.8
Unimprinted**	0.04	5.0	0.021±0.003	0.015±0.002	1.4
Unimprinted	0.25	6.0	0.475±0.071	0.287±0.043	1.6

[a]Ratio of **Rac** to the total monomers (DVB+St) [b]With 20 mg of resin for the suspension polymers and the emulsion polymers identified by an asterisk or 10 mg for the emulsion polymers except those shown with an asterisk and 50 mL of a Cu(II)-Zn(II) solution, each 5.0×10^{-5} M.

The Cu(II)/Zn(II) selectivity is 3.7 for the Cu(II)-imprinted resins compared with 1.6 for the unimprinted resins. However, the resins prepared at 1 mol % of **Rac** show a lower selectivity (Cu(II)/Zn(II) = 1.8). The Zn(II)-imprinted resins prepared by emulsion polymerization exhibit a slight Cu(II) selectivity at pH 5.0 (Cu(II)/Zn(II) = 1.1).

The metal ion-imprinted resins can be used again after treatment with an acidic solution, since the carboxyl groups are chemically bound to the resins and the metal complex should dissociate in the acidic solution. Recycled imprinted resins show similar characteristics to the original resins, *e.g.* the amount of Cu(II) adsorbed at pH 5.0 by the Cu(II)-imprinted resin with 0.25 mol % of **Rac** were 0.105 mmol/g in the first adsorption, 0.102 mmol/g in the second, and 0.110 mmol/g in the third (*16*).

The metal-imprinted resins prepared from R_nNAc and RR_nNAc also show selective adsorption from a mixture of Cu(II) and Zn(II). The ratio for competitive sorption (Cu(II)/Zn(II)) on the imprinted resins to that on the unimprinted resins is shown in Figure 7. The adsorption on the Cu(II)-imprinted resins prepared with half amount of Cu(II) is more effective than that on the unimprinted ones (1.0) and that on the Zn(II)-imprinted resins is less except for the resins of $RR_{12}NAc$. The sorption selectivity for the imprinting metal ions was $R_8NAc > R_{18}NAc > RR_6NAc > RR_{12}NAc$ which is consistent with the order of their emulsification. However, the resins prepared with twice the amount of metal ions are less selective than those prepared with half the amount. The surface-templated resins of the 1:1 complex should exhibit little recognition of Cu(II) or Zn(II) due to the lack of four-coordinate binding sites with appropriate geometry.

Thus, the facility for emulsification is found to correlate with the metal ion complexation at the surface of the resins and with the surface-template effect. The selective adsorption of surface-templated resins is apparent. Furthermore, a functional monomer having high emulsification is expected to form the excellent surface metal-imprinted resin by emulsion polymerization.

Literature Cited

1. Wulff, G.; Sarhan, A. *Angew. Chem.*, **1972**, 84, 364.
2. Wulff, G.; Sarhan, A.; Zabrocki, K. *Tetrahedron Lett.*, **1973**, 4329.
3. Wulff, G. *Angew. Chem., Int. Ed. Engl.*, **1995**, 34, 1812.
4. Robinson, D. K.; Mosbach, K. *J. Chem. Soc., Chem. Commun.*, **1989**, 969.

276

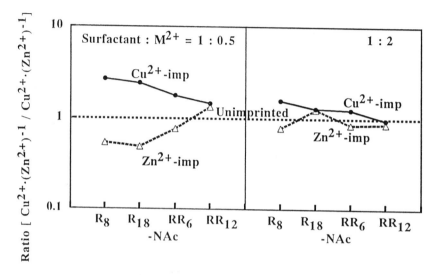

Figure 7. Competitive Sorption on the Metal Ion-Imprinted Resins from a Cu(II)-Zn(II) mixture. Ratio of the Cu(II)/Zn(II) selectivity of the metal ion-imprinted resin to that of the unimprinted resin. Initial concentration of the Cu(II)-Zn(II) mixture [Cu(II)] = [Zn(II)] = 2.5 x 10^{-4} M, 5 mL, pH 6.0, 25 °C. Resin = 0.10 g.

5. Vlatakis, G.; Andersson, L. I.; Müller, R.; Mosbach, K. *Nature*, **1993**, 361, 645.
6. Nishide, H.; Tsuchida, E. *Makromol. Chem.*, **1976**, 177, 2995.
7. Nishide, H.; Deguchi, J.; Tsuchida, E. *Chem. Lett.*, **1976**, 169.
8. Kuchen, W.; Schram, J. *Angew. Chem., Int. Ed. Engl.*, **1988**, 27, 1695.
9. Kido, H.; Miyajima, T.; Tsukagoshi, K.; Maeda, M.; Takagi, M. *Anal. Sci.*, **1992**, 8, 749.
10. Kido, H.; Sonoda, H.; Tsukagoshi, K.; Maeda, M.; Takagi, M.; Maki, H.; Miyajima, T. *Kobunshi Ronbunshu*, **1993**, 50, 403.
11. Maeda, M.; Murata, M.; Tsukagoshi, K.; Takagi, M. *Anal. Sci.*, **1994**, 10, 113.
12. Murata, M.; Hijiya, S.; Maeda, M.; Takagi, M. *Bull.Chem. Soc. Jpn.*, **1996**, 69, 637.
13. Yu, K. Y.; Tsukagoshi, K.; Maeda, M.; Takagi, M. *Anal. Sci.*, **1992**, 8, 701.
14. Tsukagoshi, K.; Yu, K. Y.; Maeda, M.; Takagi, M. *Bull. Chem. Soc. Jpn.*, **1993**, 66, 114.
15. Tsukagoshi, K.; Yu, K. Y.; Maeda, M.; Takagi, M.; Miyajima, T. *Bull.Chem Soc. Jpn.*, **1995**, 68, 3095.
16. Koide, Y.; Senba, H.; Shosenji, H.; Maeda, M.; Takagi, M. *Bull. Chem. Soc. Jpn.*, **1996**, 69, 125.
17. Yamazaki, S.; Hattori, S. *Hyomen*, **1987**, 25, 86.
18. Freedman, H. H.; Mason, J. P.; Medalia, A. I. *J. Org. Chem.*, **1958**, 23, 76.
19. Hill, J. W. *J. Am. Chem. Soc.*, **1932**, 54, 4105.
20. Clemmensen, E. *Ber.*, **1913**, 46, 1838.
21. Welch, S.C.; Wong, R.Y. *Tetrahedron Lett.*, **1972**, 1853.
22. Allen, C. F.; Kalm, M. J. *Org. Syn.*, **1963**, 4, 398.
23. Allen, C. F.; Kalm, M. J. *Org. Syn.*, **1967**, 1, 48.
24. Morris, L. R.; Rock, R. A.; Marshll, C. A.; Home, J. H. *J. Am. Chem. Soc.*, **1959**, 81, 377.
25. Takahashi, K.; Koike, M.; Nanba, Y.; Kobayashi, M. *Kaimen-kasseizai Hando-bukku*; Kougaku Tosyo: Tokyo, 1987; p.102.
26. Rosen, M. J. *Surfactants and Interfacial Phenomena*, second edition; Wiley: New York, 1989; p.121.
27. Martell, A. E.; Smith, R. M. *Kagaku-binran, Kisohen II*; 3rd ed by Nihon-kagaku Kai; Maruzen: Tokyo, 1984; p. II-342.

Chapter 19

Metal Ion-Imprinted Polymers Prepared by Surface Template Polymerization with Water-in-Oil Emulsions

Kazuya Uezu, Masahiro Goto, and Fumiyuki Nakashio

Department of Chemical Science and Technology, Faculty of Engineering, Kyushu University, Hakozaki, Fukuoka 812-81, Japan

Metal ion-imprinted polymers are prepared by surface template polymerization, an emulsion polymerization process which utilizes a functional monomer, an emulsion stabilizer, a polymer matrix-forming co-monomer, and a print molecule. Such surface-templated polymers offer high sorption rates for metal ions as well as high selectivity. For this technique, the interfacial activity of the functional monomer is a vital factor for the selection of a suitable metal ion-binding amphiphile. It is also an important factor for the firm attachment of the functional monomer onto the polymer matrix. We have succeeded in preparing highly selective ion-imprinted polymers by two approaches: a) the design of functional monomers; and b) post-irradiation with gamma rays to make the polymer matrix more rigid. Adsorption behavior of Zn(II) and Cu(II) by the newly synthesized polymers has been evaluated. The imprinting effect on metal-ion adsorption efficiency and selectivity is discussed.

Molecular imprinting is a technique used to prepare polymeric materials for applications in molecular recognition. The synthetic approach to preparing imprinted polymers involves pre-organization of the functional monomer by specific interaction between the functional monomer and the "print" molecule or ion followed by polymerization in the presence of an excess of crosslinking agent. The resulting polymer contains specific binding sites that recognize the print molecule. The polymer may exhibit high selectivity for rebinding of the print molecule or ion with which it was prepared. Since the print molecule itself directs the organization of the functional groups, specific knowledge of the imprinted structure is not necessary. Wulff et al. (1) demonstrated this approach by designing a polymeric receptor which utilized reversible covalent bonding for the monomer-print interactions. Mosbach et al. (2), Shea et al. (3), and Arnold et al. (4) have reported the preparation of a variety of imprinted polymers which employ non-covalent interactions, such as hydrogen bonding and/or electrostatic interactions, to form the monomer-print complex.

Although these imprinting techniques are conceptually attractive, few practical imprinted polymers have been reported. Problems encountered include low binding selectivities, loss of selectivity with time, and slow rebinding kinetics. Fundamental problems which are common to these conventional imprinting techniques stem from recognition sites that are usually formed within the hydrophobic polymer matrix. Also, particular difficulties arise when the template is water-soluble, such as in biological components like proteins.

We have developed a novel molecular imprinting technique called "surface template polymerization" (Figure 1) to overcome the drawbacks of conventional imprinted polymers (5-8). Surface-templated polymers are prepared by emulsion polymerization utilizing a functional monomer, an emulsion stabilizer, a polymer matrix-forming co-monomer, and a print molecule. A functional monomer, which is amphiphilic in nature, forms a complex with the print molecule at the interface of the emulsion and the thus-formed complex remains at the reaction surface. After the matrix is polymerized, the coordination structure is "imprinted" at the polymer surface. Uezu et al. (6) previously demonstrated that a Zn(II)-imprinted polymer prepared by this technique exhibits an imprinting effect as far as the amount of metal ions sorbed is concerned. The sorption results showed that the interfacial activity of the functional monomer was a vital factor in the assessment of a suitable metal ion-binding amphiphile. However, the Zn(II)-imprinted polymer showed poor selectivity for Zn(II) in competitive Cu(II)-Zn(II) sorption from aqueous solutions. This is probably due to insufficient rigidity of the polymer matrix which causes increased swelling of the imprinted polymers. In comparison, divinylbenzene polymers with no functional monomers possess stronger matrix rigidity and exhibit insignificant swelling. Therefore, the structure of the functional monomer affects the rigidity of the polymeric matrix. It has been established that monomers which possess long alkyl chains in their hydrophobic part affect the matrix rigidity.

To fix the recognition sites more rigidly and create stronger interactions between the functional monomers and imprint molecules, we have designed functional monomers which have two phosphonic acid groups and two benzene rings in the molecular structure. Zn(II)-imprinted polymers formed from such a multi-functional monomer are expected to exhibit very high selectivity towards Zn(II) over Cu(II) because the polymer combines a rigid matrix and a strong binding ability owing to the specificity of the multi-functional monomer. In general, it is difficult to realize such properties even if the uncomplexed functional monomer possesses an appropriate interfacial activity and strong binding characteristics. Therefore, it is necessary to develop a technique for creating a rigid polymeric matrix together with the development of highly stable and interactive functional monomers.

In the present study, we prepare Zn(II)-imprinted polymers by two approaches: a) the design of functional monomers; and b) post-irradiation with gamma rays to make the polymer matrix more rigid. Separations of Zn(II) and Cu(II) with the imprinted polymers are performed and the template effect of the polymer is evaluated by comparison with the unimprinted polymer and also with a conventional solvent extraction method for the same metal ions.

Experimental

Materials. The syntheses of dioleyl phosphoric acid (DOLPA) and L-glutamic acid dioleylester ribitol ($2C_{18}\Delta^9 GE$) have been reported previously (9-11). Reagent-grade di(2-ethylhexyl)phosphoric acid (D2EHPA) was purchased from Tokyo Chemical Industry Co., Ltd. and used without purification. Divinylbenzene (DVB, Wako Pure Chemical Industries, Ltd.) was used after treatment with silica gel to remove the inhibitor. Other reagents were of commercially available grades. Structures of the functional monomers and surfactant used are shown in Figure 2.

The 1,12-dodecanediol-O,O'-diphenyl phosphonic acid (DDDPA) was synthesized according to Scheme 1. A mixture of pyridine (80 mL) and diphenylphosphonic acid (56.2 g, 0.288 mol) was added to 60 mL of dry tetrahydrofuran and the solution was

280

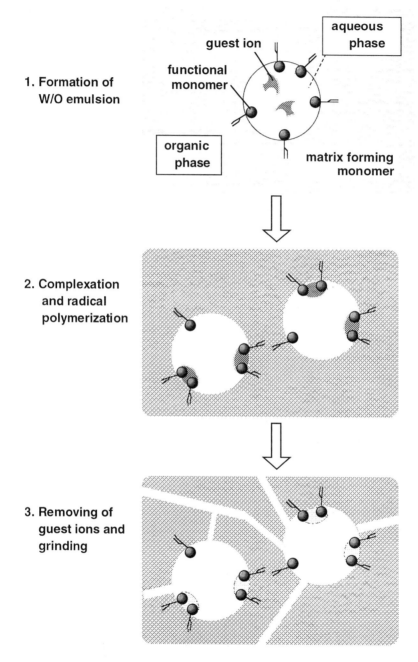

Figure 1. Schematic illustration of surface template polymerization with a water-in-oil (W/O) emulsion.

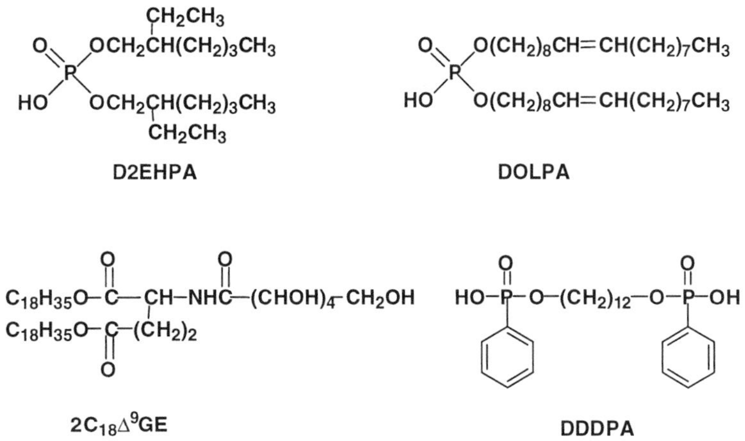

Figure 2. Structures of the functional monomers and the surfactant.

placed in an ice bath. Subsequently, a solution of 1,12-dodecanediol (24.3 g, 0.120 mol) and dry tetrahydrofuran (130 mL) was added dropwise and the mixture was stirred in the ice bath for 48 hours. Ice-water (350 mL) was added and the mixture was stirred for 5 hours. The pH of the aqueous solution was adjusted to 1-2 by addition of 30 mL of concentrated hydrochloric acid and stirring was continued for another 1 hour. Finally, chloroform (300 mL) was added and the mixture was stirred for 2 hours. The organic phase was separated and washed twice with 250 mL of 1.0 M hydrochloric acid, then dried over anhydrous magnesium sulfate. The organic solution was evaporated *in vacuo* and the residue was purified by chromatography on silica gel column with chloroform-methanol (9:1) as eluent to produce a yellowish oil, in 95 % yield. IR (neat): 2650 (OH), 2276 (OH and P=O), 1684 (P-O-R) cm^{-1}; ^{1}H NMR (250 MHz, CDCl$_3$, TMS, 30 °C): δ 1.25 (16H, m, -O-CH$_2$-CH$_2$-(CH$_2$)$_8$-CH$_2$-CH$_2$-O-), 1.76 (4H, m, -O-CH$_2$-CH$_2$-(CH$_2$)$_8$-CH$_2$-CH$_2$-O-), 4.00 (4H, m, -O-CH$_2$-CH$_2$-(CH$_2$)$_8$-CH$_2$-CH$_2$-O-), 7.47 (6H, m, ArH (para-, meta-)), 7.80 (4H, m, ArH (ortho-)), 8.32 (2H, s, OH). Elemental analysis, calculated for C$_{24}$H$_{36}$P$_2$O$_6$·0.36CHCl$_3$: C, 57.13; H, 7.09. Found: C, 56.89; H, 7.16% .

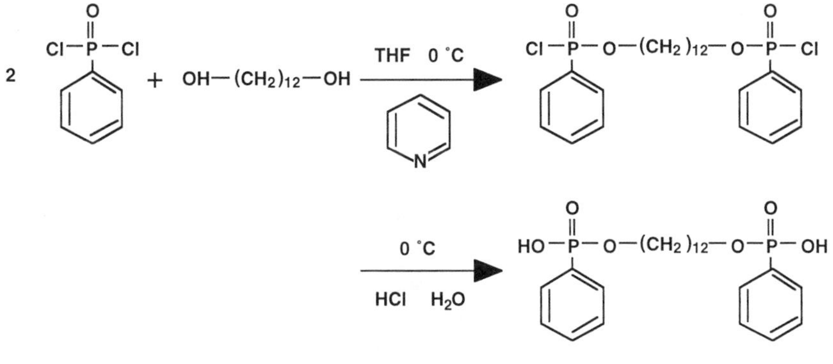

Scheme I. Synthesis of the functional monomer DDDPA

Preparation of the Zn(II)-Imprinted Polymer. DOLPA (2.9 g, 4.8 mmol) or DDDPA (1.7 g, 3.6 mmol) and $2C_{18}D^9GE$ (0.25 g, 0.30 mmol) were dissolved in 60 mL of toluene-DVB (1:2, v/v). An aqueous solution (30 mL) of 0.010 M Zn(II), which was buffered with acetic acid-sodium acetate and maintained at pH 3.8, was added. The mixture was sonicated for 3 minutes to give a water-in-oil (W/O) emulsion. After addition of 0.36 g (1.4 mmol) of 2,2'-azobis(2,4-dimethylvalero-nitrile) and 36.56 g (28 mmol) of DVB, polymerization was carried out at 55 °C for 2 hours under a flow of nitrogen. The bulk polymer was dried *in vacuo* and ground into particles which were washed with 1.0 M hydrochloric acid to remove the Zn(II) and filtered. This procedure was repeated until the Zn(II) concentration in the filtrate became negligible. The Zn(II)-imprinted polymer was dried *in vacuo*. A reference polymer was prepared similarly, but in the absence of Zn(II).

Irradiation of the Zn(II)-Imprinted Polymer with Gamma Rays. Before removal of the Zn(II) by washing with hydrochloric acid, the Zn(II)-imprinted polymer was irradiated with gamma rays at ambient temperature in a nitrogen atmosphere to induce crosslinking. The dose for crosslinking was 1820 kGy.

Interfacial Tension of the Functional Monomers. An aqueous solution of 0.10 M nitric acid-sodium nitrate was prepared. An organic solution was prepared by dissolving a weighed amount of the functional monomer in toluene. Equal volumes of the aqueous and organic solutions were shaken in a flask and allowed to attain equilibrium in a thermostatted bath at 30 °C. The layers were separated and the interfacial tensions of the functional monomer at different concentrations in the organic solution and in the aqueous nitric acid-sodium nitrate solution were measured by the drop volume method at 30 °C. The specific gravity of both solutions was measured with a digital densimeter (Paar DMA35).

Analysis of the Polymers. FT-IR spectra of polymer samples dispersed in KBr in the range of 2000-400 cm^{-1} were recorded with a Perkin-Elmer PARAGON 1000 FTIR spectrophotometer. The swelling percentage was determined by volumetric measurement (*12*). Tetrahydrofuran (THF) was employed as a swelling solvent. To assess how rigidly the DOLPA was fixed onto the polymer matrices, the amount of phosphorus contained in the DOLPA released from the matrices into the acid washing solution was measured with an ICP atomic emission spectrometer (ICPS-5000, Shimadzu Corporation).

Competitive Adsorption of Zn(II) and Cu(II). The equilibrium sorption of Zn(II) and Cu(II) by an imprinted polymer was determined batchwise. To the polymer (0.10 g) was added 5.0 mL of a solution which was of 0.10 mM in $Zn(CH_3COO)_2$ and 0.10 mM in $Cu(NO_3)_2$. The pH was adjusted to a desired value between 1 and 5 with 50 mM acetic acid-sodium acetate and 1.0 M nitric acid. The mixture was shaken in a thermostatted water bath at 30 °C for 24 hours. The polymer was filtered with a cellulose nitrate membrane (DISMIC-25, Toyo Roshi Kaisha, Ltd.). The amounts of Zn(II) and Cu(II) adsorbed on the polymers were calculated from the decrease in Zn(II) and Cu(II) concentration in the aqueous solution, as measured by atomic absorption spectrophotometry (SAS 760, Seiko Instruments Inc.).

Competitive Solvent Extraction of Zn(II) and Cu(II). The organic solution was 10 mM DDDPA or DOLPA in toluene. The aqueous solution was 0.10 mM in $Zn(CH_3COO)_2$ and 0.10 mM in $Cu(NO_3)_2$ with the pH was adjusted to desired values between 1 and 5 with 50 mM acetic acid-sodium acetate and 1.0 M nitric acid. A mixture of 10.0 mL of the organic solution and 5.0 mL of the aqueous solution was shaken in a thermostatted water bath at 30 °C for 24 hours. After phase separation, the equilibrium concentration of Zn(II) or Cu(II) in the aqueous solution was analyzed determined by atomic absorption spectrophotometry.

Results and Discussion

Design of Functional Monomers. Design requirements for the functional monomers are: a) strong binding ability for the target metal ions; b) high interfacial activity; and c) no detrimental effect on the polymer matrix. To fulfill these conditions, we designed the functional monomer DDDPA (Figure 2) which has two phosphonic acid groups and two benzene rings in the molecular structure. The two phosphonic acid groups are present to produce strong binding for Zn(II), the target metal ion. The two benzene rings are included in the hydrophobic portion of DDDPA to prevent a destructive influence on the polymer matrix, which is formed by divinylbenzene. Alkyl chains, which linked the two phenylphosphonic acid units, control the interfacial activity of the functional monomer and its solubility in organic solvents. Zn(II)-imprinted polymers derived from this multi-functional monomer are expected to exhibit high selectivity towards Zn(II) over Cu(II) because the polymers combine both rigid polymer matrices and the strong binding ability owing to the specificity of the multi-functional monomer.

Interfacial Tension of DDDPA. The relationship between the interfacial tension and concentration of DDDPA is shown in Figure 3.

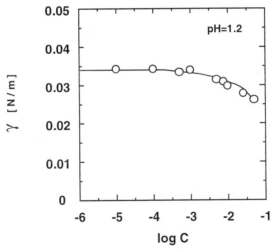

Figure 3. Relationship between the interfacial tension and concentration of DDDPA.

The adsorption equilibrium of a functional monomer is expressed as:

$$F_{org} \underset{\longleftarrow}{\overset{K_F}{\longrightarrow}} F_{ad} \tag{1}$$

where K_F is the interfacial adsorption equilibrium constant of the functional monomer, while F and the subscript ad denote the functional monomer and the adsorption state at the interface, respectively. The relation between interfacial tension and the amount of functional monomer adsorbed at the interface is expressed by the Gibbs adsorption equation. Assuming a Langmuir adsorption isotherm between the amount of functional monomer adsorbed and the bulk concentration of functional monomer, C_F, the relation between interfacial tension, γ, and C_F at temperature T is expressed as:

$$\gamma = \gamma_0 - \left(\frac{RT}{S_F}\right) \ln \left(1 + K_F C_F\right) \qquad (2)$$

where γ_0 is the interfacial tension between the organic solvent and the aqueous solution, R is the gas constant, and S_F is the interfacial area occupied by a unit mole of the functional monomer. The values of K_F and S_F can be obtained from the experimental results for the interfacial tension and Equation 2 by a nonlinear regression method. The results are shown in Table I. The solid curves in Figure 3 are calculated by Equation 2 using the constants obtained above. The interfacial activity of DDDPA is very high. The K_F value for DDDPA was 100 times larger than that for D2EHPA and 5 times larger than that for DOLPA. Therefore, DDDPA satisfied the first requirement listed above as a criterion in the Design of Functional Monomers section.

Table I. Interfacial Adsorption Equilibrium Constant, K_F, and Interfacial Areas Occupied by a Unit Mole of the Functional Monomer, S_F, in Toluene.

Functional monomer	K_F [m³/mol]	S_F [m²/mol]
D2EHPA	2.06	6.02 X 10⁵
DOLPA	4.84 X 10¹	6.35 X 10⁵
DDDPA	2.27 X 10²	1.85 X 10⁶

Adsorption Efficiency and Selectivity for Zn(II)-Imprinted Polymers Prepared with DDDPA. Zn(II)-imprinted polymers incorporating the novel bifunctional monomer DDDPA were prepared by surface template polymerization with W/O emulsions (Figure 1). The imprinted polymers were ground into particles, whose volume-average diameters were ca. 40 μm. The yield was ca. 80 %.

Figure 4 shows the pH dependence for sorption of Zn(II) and Cu(II) on a Zn(II)-imprinted polymer prepared with DDDPA. The percent sorption is enhanced with increasing pH for both ions. However, the imprinted polymer sorbs Zn(II) much more effectively than Cu(II) over the entire pH range. The ability of the imprinted polymers to sorb Zn(II) is significantly higher than for Cu(II), which is attributed to the interfacial activity of DDDPA and the strong interaction between Zn(II) and DDDPA. It should be noted that Zn(II) is completely separated from Cu(II) for aqueous solutions with pH ~3. This high selectivity is considered to be produced by Zn(II)-imprintied cavities on the surface of the polymers.

The pH dependence for competitive liquid-liquid extraction of Zn(II) and Cu(II) from aqueous solutions by toluene solutions of DDDPA is shown in Figure 5. In the solvent extraction system, much lower selectivity for Cu(II) is observed than in sorption by the imprinted polymer (Figure 4). Since the functional monomer (extract-

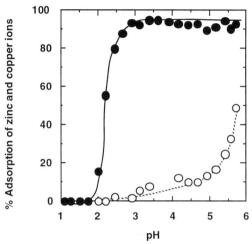

Figure 4. pH dependence for sorption of Zn(II) (●) and Cu(II) (○) with a Zn(II)-imprinted polymer prepared with DDDPA.

ant) in the solvent extraction may assume both a tetrahedral configuration for Zn(II) and a square planar configuration for Cu(II) due to its flexibility in the organic solvent, the metal ions selectivity is controlled by the stability of the complexes between the extractant and the ions. On the other hand, the DDDPA functional monomer from which the Zn(II)-imprinted polymer is formed is restricted to the tetrahedral configuration in which it complexes the Zn(II) template. Therefore its ability to recognize Zn(II) is considerably higher.

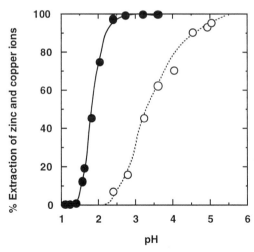

Figure 5. pH dependence for solvent extraction of Zn(II) (●) and Cu(II) (○) from aqueous solutions into a toluene solution of DDDPA.

Post-Irradiation of Zn(II)-Imprinted Polymers with Gamma Rays. A Zn(II)-imprinted polymer was prepared by surface template polymerization with a

W/O emulsion. DVB, toluene, DOLPA, and $2C_{18}D^9GE$ were used as the cross-linking agent, diluent, Zn(II)-binding amphiphile, and an emulsion stabilizer, respectively. After polymerization, the bulk polymer was ground into particles whose diameters were *ca.* 10 µm. After drying *in vacuo*, the Zn(II)-imprinted particles were subjected to irradiation by gamma rays to bind the DOLPA thoroughly and to crosslink the polymer matrix. The Zn(II) template was completely removed from the polymer particles with 1.0 M hydrochloric acid. The particles were then dried *in vacuo*. The yield was *ca.* 70 %.

Figure 6 shows the pH dependence for competitive sorption of Zn(II) and Cu(II) by the Zn(II)-imprinted polymer prepared with DOLPA without gamma-ray irradiation after the initial polymerization. The amounts of metal ions in the rebinding study were set below the maximum capacity of the Zn(II)-imprinted sites (1.1×10^{-6} mol for 0.10 g of polymer) to evaluate the binding affinity of the polymer. The Zn(II)-imprinted polymer sorbed both Zn(II) and Cu(II) over the entire pH range. However, this imprinted polymer shows a comparatively low selectivity for sorption of Zn(II) over Cu(II).

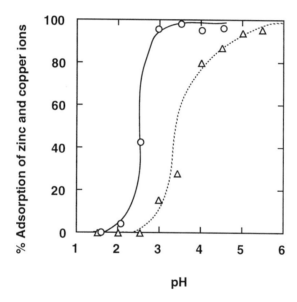

Figure 6. pH dependence for sorption of Zn(II) (O) and Cu(II) (Δ) with a Zn(II)-imprinted polymer prepared with DOLPA without gamma-ray irradiation.

In a solvent extraction system (Figure 7), only low selectivity for Zn(II) is observed because the functional monomer (extractant) can assume both a tetrahedral configuration for Zn(II) and a square planar configuration for Cu(II) due to its flexibility in the organic solvent. It is noted that the Zn(II) selectivities of the Zn(II)-imprinted polymer prepared from DOLPA without gamma-ray irradiation (Figure 6) and in solvent extraction by DOLPA (Figure 7) are very similar. Thus the Zn(II)-imprinted polymer has a metal ion selectivity equal only to that of the functional monomer as is the case for other conventional imprinted polymers (*13-15*).

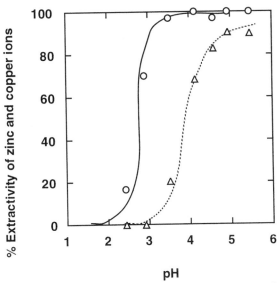

Figure 7. pH dependence for solvent extraction of Zn(II) (O) and Cu(II) (Δ) from aqueous solutions into toluene solutions of DOLPA.

Figure 8 shows the pH dependence for competitive sorption of Zn(II) and Cu(II) by the Zn(II)-imprinted polymers prepared with DOLPA with post-irradiation by gamma rays. No difference is noted for Zn(II) sorption by the imprinted polymers prepared with DOLPA without (Figure 6) and with post-irradition (Figure 8) by gamma rays. An irradiated blank polymer, which did not include DOLPA, adsorbed no metal ions. These results clearly show that irradiation with gamma rays does not destroy the Zn(II)-DOLPA complex and produces few functional groups, such as carbonyl groups, within the dose range used. In contrast, the Cu(II) binding ability of the gamma-ray-irradiated, imprinted polymer was markedly decreased. Thus the gamma-ray-irradiated, imprinted polymers can distinguish the coordination of Zn(II) from that of Cu(II).

Comparison of the IR spectra for both imprinted polymers with and without gamma-ray irradiation shows that the peak at 1630 cm^{-1}, associated with RHC=CH$_2$ ($n_{C=C}$), was diminished owing to crosslinking induced by irradiation. From the swelling percentage in THF for all of the polymers (Table II), the irradiated polymers were found to swell less than non-irradiated polymers. The ratio of DOLPA released from each polymer matrix during acid-washing operations (Table III) was remarkably reduced by irradiation with gamma rays. Improvement in the selectivity of gamma-ray-irradiated imprinted polymers is attributed to induced crosslinking in the polymer matrix by the irradiation which renders the polymer matrix rigid, and thereby enhances the stability of the binding sites toward recognition of Zn(II). Furthermore, the double bond in the oleyl chains of DOLPA can form a single bond with a free radical on each end to combine DOLPA rigidly with the polymer matrix. Therefore, the application of gamma-ray irradiation is shown to be a very convenient method for fixing the binding sites of target metal ions.

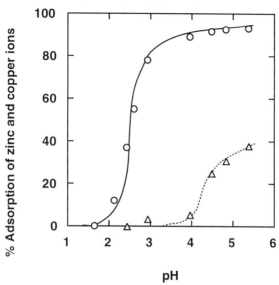

Figure 8. pH dependence for sorption of Zn(II) (O) and Cu(II) (Δ) by a Zn(II) -
imprinted polymer prepared from DOLPA with post-irradiation by gamma rays.

Table II. Swelling of the Polymers in THF.

polymer	swelling, %
unimprinted polymer	27
gamma-ray-irradiated unimprinted polymer	22
imprinted polymer	37
gamma-ray-irradiated imprinted polymer	27

**Table III. Percent of the Functional Monomer (DOLPA) Released
from the Polymer Matrix During the Acid-Washing Operation**

polymer	$\dfrac{amount\ of\ DOLPA\ released}{total\ amount\ of\ DOLPA\ added} \times 100\%$
unimprinted polymer	1.5
gamma-ray-irradiated unimprinted polymer	1.3
imprinted polymer	3.1
gamma-ray-irradiated imprinted polymer	1.5

These results show that rigid and dimensionally stable metal ion-imprinted polymers which recognize metal coordination by anchoring a functional monomer, such as DOLPA, to the polymer surface can be prepared. It is well known that properties such as matrix rigidity bring about poor mass transfer in conventional imprinted polymers. The ability to enhance the matrix rigidity without decreasing mass transfer is therefore an important advantage in surface template polymerization. Furthermore, the combination of surface template polymerization with irradiation by gamma rays offers a potential technique for the construction of highly selective, molecule-recognizing polymers applicable to the sorption of various water-soluble substances.

Acknowledgments

This work was supported in part by the Arai Science and Technology Foundation and a research fund from the Power Reactor and Nuclear Fuel Development Corporation. Financial support by a Grant-in-Aid for Scientific Research (No. 09750842) from the Ministry of Education, Science and Culture of Japan is acknowledged. We are also grateful to Henry Kasaini for editing the manuscript.

Literature Cited.

1. Wulff, G.; Sarhan, A. *Angew. Chem., Int. Ed. Engl.* **1972**, *11*, 341.
2. Kempe, M.; Mosbach, K. *Tetrahedron Lett.* **1995**, *36*, 3563.
3. Sellergren, B.; Shea, K. J. *J. Chromatogr., A* **1995**, *690*, 29.
4. Dhal, P. K.; Arnold, F. H. *Macromolecules* **1992**, *25*, 7051.
5. Tsukagoshi, K.; Yu, K. Y.; Maeda, M.; Takagi, M. *Bull. Chem. Soc. Jpn.* **1993**, *66*, 114.
6. Uezu, K.; Nakamura, H.; Goto, M.; Murata, M.; Maeda, M.; Takagi, M.; Nakashio, F. *J. Chem. Eng. Jpn.* **1994**, *27*, 436.
7. Yoshida, M.; Uezu, K.; Goto, M.; Nakashio, F. *J. Chem. Eng. Jpn.* **1996**, *29*, 174.
8. Uezu, K.; Nakamura, H.; Kanno, J.; Sugo, T.; Goto, M.; Nakashio, F. *Macromolecules* **1997**, *30*, 3888.
9. Goto, M.; Kondo, K.; Nakashio, F. *J. Chem. Eng. Jpn.* **1987**, *20*, 157.
10. Goto, M.; Kondo, K.; Nakashio, F. *J. Chem. Eng. Jpn.* **1989**, *22*, 79.
11. Goto, M.; Matsumoto, S.; Nakashio, F.; Yoshizuka, K.; Inoue, K. *Proc. ISEC'96* **1996**.
12. Green, T. K.; Kovac, J.; Larsen, J. W. *Fuel* **1984**, *63*, 935.
13. Nishide, H.; Tsuchida, E. *Makromol. Chem.* **1976**, *177*, 2295.
14. Efendiev, A. A.; Kabanov, V. A. *Pure Appl. Chem.* **1982**, *54*, 2077.
15. Kuchen, W.; Schram, J. *Angew. Chem., Int. Ed. Engl.* **1988**, *27*, 1695.

Chapter 20

A Physicochemical Study on the Origin of the Imprinting Effect

Tohru Miyajima[1], Kyoko Sohma[1], Shin-ichi Ishiguro[1], Masaki Ando[2], Shigeo Nakamura[2], Mizuo Maeda[2], and Makoto Takagi[2]

[1]Department of Chemistry, Faculty of Science, Kyushu University, Hakozaki, Fukuoka 812-81, Japan
[2]Department of Chemical Science and Technology, Faculty of Engineering, Kyushu University, Hakozaki, Fukuoka 812-81, Japan

The origin of the imprinting effects which have been reported for metal ion-templated microspheres with acidic functionalities is probed by examining the acid dissociation properties of non-templated carboxylated resins prepared by emulsion polymerization. The equilibria are compared with those of polyacrylic acid, a linear polymer analog. It is found that: a) a strong and variable electric field is formed at the resin surface due to two-dimensional distribution of the carboxyl groups; b) the salt concentration effect on the acid dissociation equilibria for the resin is much smaller than that for polyacrylic acid, probably because of restricted mobility of the carboxylate groups at the water/polymer interface; and c) the intrinsic acid dissociation reaction at the interface proceeds by a two-step reaction in which hydrogen bonds are formed between two adjacent carboxyl groups. These findings indicate a carboxyl group arrangement which is favorable for the metal ion imprinting.

Despite rapid advances in the synthesis of templated polymers, fundamental studies on the origin of the imprinting effect are limited. A combination of several physicochemical factors, such as electrostatic and hydrophobic interactions as well as specific chemical bond formation between the host and guest molecules, *e.g.* in metal ion coordination, are believed to be of primary importance. To obtain microscopic information concerning such factors, a metal ion-templated resin system was selected for investigation. This system allows several experimental methods to be utilized in evaluating the complexation behavior and enables comparisons to be made with relevant linear polymer analogs which contain the same functionalities.

In recent years, complexation by metal ion-templated resins has been extensively studies by our research group (*1-9*). Kido *et al.* (*1,2*) first reported a template polymerization technique using oleic acid as the host monomer, divinylbenzene as the matrix-forming monomer, and Cu^{2+} as the metal ion. Rebinding of Cu^{2+} by the templated resin was studied by potentiometric titration. It was found that the complexation reaction was rapid and reversible and high selectivity was achieved with the Cu^{2+}-templated resin compared to an analogous non-templated resin. Subsequently Tsukagoshi *et al.* (*3-5*) prepared carboxylated microspheres from methacrylic acid, butyl acrylate, and styrene and demonstrated a surface imprint effect for several metal ion species including Cu^{2+}. Koide *et al.* (6) utilized 10-(*p*-vinylphenyl)decanoic acid as a host monomer instead of oleic acid and prepared Cu^{2+}-imprinted resin which exhibited a surface template effect. Metal ion-templated resins with phosphate groups were prepared by Maeda *et al.* (7-9). In this case, the host monomers were phosphate surfactants, such as oleylphosphoric acid, oleyl phenylphosphoric acid, and dioleyl-phosphoric acid. The surface template effect observed for these metal ion-selective resins was attributed to multidentate coordination of surface-anchored metallophilic groups to the template metal ions.

Since these surface-template resins may be classified as chelating resins, their complexation equilibria can be analyzed on the basis of conventional ion-exchange and polyelectrolyte theories. The major difference between ordinary crosslinked chelating resins and the present surface template resins is the dimensionalitiy of the polyelectrolytes. Polyion gels have a three-dimensional distribution of the functional groups, whereas the functional groups of a surface template resin are located in a two dimensional plane. In addition, the mobility of the functional groups on the two-dimensional surface is expected to be considerably lower than that in a three-dimensional resin which may result in higher selectivity.

The present study was undertaken to obtain fundamental information on the origin of the imprinting effect by a careful examination of the acid dissociation behavior of a carboxylate resin prepared by emulsion polymerization. A non-templated resin was prepared since we believe that the most characteristic features for the imprinting effect are a high surface charge density and multidentate complexation behavior. The importance of these factors can be evaluated from equilibrium studies conducted with the non-templated resin.

Experimental.

Resin Synthesis. Purified divinylbenzene (20.0 g, Sankyo Chemical Industries, Ltd.), 2.0 g of potassium oleate, 0.20 g of the initiator 2,2'azobis(2,4-dimethylvaleronitrile), and 40 mL of distilled water were mixed and stirred under nitrogen for eight hours at 25 °C. The polymer was filtered and washed with 1.0 M hydrochloric acid. This resultant H^+-form resin was washed with distilled water. The density and average diameter of the particles thus obtained were 1.019 g/mL and 1.3×10^{-4} cm, respectively.

Potentiometric Titration. A mixture of 2.0 g of the resin and 30 mL of 0.10 M sodium hydroxide + Cs M sodium chloride solution was titrated with a 0.10 M hydrochloric acid + Cs M sodium chloride solution under nitrogen at 25 °C. The equilibrium pH values were measured potentiometrically with an Orion 91-01 glass electrode and an Orion single junction reference electrode equipped with an Orion Ionalyzer 720A. Just before and after the resin titration, the electrochemical cell was calibrated by a Gran's plot (10). The titrations were conducted automatically with an auto-Buret (APB-118, Kyoto Electronics Manufacturing Company, Ltd., Tokyo) which was controlled by a personal computer.

Results and Discussion

Surface Charge Density. Due to the difficulty in dispersing the H^+-form resin in the aqueous solution, a back titration procedure was utilized. Since it was expected that a strong electric field would be formed on the resin surface, the titrations were carried out at quite high salt concentrations, *i.e.* $Cs > 1$ M, for determination of inflection points in the titration curves. A typical titration curve together with its first-derivative curve, *i.e.* $d(pH)/d(V_{HCl})$ *vs.* V_{HCl}, obtained at $Cs = 3.00$ M are shown in Figure 1. The pH decreased with V_{HCl} (the volume of added HCl solution) and shows two inflection points, V_1 and V_2. The pH decrease in the region $0 < V_{HCl} < V_1$ is attributed to neutralization of the excess sodium hydroxide. The region $V_1 < V_{HCl} < V_2$ corresponds to neutralization of the Na^+-form resin and the pH decrease when

$V_{HCl} > V_2$ is due to the excess of hydrochloric acid. The difference between V_2 and V_1 represents the total exchange capacity of the resin. The specific exchange capacity, n_{ex} (milliequivalents/g) thus evaluated is 0.35 ± 0.02 milliquiv/g. This value is somewhat larger than the 0.30 milliequiv/g which is estimated from the amounts of oleic acid and divinylbenzene which were used to prepare the resin. This discrepancy may indicate the incomplete polymerization of the divinylbenzene in the present polymerization procedure.

The average particle size was determined by scanning electron microscopy to be 1.3×10^{-4} cm. This gives an apparent particle surface area of 5.3×10^{-8} cm^{-2} and a particle volume of 1.2×10^{-12} cm^3. Using the resin density, the number of oleic acid molecules (carboxyl groups) fixed per particle is calculated to be 2.0×10^8. Thus one carboxyl group occupies 1.62 Å^2 on the spherical surface of the resin particle. This value of surface area for a single functionality is unexpectedly small and indicates an uneven resin surface and highly crowded carboxyl groups.

Acid Dissociation Equilibria. Acid dissociation equilibria of weakly acidic linear and crosslinked polyelectrolytes are usually described by the Henderson-Hasselbach equation:

$$pK_{app} = pH - \log\{\alpha/1 - \alpha)\} \tag{1}$$

where pK_{app} and α are the apparent acid dissociation constant and the degree of dissociation of the polyacid, respectively (11-13). In the present study, Equation 1 was used to express the equilibria. The value of α was calculated by the equation:

$$1 - \alpha = \{C_{HCl}(V_{HCl} - V_1 - [H](V_0 + V_{HCl})\}n_{ex}W_{resin} \tag{2}$$

where V_0 is the initial volume of the sample solution, C_{HCl} is the concentration of the hydrochloric acid solution (i.e. 0.100 M), and W_{resin} is the weight of resin used in the titration. Titration and first-derivative curves obtained at $Cs = 0.1$ and 1.0 M are shown in Figures 2 and 3, respectively. It is apparent that the heights of the first peaks for the first-derivative curves are much lower than that obtained for $Cs = 3.0$ M (Figure 1), even though the peak heights for the second peaks at V_2 are of the same magnitude.

Figure 4 shows a plot of the pK_{app} values thus calculated against α. It is readily apparent that the pK_{app} value is highly dependent upon α. For the carboxylic acid monomer, the calculated value of the right-hand side of Equation 1 remains constant and equals the pK_a value of the monomer acid. The remarkable enhancement in pK_{app} with increase in α is primarily attributable to the polyelectrolytic nature of the resin, i.e. enhancement of the electrostatic field formed at the particle surface leads to attraction of mobile H$^+$ ions in the vicinity of the resin carboxyl sites. At $\alpha = 0$, where no electrostatic attraction is expected, all the pK_{app} vs. α curves converge to a pK value of ca. 4.5 which is quite close to the acid dissociation constant of acetic acid, 4.56 (14).

Since the carboxyl groups of the resin are expected to be distributed over the surface of the polymerized divinylbenzene particle with a high charge density, it is of special interest to compare the pK_{app} vs. α plots determined for linear polyacrylic acid, PAA in which the average separation of the carboxyl groups fixed on the backbone is ca. 2.5Å. The pK_{app} vs. α plots for PAA are shown in Figure 5. Although the intrinsic acid dissociation constants for the resin and PAA are the same within

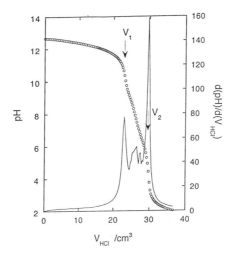

Figure 1. Titration and First Derivative (Solid Line) Curves for the Carboxylic Acid Resin. The H^+-form resin (2.00 g) dispersed in 30.0 mL of 0.10 M sodium hydroxide and 3.00 M sodium chloride was titrated with a solution of 0.10 M hydrochloric acid and 3.00 M sodium chloride under nitrogen at 25 °C.

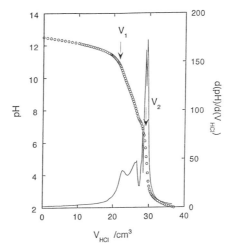

Figure 2. Titration and First Derivative (Solid Line) Curves for the Carboxylic Acid Resin. The H^+-form resin (2.00 g) dispersed in 30.0 mL of 0.10 M sodium hydroxide and 0.10 M sodium chloride was titrated with a solution of 0.10 M hydrochloric acid and 0.10 M sodium chloride under nitrogen at 25 °C.

294

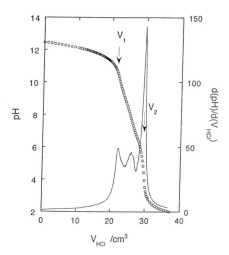

Figure 3. Titration and First Derivative (Solid Line) Curves for the Carboxylic Acid Resin. The H$^+$-form resin (2.00 g) dispersed in 30.0 mL of 0.10 M sodium hydroxide and 1.00 M sodium chloride was titrated with a solution of 0.10 M hydrochloric acid and 1.00 M sodium chloride under nitrogen at 25 °C.

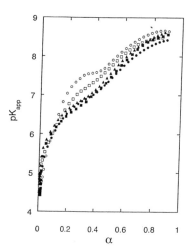

Figure 4. Effect of α and Cs on pK$_{app}$ for the Carboxylic Acid Resin. Cs = 0.10 (O), 0.50 (□), 1.00 (●), 2.00 (▲), 3.00 (■) M.

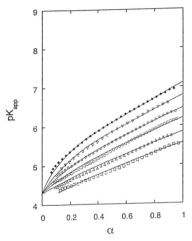

Figure 5. Effect of α and Cs on pK_{app} of Polyacrylic Acid. $Cs = 0.01$ ($\blacklozenge$), 0.02 (∇), 0.05 (O), 0.20 (Δ), 0.50 ($\square$) M.

experimental error, the magnitude of the electrostatic effect for PAA appears to be much smaller than that of the resin, when compared at the same concentration. Assuming that the difference between pK_{app} and the intrinsic acid dissociation constant (ΔpK), reflects only the surface potential (Ψ_p) of the resin, Ψ_p can be related to ΔpK as follows:

$$\Delta pK = pK_{app} - 4.50 = -0.4343e\Psi_p/kT \tag{3}$$

where e is the protonic charge and kT has its usual meaning. The Ψ_p value of the resin in its fully dissociated state is calculated to be as high as *ca.* -250 mV.

With regard to the magnitude of ΔpK, it is notable that the Cs effect in the resin system was much smaller than that in the PAA system. In our previous studies of the acid dissociation equilibria for linear and crosslinked polyelectrolytes, high sensitivity of the equilibria to the added salt concentration level has always been observed (*15,16*) This response has been ascribed to the thermal motion of the functionalities as well as of the polymer backbone and the equilibria have been quantified by the Gibbs-Donnan concept (*15,16*). The quite small Cs dependence in the pK_{app} *vs.* α plot for the resin system is quite characteristic of thermal motion of carboxyl groups fixed on the resin surface. It is assumed that mobility of the carboxyl group on the resin surface is highly restricted due to surface fixation and the hydrophobic nature of the water/resin surface interface.

Another differentiating feature for the acid dissociation equilibria of the resin the discontinuities observed in the pK_{app} *vs.* α curves (Figure 4). Each curve appears to be composed of at least two portions with an inflection point at $\alpha = 0.5$. Inflection points are not observed for potentiometric titrations of PAA (Figure 5 and References 11-13). Since no conformational change upon acid dissociation is expected for the polymer surface, two different protonation processes are postulated. At $0<\alpha<0.5$, a normal protonation reaction occurs; whereas at $0.5<\alpha<1.0$, two adjacent carboxyl groups may participate in the protonation simultaneously due to their close proximity on the resin surface.

Since it is anticipated that a slow acid dissociation process at the water/resin-surface interface would affect the pK_{app} *vs.* α plots, the time dependence of the electrochemical response for each titration point was examined carefully. No drift was observed for any titration point and constant emf values were always obtained within 10 minutes. This demonstrates a rapid and reversible acid dissociation reaction at the interface.

Implications for the Imprinting Effect. Although the present work is limited to a study of acid dissociation equilibria for non-templated resins prepared by emulsion polymerization with oleic acid as the host surfactant, several characteristic features of the origin of reported imprinting effects in metal ion-templated resins can be elucidated. The first is the availability of all of the carboxylate groups at the water/resin-surface interface. A negligible number of the carboxyl groups are buried in the resin matrix. Second is the high electric field formed at the resin surface, which strongly enhances the interaction between a metal cation and the anionic polymer surface. Similar carboxylate surface polyelectrolytes have been reported recently by Winnik *et al.* (*17*) and their acid dissociation properties have also been examined potentiometrically. However, the surface charge density of the present carboxylic acid resin is estimated to be much higher than that for the resin prepared by Winnink. Thirdly, the quite small salt concentration dependence of the acid dissociation equilibria reveals the hydrophobic nature of the water resin-surface. This property of the interface may contribute to the rapidity and reversibility of metal ion binding by the resin. Finally, the possibility of hydrogen bonding between adjacent carboxyl groups must be pointed out. Such hydrogen bonding effects reveal the highly concentrated functionalities at the interfacial region. Such high concentrations should enable multidentate coordination of metal ions, which is indispensable for the imprinting effect.

Quite recently a resin with amine groups has been prepared by Fujiwara et al. (18). Our studies of ion binding by basic linear polyelectrolytes (19) showed that higher ion selectivity is expected for such species compared to acidic linear polyelectrolytes due to the hydrophobic nature of the interactions. Since hydrophobic interactions at a water/resin-surface interface are enhanced relative to those of linear polyelectrolytes, it of special interest to probe the imprinting effect in positively charged resin particles.

Literature Cited.

1. Kido, H,; Miyajima, T.; Tsukagoshi, K.; Maeda, M.; Takagi, M. *Anal. Sci.* **1992**, *8* , 749.
2. Kido, H,; Sonoda, H.; Tsukagoshi, K.; Maeda, M. Takagi, M.; Miyajima, T. *Kobunshi Ronbunshu* **1993**, *50*, 403.
3. Yu, K. Y.; Tsukagoshi, K.; Maeda, M.; Takagi, M. *Anal. Sci.* **1992**, *8*, 70.
4. Tsukagoshi, K.; Yu, K. Y..; Maeda, M.; Takagi, M. *Bull. Chem. Soc. Jpn.* **1993**, *66*, 114.
5. Tsukagoshi, K.; Yu, K. Y.; Maeda, M.; Takagi, M. *Kobunshi Ronbunshu* **1993**, *50*, 455.
6. Koide Y.; Senba, H.; Shosenji, H.; Maeda, M., Takagi, M. *Bull. Chem. Soc. Jpn.* **1993**, *69*, 125.
7. Maeda, M.; Murata, M.; Tsukagoshi, K.; Takagi, M. *Anal. Sci.* **1994**, *10*, 113.
8. Murata, M.; Maeda, M.; Takagi, M. *Anal. Sci. Technol.* **1995**, *8*, 529.
9. Murata, M.; Hijiya, S.; Maeda, M.; Takagi, M. *Bull. Chem. Soc. Jpn.*, **1996**, *69*, 637.
10. Gran, G. *Analyst* **1951**, *77*, 661.
11. Nagasawa, M. *Pure Appl. Chem.* **1971**, *26*, 519.
12. Miyajima, T.; Mori, M.; Ishiguro, S.; Chung, K. H.; Moon, C. H. *J. Colloid Interface Sci.* **1996**, *184*, 279.
13. Miyajima, T.; Mori, M.; Ishiguro, S. *J. Colloid Interface Sci.* **1997**, *187*, 259.
14. Portnanova, R.; DiBernardo, P.; Cassol, A.; Tondello, E.; Magon, L. *Inorg Chim. Acta* **1974**, *8*, 233.
15. Marinsky, J. A. In *Ion Exchange and Solvent Extraction*; Marinsky, J. A.; Marcus, Y. Eds.; Dekker: New York, 1993, volume 11; p. 237.
16. Miyajima, T. In *Ion Exchange and Solvent Extraction*; Marinsky, J. A.; Marcus, Y. Eds.; Dekker: New York, 1995, volume 13; p. 279.
17. Kawaguchi, S.; Yekta, A.; Winnick, M. A. *J. Colloid Interface Sci.* **1995**, *176*. 362.
18. Fujiwara, I.; Maeda, M.; Takagi, M. *Anal. Sci.* **1996**, *12*, 545.
19. Kodama, H.; Miyajima, T.; Mori, M.; Takahashi, H.; Nishimura, H.; Ishiguro, S. *Colloid Poly. Sci.*, in press.

RECOGNITION
WITH INORGANIC-BASED POLYMERS

Chapter 21

Recognition over Footprint Cavities

Kensaku Morihara

Department of Chemistry, Faculty of Science, Nara Women's University, Nishiuoya-kita, Nara 630, Japan

"Footprint cavities" are molecular-imprinted sites on an aluminum ion-doped silica gel surface with complementary cavities which also possess Lewis acid sites. Like catalytic antibodies, these sites exhibit catalytic behavior with tailorable selectivity. This distinguishes the footprint cavities from the imprinted adsorption sites in vinyl polymers. These sites catalyze substrate-specific and stereo-selective acyl transfer reactions, condensations, racemizations, and reductions. Our studies of these reactions reveal that the configuration of the template bound during imprinting predetermines the structure of the cavities and the orientation of substrate molecules bound within the cavities determine the mechanisms of the catalyzed reactions. Thus, our investigation of these phenomena provides useful information for the design of templates to allow construction of cavities capable of fine molecular recognition and specific catalysis.

Recently, remarkable developments have occurred in the field of recognition and separation using molecular-imprinted polymers (MIPs) (1,2). However, very little imprinting of inorganic materials has been reported, even though the concept of molecular imprinting itself originated with Dickey's experiments using silica gel (3). We have developed a somewhat different method for molecular imprinting on silica gel. It involves the surface modification of an aluminum ion-doped silica gel in the presence of a template (4-10). The molecular-imprinted sites are shallow cavities, which consist of a Lewis acid site (the aluminum atom of the aluminum species) and a surrounding matrix which is complementary to the template molecule. We call these sites "footprint cavities" because they can be considered to be the footprint of the template molecule. A variety of compounds can serve as templates for the imprinting. Their tailor-made preparation is an important feature of the footprint cavities, just as for MIPs. The term footprint cavity is, however, only an expression. Smooth boundaries are absent, since the dimensions of the silicate matrix are of the same order as those of the template. A footprint cavity may be considered as a Lewis acid site surrounded by properly oriented silanol functions and space for binding of a guest molecule. These factors cooperate to maximize interactions of a molecule bound within the footprint cavity. This well-defined arrangement of functional groups provides a unique micro-environment with fine molecular recognition capabilities.

When a proper transition state analogue is chosen as a template, this micro-environments can function as a specific catalytic site. These tailor-made footprint catalytic cavities can be referred to as artificial catalytic antibodies. We have investigated this concept of catalyst design to determine its scope and limitations (*4-18*).

The Imprinting Procedure

The imprinting procedure involves five steps, as shown in Figure 1.

Surface Activation of the Silica Gel. This process releases free silanol groups from the surface of commercial silica gel (Merck Kiesel Gel) by partial hydrolysis with acid.

Doping with Al(III). This process incorporates Al(III) into the matrix by isomorphic substitution of silicate with aluminate. The substitution generates Lewis acid sites on the surface. These are "native" Lewis acid sites on the surface which are converted to acid sites within complementary cavity structures in subsequent steps.

Loading of the Template on the Al(III)-Doped Silica Gel. This process forms a Lewis acid-base coordination complex between the acid sites and the template on the surface, which are precursors of the footprint cavities.

Aging and Drying of the Treated Gel. The objective of this process is not to simply dry the gel, but to rearrange the silicate matrix around the acid-base complexes. A continuous and reversible rearrangement of the silicate matrix occurs during this step by depolymerization of the silicate matrix and repolymerization of the released silicic acid in a state of equilibrium around the acid-base complexes. The strong coordinate bonding anchors the template in the cavity. Drying gradually increases the concentration of silicic acid, which is favorable to the polymerization in the equilibrium. Such reorganization must proceed under thermodynamic control. The acid-base complexes are stabilized by forming a maximum number of interactions between the acid-base complexes and the surrounding silicate matrix. This process is the key to the imprinting process. After this step, trimethylsilylation of the free silanol groups outside the cavities can be performed, if desired.

Removal of the Template by Extraction with Methanol. Contact with refluxing methanol gradually decomposes the complexes. The released template molecules are removed by the continuous extraction with methanol. This leaves "footprint cavities" containing Lewis acid sites on the silica gel surface.

Template Molecules and Corresponding Substrates

Theoretically, any compound with Lewis base character could be used as a template. However the template must be stable and moderately insoluble in the reaction mixture for template loading (usually a mixture of acetone and water adjusted to pH 4.0). Carboxylic acids, simple hydrocarbons, and highly water-soluble compounds cannot serve as templates. Figure 2 shows some typical templates with arrows indicating sites which bind to the Lewis acid site. Most template molecules possess a hydrogen bond acceptor close to the coordinating site. Templates with a chiral center in the molecule can be used to imprint cavities that exhibit chiral recognition and enantioselective catalysis (*9,12,14-16*). Reactive compounds which can be embedded in the cavities may serve as substrates in catalyzed reactions. We used acid anhydride substrates which resemble the template molecules for catalyzed acyl transfer reactions

302

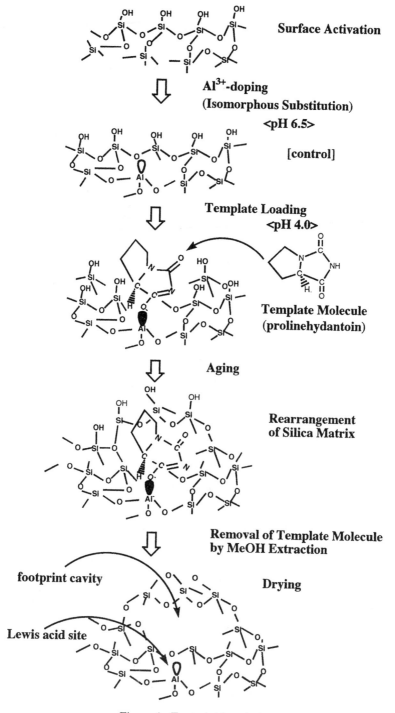

Figure 1. Footprint Imprinting.

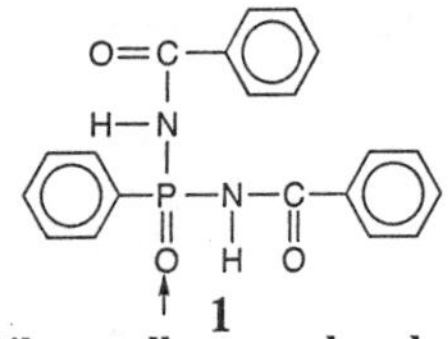

1

N, N'-Dibenzoylbenzenephosphonediamide

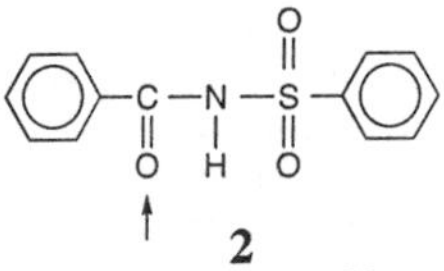

2

N-Benzoylbenzenesulfonamide

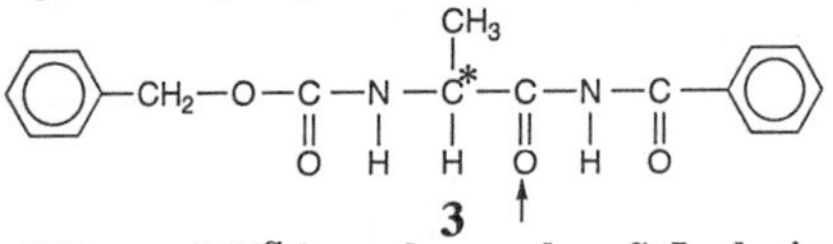

3

N-Benzoyl-(N^{α}-benzyloxycarbonyl)-L-alaninamide

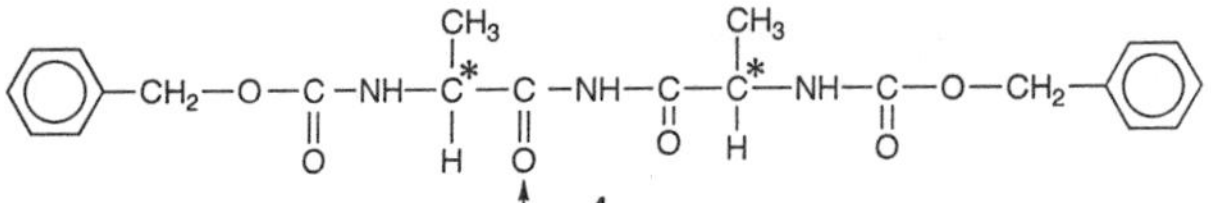

4

Bis-(*N*-benzyloxycarbonyl-L-alanyl)amine

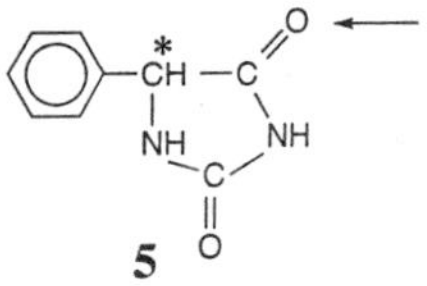

5

5-Phenyl-2,4-imidazolinedione
(5-Phenylhydantoin)

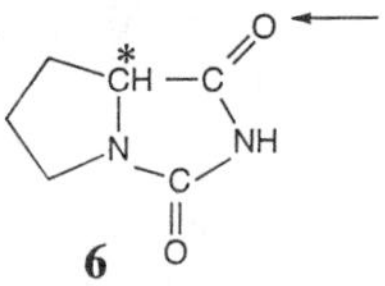

6

Prolinehydantoin

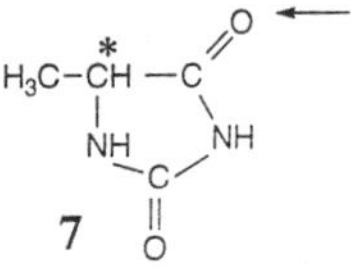

7

5-methyl-2,4-imidazolinedione
(5-Methylhydantoin)

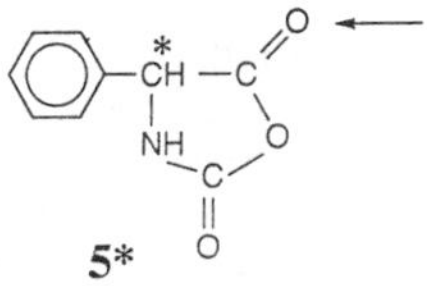

5*

(*N*-carboxy)-phenylglycine anhydride

Figure 2. Templates for Imprinting. (Arrows indicate coordination sites.)

in the footprint cavities. Hydantoins are unique compounds that can serve as a template for imprinting and also as a substrate in catalytic acyl transfer (*12*).

Characterization and Homogeneity of Footprint Cavities

Formation of imprinted cavities can be verified by virtue of the strong competitive inhibition which is observed upon addition of template molecules to the catalyzed reaction media. This can be used as a measure of the formation of footprint cavities. Homogeneity of the imprinted cavities is important in discussing the molecular recognition capability of the cavities. In all of the cases which we investigated, catalytic reactions over footprint cavities obeyed Michaelis-Menten kinetics with regard to the substrate and were first order with respect to the nucleophile concentration. Also linear Lineweaver-Burk plots were obtained under pseudo-first order conditions. This linearity demonstrates the homogeneity of the footprint cavities. Assuming such homogeneity, molecular recognition of the cavities can be simply defined as the free energy change, $-\Delta G = RT \ln(1/K_d)$, where K_d is the dissociation constant of the template molecule from the cavity. K_d is practically equal to the competitive inhibition constant, K_i, for Michaelis-Menten kinetics. Footprint cavities show maximal molecular recognition for the templates used in the imprinting, as expected. Structurally related compounds also show affinities for the cavities according to their resemblance to the template. Quantitative comparison of these affinities suggested that footprint cavities are comprised of subsites that recognize partial structures of the binding molecules. Therefore molecular recognition is the sum of the partial recognitions (*5,8*) by these subsites. This finding is the basis of our catalytic site design.

Features of the Footprint Cavities

Both the Lewis acid sites present in the cavities and the chemical nature of the silicate matrix cause footprint cavities to be remarkably different than MIPs. The footprints are shallow cavities; they cannot recognize the bound molecules in entirety, only the lower parts (*17,18*). The acid sites can coordinate Lewis bases and function as catalysts. Water molecules can bind to the Lewis acid sites to convert them into Brönsted acid sites. Free silanol and siloxane groups of the neighboring silicate matrix can act as hydrogen bond donors and acceptors. Furthermore, the polar surface of the matrix can provide dipole-dipole interactions with the bound molecules. Highly complementary structures formed under thermodynamically controlled conditions can further strengthen these interactions. Coordination provides the primary binding force, and the cooperation of other functional groups contributes to the fine molecular recognition capabilities of the footprint cavities. The strong coordination and the polar nature of the cavities are better suitable to the creation of catalytic sites rather than to adsorption sites for separations.

Molecular Recognition and Catalysis over Footprint Cavities

If the footprint cavities can differentiate species in their transition state ($S^{\ddagger}$) or reactive intermediate (I) structures from those in their reactant states (S), the cavities can function as catalytic sites. The cavities can bind $S^{\ddagger}$ to stabilize the transition state and thereby lower the activation energy. When a transition state analogue (S') or an intermediate analogue (I') is utilized as the template, the cavities can recognize $S^{\ddagger}$ and I according to their resemblance to S' and I'. This is a reverse application of the inhibition mechanism for catalysis with S' described by Wolfenden (*19*) , who demonstrated that the more closely a compound resembles the transition state the better the best competitive inhibitor it is in enzymatic reactions. According to this concept, the cavities imprinted with and complementary to the most closely resembling analogue should function as a highly effective catalyst in the reaction of a substrate.

Thus, how to design highly catalytic sites for a given reaction leads to how to design a proper template of the transition state analogue. This is a similar concept to those used for production of catalytic antibodies (20,21). Therefore, catalytic footprint cavities can be referred to as artificial catalytic antibodies.

Catalyzed Reactions over Footprint Cavities

In principle, the footprint cavities can catalyze any reaction that is subject to Lewis acids catalysis. We have investigated mainly acyl transfer reactions, since the substrates and transition state analogues were readily available. The substrates were acid anhydrides with structures that resemble those of the templates and the nucleophile was 2,4-dinitrophenolate, the consumption of which was followed photometrically at 430 nm. Of greater interest is stereoselective catalyses over chirally imprinted cavities which could exhibit marked selectivity and thereby provide a great deal of information about recognition and catalysis over cavities. Typical examples of specific catalysis over cavities are described below.

Substrate-Specificity over Footprint Cavities. As shown by the data in Table I, cavities imprinted with templates **1** and **2**, (hereafter referred to as {**1**} and{**2**}) catalyzed the acyl transfer from benzoic anhydride to 2,4-dinitrophenolate in acetonitrile at 30 °C. Cavities {**1**} displayed more than 10 times the catalytic activity, (k_{cat}), and 9.5 times the substrate selectivity (k_{cat}/K_m) of {**2**} (7). Cavities {**1**} might be comprised of two subsites. One subsite, involving the Lewis acid, can bind a tetrahedral phenylphosphonyl moiety; while the other can bind a trigonal benzoyl moiety. This is because the phosphonyl oxygen coordinates with the Lewis acid site in the imprinting process to form the cavity structures. On the other hand, one subsite of {**2**} involves the Lewis acid and structures complementary to the trigonal benzoyl moiety; the other subsite has a structure complementary to the tetrahedral phenylsulfonyl moiety (see Figure 3). Therefore, the cavities in {**1**} can stabilize the tetrahedral intermediate in the acyl transfer reaction better than those in {**2**} and therefore show a distinct difference in the rate enhancement (k_{cat}).

Table I. Kinetic Parameters for the Catalyzed Reactions

Reaction	Catalyst	k_{cat} $M^{-1}s^{-1}$	K_m M	k_{cat}/K_m $M^{-2}s^{-1}$	Ratio
Acyl transfer[a]	Control[b]	3.41	11.38	3.0	(1.0)
	{**1**}	33.36	4.06	82.2	27.4
	{**2**}	3.27	3.78	8.6	2.9
Crossed Aldol Condensation	Control	not determined			
	{**1**}	48.99	4.86	10.0	(1,000)[c]
	{**2**}	not determined			

[a]The substrate was benzoic anhydride and the nucleophile was 2,4-dinitrophenolate in acetonitrile at 30 °C. [b]The control is Al(III)-doped, but non-imprinted, silica gel. [c]Ratio of the yield with {**1**} to the yield with {**2**} after 234 hours.

In addition to acyl transfer reactions, cavities {**1**} and {**2**} also exhibit substrate-selective catalytic behavior in the Claisen-Schmidt condensation (a crossed aldol condensation) between benzaldehyde and acetophenone. Two reaction mechanisms can be proposed for this condensation, as shown in Figure 3 (22): a) a Lewis acid site in a cavity activates the carbonyl group of the bound benzaldehyde, with rate limiting nucleophilic attack of the conjugate base (formed by proton abstraction by a base, e.g., Proton Sponge) of acetophenone; and b) the conjugate base (formed by the action of the Lewis acid site and the base) of acetophenone attacks the bound benzaldehyde. Both cavities have a trigonal and a tetrahedral subsite and can

306

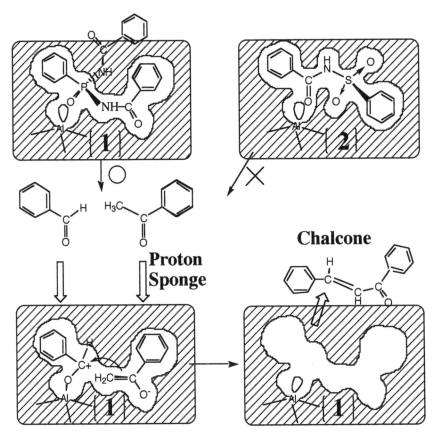

Figure 3. A Mechanism for the Crossed Aldol Condensation Over Footprint Cavities {1}.

accommodate the proposed tetrahedral transition states. Figure 4 and the data in Table I show that the reaction proceeds very poorly over {2} as does a control catalyst with Lewis acid sites but no cavity structures. On the other hand, {1} appears to catalyze the condensation. So the first proposed mechanism is more likely. This finding shows the importance of the design of cavities which are complementary to the transition state for a given reaction.

Enantioselective Catalysis over Footprint Cavities. Since all of the catalytic reactions obeyed Michaelis-Menten kinetics, enantioselectivity was determined using the $(S)/(R)$ ratio of k_{cat}/K_m. For all of the cases examined, the enantioselective cavities could be divided into two groups. One group shows rather low enantioselectivity by virtue of a "productive and non-productive binding" mechanism and the other group exhibits high selectivity due to "lock-and-key" binding. The prevailing mechanism is determined by the structure of the cavities, and, therefore, of the template.

Enantioselective Catalysis Based on Productive and Non-productive Binding. Imprinting with chiral amino acid derivatives **3**, **4**, and **5** produces cavities which consist of three subsites (see Figure 5). Thus, the carboxyl moiety of the amino acid coordinates to the Lewis acid site, and the acyl amide and the side chain moieties are bound by the silicate matrix during the imprinting process. Thus complementary structures are formed around these three moieties. Kinetic investigations showed that cavities of this kind allow two binding conformations for substrate molecules (*12,15*). The cavities can hold both enantiomers in either productive or non-productive binding orientations (Figure 5), the former being more reactive than the latter. However, this mechanism results in low selectivity because it depends on small differences between the two orientations and their steric effect on the attacking nucleophile. Table II shows that the enantioselectivities (the $(S)/(R)$ ratio of k_{cat}/K_m values) in the acyl transfers (nucleophilic ring opening reactions) of **5** and **5*** are only 1.3-2.3 (*12*).

Table II.   Kinetic Parameters for Enantioselective Catalyses 1

Run	Cavity	Substrate	k_{cat} $10^2 s^{-1}$	K_m $10^{-4} M^{-3}$	k_{cat}/K_m $10^5 Ms^{-1}$	Selectivity
1	{(S)-**5**}	(S)-**5**	3.49	4.79	7.3	2.3
	{(S)-**5**}	(R)-**5**	2.66	8.37	3.2	(1.0)
	{(S)-**5**}	(RS)-**5**	2.77	6.37	4.4	1.4
2	{(R)-**5**}	(R)-**5**	5.11	12.80	4.0	2.1
	{(R)-**5**}	(S)-**5**	3.84	20.40	1.9	(1.0)
	{(R)-**5**}	(RS)-**5**	4.14	13.04	3.2	1.7
3	{(R)-**5**}	(R)-**5***[a]	4.29	1.93	22.2	1.3
	{(R)-**5**}	(S)-**5***[a]	3.59	2.14	16.6	(1.0)
4	{(S)-**5**}	(R)-**5***[a]	4.87	5.26	9.3	2.0
	{(S)-**5**}	(R)-**5** *[a]	3.29	7.16	4.6	(1.0)

[a]The substrates (R)-**5** and (S)-**5** are N-carboxy-(R and S)-phenylglycine anhydrides, respectively.

Enantioselective Catalysis Based on the Lock-and-Key Principle. This occurs over highly selective cavities which strictly specify the configuration of a bound substrate molecule, so that only one enantiomer can bind and react. Therefore, very high enantioselectivity is expected. However, with amino acid derivatives as templates, such highly selective cavities cannot be formed. This is because the α-hydrogens of amino acids have minimal affinity for the silicate matrix. Thus, the other three groups attached to the chiral carbon preferentially bind to form three subsites, thereby allowing non-productive binding. We overcame this difficulty by finding a bicyclic template, (S)-prolinehydanto **6**, based on the examination of CPK models and

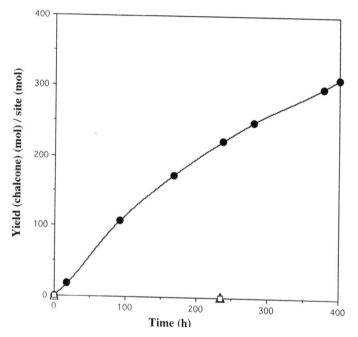

Figure 4. Crossed Aldol Condensations Over Cavities {1}. (● = PhPO(NHCOPH)$_2$, □ = PhCONHSO$_2$Ph, and Δ = Control)

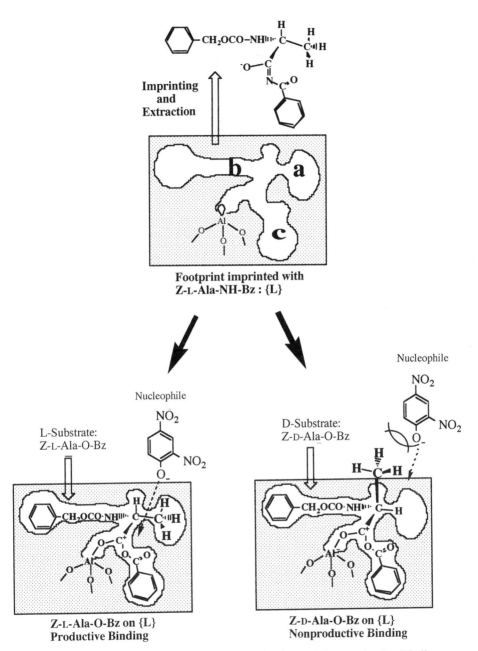

Figure 5. Enantioselective Catalysis by Productive and Non-productive Binding.

from the results of MOPAC calculations. The dihedral angle between the pyrolidine and hydantoin rings is 126° which indicates that the bridgehead nitrogen is sp³ hybridized with a lone pair and can function as a hydrogen bond acceptor. Template **6** can imprint cavities which exclude the pyrolidine ring and include the hydantoin ring, as shown in Figure 6. The cavities may consist of three chemoselective recognition sites (a, b, and c) for three points in the hydantoin ring: the a-site for a carbonyl group at the 3-position of the hydantoin ring as a coordination donor; and the b-site for the other carbonyl group at the 5-position and the c-site for the nitrogen at the 1-position as hydrogen-bond acceptors. Judging from CPK models and MOPAC calculation considerations, the imide function at the 4-position is unlikely to be involved in the recognition site. Such cavities could discriminate between the enantiomers by placement of the recognition sites for the three points. That is, cavities imprinted with (S)-prolinehydantoin place the three recognition subsites (a, b, and c) clockwise, and those imprinted with (R)-prolinehydantoin position the three subsites counterclockwise. According to this scheme for the recognition of the hydantoin ring, the side chains of amino acids should be independent of chiral recognition. As shown by the data in Table III, {**6**} displays remarkable enantioselectivities ranging from 14 to 21 in acyl transfers from **5**, **6**, **7**, and **5*** to 2,4-dinitrophenolate. These high selectivities of {**6**} compared with those of {**5**} shown in Table III must arise from the K_m term, not from the V_{max} term. These findings showed that the {(S)-**6**} strictly discriminates between the enantiomers in the binding step and rejects the (R)-substrates, and vice versa, as expected (*17,18*). Thus, the chiral recognition mechanism is reasonable and the template design is effective. Since no significant influence of the side chain residues was observed on the selectivities, kinetic optical resolution of any racemic amino acids by use of {**6**} should be possible via their hydantoin derivatives.

Table III. Kinetic Parameters for Enantioselective Catalyses 2

Run	Cavity	Substrate	k_{cat} $10^2 s^{-1}$	K_m $10^{-4}M^{-3}$	k_{cat}/K_m $10^5 Ms^{-1}$	Selectivity[a]
1	{(S)-**6**}	(S)-**5**	5.51	5.19×10^{-5}	106.0	21.2
	{(S)-**6**}	(R)-**5**	5.64	1.13×10^{-4}	5.0	(1.0)
2	{(S)-**6**}	(S)-**6**	9.23	2.02×10^{-4}	18.6	17.2
	{(S)-**6**}	(R)-**6**	13.60	1.25×10^{-3}	1.1	(1.0)
3	{(S)-**6**}	(S)-**7**	8.16	3.65×10^{-5}	22.4	16.1
	{(S)-**6**}	(R)-**7**	9.61	6.92×10^{-4}	1.4	(1.0)
4	{(S)-**6**}	(S)-**5***	10.80	6.69×10^{-4}	16.2	17.4
	{(S)-**6**}	(R)-**5***	16.20	1.74×10^{-3}	0.9	(1.0)
5	{(R)-**6**}	(R)-**5**	8.37	7.01×10^{-5}	11.9	14.4
	{(R)-**6**}	(S)-**5**	12.90	1.53×10^{-3}	0.8	(1.0)
6	{(R)-**6**}	(R)-**6**	8.62	5.31×10^{-5}	16.2	19.9
	{(R)-**6**}	(S)-**6**	10.05	1.29×10^{-3}	0.8	(1.0)
7	{(R)-**6**}	(R)-**7**	6.82	8.85×10^{-5}	7.7	14.4
	{(R)-**6**}	(S)-**7**	2.15	4.03×10^{-4}	0.5	(1.0)
8	{(R)-**6**}	(R)-**5***	8.24	6.10×10^{-5}	13.5	19.5
	{(R)-**6**}	(S)-**5***	13.20	1.90×10^{-4}	0.7	(1.0)

[a] The selectivities for the catalytic sites in the control material doped with Al(II), but not imprinted are negligible (0.97-1.01).

Enantioselective Racemization over Chirally Imprinted {6}. Using pyridine, {(R)-6} catalyzed the racemization of (R)-5-phenylhydantoin 1.6-fold more effectively than did the (S)-isomer and vice versa (*23*).

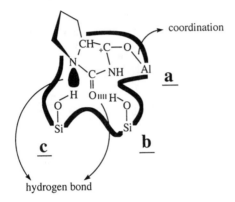

Figure 6. The Cavity Imprinted with (*S*)-**6**.

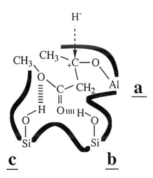

Figure 7. Asymmetric Reduction of Methyl Acetoacetate in Cavity {(*S*)-**6**}.

Asymmetric Synthesis over Chirally Imprinted Cavities.

Chirally imprinted {(S)-6} can recognize the enantio-face of a hydantoin ring using different recognition sites (a, b, and c) placed counterclockwise. When the cavities of {(S)-6} bind the β-ketoester, methyl acetoacetate, the recognition sites impose a definite configuration of the bound molecule, as shown in Figure 7. The pro-chiral ketone should coordinate to the a-site. Attack of a nucleophile is allowed only from the Re-face. Therefore, an asymmetric reaction is expected. We investigated the asymmetric reduction of methyl acetoacetate using NaHB(OCH$_3$)$_3$ over {(S)-6} in acetonitrile at room temperature. The highest enantiomeric excess observed was 80% (Morihara, unpublished results). Similar asymmetric reductions over chiral cavities are also promising. Cavities imprinted with chiral (+)/(-)-benzyl-phenyl-sulfoxide catalyzed the asymmetric reduction of acetophenone with NaBH$_3$(CN) in acetonitrile at room temperature. The observed e.e. values varied from 60 to 99% (Morihara, unpublished results). Cavities imprinted with chiral (+)/(-)-N-benzyl-methyl-p-toluyl-N-oxide catalyzed the asymmetric reduction of acetophenone with NaBH$_3$(CN) under the same reaction conditions with 30% e.e. (24). Optimization is now in progress with the objective of producing higher and more reproducible e.e. values.

Conclusions

Footprint cavities can catalyze a variety of reactions, i.e. substrate-selective acyl transfer reactions, substrate-selective Claisen-Schmidt condensations, enantioselective acyl transfer reactions with different mechanisms, enantioselective racemizations, and asymmetric reductions. The configuration of template which is bound during the imprinting process predetermines the structure and recognition capability of the cavities. Recognition of substrates by the cavities controls the orientation of the molecules, which determines the nature and mechanisms of the catalyzed reactions. These findings demonstrate that consideration of the optimal binding orientations of template molecules can assist in the enhancement of molecular recognition and catalysis by the cavities. Our strategy for cavity design is shown to be widely applicable and of promise in the fields of specific separations and specific catalysis.

Literature Cited

1. Wülff, G. Angew. Chem., Int. Ed. Engl. 1995, 34, 1812.
2. Mosbach, K.; Ramström, O. Biotechnology 1996, 14, 163.
3. Dickey; F. H. J. Phys. Chem. 1955, 59, 695.
4. Morihara, K.; Kurihara, S.; Suzuki, S. Bull. Chem. Soc. Jpn. 1988, 61, 3991.
5. Morihara, K.; Nishihata, E.; Kojima, M.; Miyake, S. Bull. Chem. Soc. Jpn. 1988, 61, 3991.
6. Morihara, K.; Tanaka, E.; Takeuchi, Y.; Miyazaki, K.; Yamamoto, N.; Sagawa, Y.; Kawamoto, E.; Shimada, T. Bull. Chem. Soc. Jpn. 1989, 62, 499.
7. Shimada, T.; Nakanishi, K.; Morihara, K. Bull. Chem. Soc. Jpn. 1992, 65, 954.
8   Shimada, T.; Kurazono, R.; Morihara, K. Bull. Chem. Soc. Jpn. 1993, 66, 836.
9. Morihara, K.; Kawasaki, S.; Kofuji, M.; Shimada, T. Bull. Chem. Soc. Jpn. 1993, 66, 906.
10. Morihara, K.; Doi, S.; Takiguchi, M., Shimada, T. Bull. Chem. Soc. Jpn. 1993, 66, 297.
11. Morihara, K.; Iijima, T.; Usui, H.; Shimada, T. Bull. Chem. Soc. Jpn . 1993, 66, 3047.
12. Matsuishi, T.; Shimada, T.; Morihara, K. Bull. Chem. Soc. Jpn. 1994, 67, 748.

13. Shimada, T.; Hirose, R.; Morihara, K. *Bull. Chem. Soc. Jpn.* **1994**, *67*, 227.

14. Morihara, K.; Takiguchi, M.; Shimada, T. *Bull. Chem. Soc. Jpn.* **1994**, *67*, 1078.

15. Morihara, K.; Kurokawa, M.; Kamata, Y.; Shimada, T. *J. Chem. Soc., Chem. Commun.* **1992**, 358.

16. Matsuishi, T.; Shimada, T.; K. Morihara, K. *Chem. Lett.* **1992**, 1921.

17. Shimada, T.; Tomita, M.; Morihara, K. *1995 International Chemical Congress of Pacific Basin Societies,* Honolulu, Hawaii, December 1996, Abstract 281.

18. Tomita, M.; Shimada, T.; Morihara, K. *Proceedings of the 70th Annual Meeting of the Chemical Society of Japan,* 1996, 3PC079.

19. Wolfenden, R. *Acc. Chem. Res.* **1972**, *5*, 10.

20. Tramontano, A.; Janda, K.; Lerner, R. *Science* **1986**, *234*, 1566.

21. Pollack, S. J.; Jacobs, J. W.; Schultz, P. G. *Science* **1986**, *234*, 1570.

22. Shimada, T.; Munekyo, H.; Takahashi, Y.; Morihara, K. *Proceedings of the 72nd Annual Meeting of the Chemical Society of Japan,* 1997, 3PA105.

23. Tomita, M.; Yamamoto, T.; Shimada, T.; Morihara, K. *Proceedings of the 72nd Annual Meeting of the Chemical Society of Japan,* 1997, 2B440.

24. Shimada, T.; Ishii, M.; Toyota, K.; Imamura, K.; Kakehi, K.; Morihara, K. *Proceedings of the 72nd Annual Meeting of the Chemical Society of Japan,* 1997, 3PA104.

Chapter 22

Molecular Imprinted Receptors in Sol-Gel Materials for Aqueous Phase Recognition of Phosphates and Phosphonates

Darryl Y. Sasaki[1], Daniel J. Rush[2], Charles E. Daitch[1], Todd M. Alam[1], Roger A. Assink[1], Carol S. Ashley[1], C. Jeffrey Brinker[1], and Kenneth J. Shea[2]

[1]Sandia National Laboratories, Albuquerque, NM 87185
[2]Department of Chemistry, University of California at Irvine, Irvine, CA 92717

Synthetic receptors for phosphates and phosphonates have been generated in SiO_2 xerogels via a surface molecular imprinting method. The monomer 3-trimethoxysilylpropyl-1-guanidinium chloride (1) was developed to prepare receptor sites capable of binding with substrates through a combination of ionic and hydrogen bond interactions. HPLC studies and adsorption isotherms performed in water have found that molecular imprinting affords a significant improvement in K_a for phosphate and phosphonate affinity over a randomly functionalized xerogel. Affinities for these materials offer about an order of magnitude improvement in affinity compared to analogous small molecule receptors reported in the literature. The xerogel matrix appears to participate in host-guest interactions through anionic charge buildup with increasing pH.

The selective and rapid detection of phosphate and phosphonate compounds, which include numerous biologically important signaling molecules (1-3) as well as pesticides and chemical warfare agents (4), is dependent upon the development of efficient host-guest systems. Various solution phase synthetic molecular host-guest systems have been developed to selectively bind phosphate esters using hydrogen bonding and electrostatics to achieve high affinity (5-9). A major drawback is the inability of these systems to achieve high complexation in the presence of water, a situation in which the detection of these compounds is desired. In contrast to these small molecule hosts, self-assembled Langmuir monolayers with guanidine functionality have produced some high binding constants for nucleotides in water through guanidine-phosphate pairing (10). It is believed that the monolayer host systems have some uniquely structured water near the interface that enhances host-guest binding, a feature non-existent for small molecule systems which operate in bulk water. Molecular receptors embedded in polymer scaffolds, like proteins, may offer interfacial reordering of water near the receptor that could enhance substrate binding (11,12). Although it is unclear how to mimic proteins in this regard, efforts into the design and preparation of molecular receptors on polymer matrices may lead to the development of molecular recognition materials with enhanced selectivity and sensitivity under aqueous conditions.

Successful preparation of synthetic receptors in crosslinked polymers via the

314

molecular imprinting methodology has generated a broad range of materials with molecular recognition properties (*13*). This method has already demonstrated that some highly selective receptor sites can be built for complex molecules, such as sugars (*14*), amino acids (*15*), peptides (*16*), nucleosides (*17*), and others, in polar organic solvents. Some strides in binding affinity for host-guest interactions in aqueous solution have been made in recent years (*18, 19*). However, selectivity and affinity in the presence of water remains a problem in these organic based materials.

We describe herein a new silica sol-gel molecular imprinted material for aqueous phase recognition of phosphate and phosphonate substrates. The receptors are functionalized with the guanidine moiety which provides specific points of interaction in the aqueous environment. The guanidine residue was chosen for its ubiquitous presence in protein receptors sites often acting as the main interaction point for phosphate compounds (*20*). The guanidinium-phosphate interaction employs a combination of electrostatic as well as bidentate hydrogen bonds to achieve high recognition properties (*21*). Guanidine alkyl silanes were used to functionalize the silica matrix formed by a sol-gel process. Previous studies of imprinting in silica gels have demonstrated that metal oxides can provide for good affinity and selectivity of guest substrates in organic and aqueous solutions (*22-26*). The sol-gel process for molecular imprinting introduced here offers the ability of designing and building the material's matrix starting at the molecular level allowing facile tailoring of the material's polarity, pore dimension, and surface functionalization. Additionally, silica sol-gels offer many desirable features as the scaffold for molecular receptor sites which include a hydrophilic surface, optical transparency (for applications as sensor materials), and a highly crosslinked, robust structure with high surface area.

Experimental

General. Silane reagents were obtained from Gelest, Inc. Reagent 3-Amino-1-trimethoxypropylsilane was distilled prior to use. Water was purified through a Barnstead Type D4700 NANOpure Analytical Deionization System. 1-H-pyrazole-1-carboxamidine hydrochloride reagent was prepared by the reported procedure (*27*). Solution phase ^{1}H, ^{13}C, and ^{31}P NMR experiments were performed on a Bruker AM300 NMR spectrometer, and infrared spectra were obtained on a Perkin-Elmer 1750 FTIR spectrophotometer. Mass spectroscopy was performed by Mass Consortium, Corp.

1-Trimethoxysilylpropyl-3-guanidinium chloride (1). A solution of 3-amino-1-trimethoxypropylsilane (10.0 g, 55.8 mmole) and 1-H-pyrazole-1-carboxamidine hydrochloride (8.20 g, 55.9 mmole) in dry methanol (30 mL) was stirred for 15 hours at room temperature. The solvent was removed in *vacuo* leaving a viscous, colorless oil which was purified by Kugelrohr distillation (150 °C, 100 μmHg) to remove the pyrazole byproduct. A clear, colorless, viscous oil of the product **1** remained (12.7 g, 88%). ^{1}H NMR (CDCl$_3$) δ 7.83 (t, J = 6.5 Hz, 1H, CH$_2$N$\underline{H}$), 7.07 (br s, 4H, N$\underline{H}_2$), 3.58 (s, 9H, OC$\underline{H}_3$), 3.20 (dt, J = 6.5, 6.5 Hz, 2H, NHC$\underline{H}_2$), 1.72 (m, 2H, C$\underline{H}_2$), 0.72 (t, J = 7.8 Hz, 2H, Si-C$\underline{H}_2$). ^{13}C NMR (CDCl$_3$) δ 157.75, 50.59, 43.31, 22.42, 5.83. IR (NaCl) 3330, 3166, 2946, 2850, 1652, 1469, 1192, 1083, 819 cm^{-1}. High resolution MS calcd for C$_7$H$_{20}$N$_3$O$_3$Si: 222.1274. Found: 222.1281.

Xerogel Preparation. Ethanol (305 mL) and tetraethoxysilane (305 mL) were stirred in a 1 L vessel followed by the addition of water (24.5 mL) and 1N aqueous HCl (1 mL). The mixture was warmed to 60 °C with stirring for 1.5 hours, then cooled to room temperature to afford the homogeneous sol solution. To the sol (91 mL) was added 0.1 M aqueous NH$_4$OH (9.1 mL). The solution gelled overnight and was aged in a closed container at 50 °C for one day. The gel was then crushed, washed with ethanol twice, collected, placed in fresh ethanol (200 mL), and kept at 50 °C overnight. For molecular imprinting the gel was collected and placed in a 100 mL solution of ethanol containing **1** (8.0 mmole) and phenylphosphonic acid (4.0 mmole) template and the mixture incubated for another day at 50 °C. The solvent was subsequently evaporated at 50 °C over a period of 12 hours. Blank and randomly functionalized xerogels were prepared identically with the exclusion of **1** and phenylphosphonic acid (PPA) for the blank gel and of PPA for the randomly functionalized gel. The xerogel was then crushed to a 75-250 micron particle size, washed with ethanol, and dried under vacuum at 60 °C for a day. Prior to rebinding or HPLC studies, all materials were washed three times with 1 N aqueous HCl solution (100 mL/g xerogel) by swirling one hour for each wash at room temperature. The gels were then washed liberally with pure water followed by drying under vacuum at 60 °C overnight. Quantitative removal of the PPA template was determined by UV analysis (PPA in 50% methanol/1 N aqueous HCl, $\varepsilon = 8332$ at 210 nm).

HPLC Analysis. Xerogels processed to a particle size of 25-38 μm were slurry packed into a 4.6 mm i.d. x 100 mm HPLC column which was connected to a Waters 600-MS HPLC fitted with a Waters 484 tunable wavelength UV detector set at 260 nm. All studies used a mobile phase consisting of 95% aqueous buffer solution/5% acetonitrile. The ionic strength dependence study used an aqueous buffer composed of 0.01 M potassium phosphate adjusted to pH 6.0 with either HCl or KOH, and the ionic strength was adjusted to the appropriate value by the addition of KCl. The pH dependence study used an aqueous buffer composed of 0.01 M potassium phosphate adjusted to the appropriate pH with either HCl or KOH, where the ionic strength was maintained at 0.05 M with the addition of KCl. Capacity factors for Figure 5 were determined using acetone as the void volume marker, an aqueous phase of 0.01 M potassium phosphate buffer adjusted to pH 6.0 or 8.0, and ionic strength of 0.05 M.

Rebinding Studies. Xerogel (50 mg) at 75-150 μm particle size was placed in a solution containing a phosphate or phosphonate substrate at various concentrations in a volume such that the total bound substrate in the xerogel amounted to less than 10 mole % of the total substrate in solution. The pH of the solution was adjusted to ~5 using the substrates as the buffering medium. Although the solution equilibrates with the xerogel in minutes the solution was swirled overnight prior to analysis. An aliquot was taken from the solution and analyzed for depletion by an acid molybdate/Fiske & SubbaRow reducer assay (Sigma Chemicals) for PO_4 or by ^{31}P NMR for PPA. The binding data were evaluated by Scatchard analysis giving linear plots over the concentration range of 1 - 10 mM. Binding constants (K_a) and binding capacities (guanidine:substrate) are reported in Table II.

Solid State NMR. Direct polarization ^{29}Si solid state MAS NMR spectra were recorded on a Bruker AMX 400 spectrometer at a resonant frequency of 79.500 MHz using a 7 mm bb MAS probe and a 4 kHz spinning speed. The reference was external Q_8M_8 ($\delta = 11.5$ ppm). These spectra were recorded with a delay time of 300 seconds which is several times longer than the relaxation times of the sample to ensure quantitative spectra of the Q silicon species, where $Q^n = Si(OSi)_n(OR)_{4-n}$. Cross-polarization ^{29}Si spectra of the 4% xerogel were recorded at 39.7 MHz on a Tecmag console interfaced to a Chemagnetics spectrometer. A 7 mm MAS probe, a 3.5 kHz spinning speed, a contact time of 6 ms and a delay time of 4 seconds were used. The cross-polarization time was chosen so that the T resonances and the Q^2 and Q^3 resonances were expected to be quantitative. The resonances of the Q and T silicon species were deconvoluted using Gaussian components to calculate the extents of reaction and the relative amounts of Q and T silicons.

Results and Discussion

The molecular imprinting technique employed for the sol-gel materials allowed surface functionalization of the matrix to maximize accessibility of receptor sites while minimizing structure modification of the gel matrix. A SiO_2 xerogel was first prepared from tetraethoxysilane using a standard literature procedure for sol-gel processing (28). The xerogel was functionalized with the guanidine siloxane monomer, 3-trimethoxysilylpropyl-1-guanidinium chloride (1), prepared by reaction of 1-trimethoxysilylpropyl-3-amine with 1-H-pyrazole-1-carboxamidine hydrochloride. The siloxane monomer 1 was condensed onto the silica surface in ethanol solution in the presence of the template molecule, phenylphosphonic acid (PPA), in a 2:1 ratio to generate a receptor site as ideally illustrated in Figure 1. The finished material was dried, gently ground, and sieved to appropriate size.

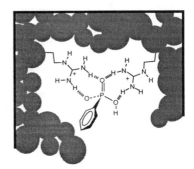

Figure 1. Idealized Bis-guanidine Receptor Site for Phosphonate.

Materials Characterization. Structural information for the xerogels was obtained through ^{29}Si solid state MAS NMR providing quantitative determination of the extent of condensation of the matrix and functionalization with 1. Direct polarization ^{29}Si NMR was used to quantify the Q, or $Si(OR)_4$, species in the silica gel, where R = Si, H, or ethyl (29). From the abundance of various Q species the extent of condensation of matrix silica could be determined. The blank xerogel was found to have $90 \pm 1\%$ condensation indicating high crosslinking. A comparison of blank xerogel and xerogel functionalized with 1 showed that 1 to 4 mole % functionalization does not affect condensation of the gel matrix. Cross-polarization experiments measuring the T silicon ($SiR'(OR)_3$, where R = Si, H, or ethyl and R' = propyl-3-guanidinium) of covalently bound 1 determined an incorporation of 1 into the gel at $93 \pm 10\%$ with an extent of condensation of $91 \pm 2\%$. Thus, the procedure used for surface functionalization was successful with near quantitative coupling of siloxane monomer to the matrix and an approximate three point covalent attachment.

Surface area analyses were performed to determine any gross structural changes that might occur upon guanidine functionalization of the xerogel surface. Table I shows nitrogen BET analyses of blank xerogel and gels randomly functionalized with 1 at 2% and 4%, and an imprinted gel with 4% of 1 and PPA imprinting at a 2:1 ratio of guanidine to PPA. The percentages of 1 are indicated as a mole percent relative to total silicon in the gel. With 2% functionalization the xerogel shows only a slight increase in surface area to 915 m^2/g over the non-functionalized material (876 m^2/g) but a 65% increase in pore volume. Upon further functionalization to 4%, the surface area drops appreciably (836 m^2/g) along with pore volume. Addition of the PPA template brings about a further decrease in surface area while pore volume remains constant. Overall the functionalization process does not significantly alter the xerogel structure, however, some trends were observed. Most noticeable is that surface functionalization with 1 leads to an increase in pore volume of 35-65%. Since 1 can act as a capping agent for surface silanols, surface functionalization could minimize condensation across walls of collapsing pores leading to higher pore volumes for functionalized xerogels (30). We can offer no good explanation as to why the surface area decreases as functionalization increases from 2 to 4% and further with PPA imprinting. With an average calculated area per bound guanidine of 225 Å^2, site isolation is highly probable assuming random distribution.

Table I. BET Surface Area Analysis

% 1/SiO$_2$	Template molecule	Surface area (m^2/g)	Pore volume (cm^3/g)
0	none	876	0.63
2	none	915	1.01
4	none	836	0.86
4	PPA	771	0.85

Host-guest Interactions.

Ionic Strength. In an effort to understand host-guest interactions of phosphate and phosphonate substrates to these sol-gel materials, several HPLC studies were performed. Relative affinities of a phosphonate diester and a phosphonic acid to blank and randomly functionalized 4% guanidine xerogels were measured in an ionic strength study to qualitatively determine the contribution of ionic interaction between the host and guest. The eluent used was a solvent mixture of 95% 0.01 M potassium phosphate buffered water at pH 6.0 and 5% acetonitrile. The ionic strength was adjusted with KCl. Specific interaction of the guanidines with the phosphonic acid substrate should involve a combination of electrostatics and hydrogen bonding. On the other hand, binding of phosphonate diester would occur solely through hydrogen bonds. Figure 2 shows the effect of ionic strength on retention time to blank and functionalized gels with acetone, diisopropylphenylphosphonate (DIPP), and phenylphosphonic acid (PPA) substrates. Increasing ionic strength is indicated with increasing darkness of the bar. Acetone, which has no specific interaction to the materials, shows essentially no difference in affinity to both gels and no effect due to changes in ionic strength. DIPP exhibits a high non-specific affinity to the blank xerogel, but this affinity diminishes upon functionalization of the gel. Conversely, PPA shows no affinity for the blank gel, but high affinity for the 4% guanidine gel.

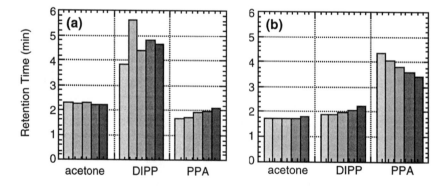

Figure 2. HPLC Study of the Effect of Ionic Strength on Substrate Affinity to Blank and Functionalized Xerogel. Graphs are of data from blank (a) and 4% guanidine (b) xerogels. The ionic strength increases from left to right above each substrate from 0.02, 0.05, 0.10, 0.20, to 0.50 M.

Functionalization of the gel with **1** effectively minimizes non-specific interaction of phosphonate esters while adding specificity for phosphonic acids. The effect of ionic strength on the affinity of PPA to 4% guanidine gel shows a trend of reduced affinity with increasing ionic strength. This is an expected observation for phosphonate-guanidine complexation where ionic pairing contributes to the binding interaction. Guanidine-phosphonate pairing has an additional hydrogen bond component which is believed to provide coordination of substrates in protein receptor sites (*11*). It is possible that hydrogen bonding also actively participates in these synthetic receptors and is responsible for the affinity observed at higher ionic strengths. From the data, however, it is evident that hydrogen bonding between DIPP and guanidine is negligible under the aqueous conditions.

Effect of Solution pH. HPLC studies on the effect of pH on substrate retention time shows a trend of diminishing affinity with increasing pH. The mobile phase in these studies was a 5% acetonitrile/95% aqueous phosphate buffer solution adjusted to the desired pH, and ionic strength held at 0.05 with added KCl. Figure 3 shows a comparison of blank and guanidine functionalized xerogel run with acetone, DIPP, and PPA at several pH values from 2 to 10. The retention time of acetone is unaffected by changes in pH or gel type. For the phosphonate substrates, however, DIPP on the blank gel and PPA on functionalized gel show an inverse relationship between retention time (affinity) and pH. Moreover, a complete loss of affinity is found at pH 8 and above. In considering the specific interaction of the guanidine functionalized gel with PPA, strongest host-guest interactions should be found near neutral pH. However, the data show that highest affinity is found at the lower pH values. The PPA affinity appears to be influenced by the extent of deprotonation of the acidic silanol groups on the silica matrix producing an increasingly anionic surface above pH 2. Although the guanidine group can specifically bind with PPA the silica matrix dominates substrate interaction with changing solution pH. Figure 4 illustrates the possible host-guest-matrix interaction at low and high pH. At low pH the silica surface will be close to neutral charge allowing the guanidinium group to bind freely with PPA. At higher pH a negatively charged surface could force the cationic guanidinium to interact with the surface and restrict binding with the PPA guest (Figure 4). Other contributions for poor affinity at high pH could arise from ionic repulsion between the matrix and substrate or surface effects which may lower the pK_a of guanidinium (13.6 in water) thereby minimizing electrostatic interactions with

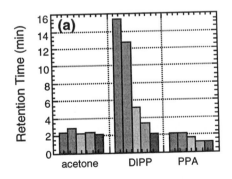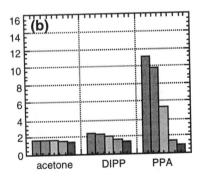

Figure 3. HPLC Study of the Effect of pH on Substrate Affinity on Blank and Functionalized Xerogels. Graphs are of data from blank (a) and 4% guanidine (b) xerogels. The pH increases from left to right in the order of 2.0, 4.0, 6.0, 8.0, and 10.0.

320

PPA at pH's lower than predicted. For the phosphonate diester DIPP it is unclear why there is a pH dependent non-specific interaction to the blank gel although reasonable explanations, which are out of the scope of this text, involving polar interactions with the surface could be made. Functionalization of the gel with **1**, however, appears to remove the non-specific affinity at all pH levels.

An HPLC study comparing imprinted *vs* randomly functionalized 4% guanidine xerogels is shown in Figure 5. The two graphs present a comparison of capacity factors for DIPP and PPA to the xerogels at two different pH values of 6 and 8. At pH 6, PPA shows a modest, yet significantly higher, affinity for the imprinted material with a 25% increase in the capacity factor over the randomly functionalized material. DIPP, on the other hand, shows a reduced affinity for the imprinted gel. The non-specific affinity for DIPP may be affected by differences in surface area which appears as a byproduct of molecular imprinting as found through BET analysis

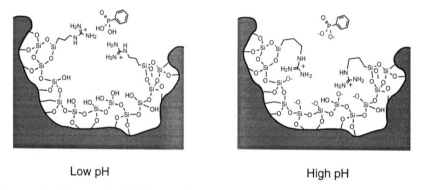

Low pH   High pH

Figure 4. Possible Scenarios for PPA Interaction with the Guanidine-Functionalized Receptor Site at Low and High pH.

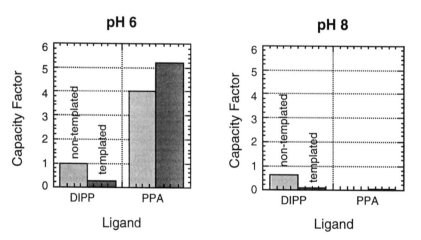

Figure 5. A Comparison of HPLC Capacity Factors of Imprinted, or Templated, *vs* Non-templated Guanidine-functionalized Xerogels at pH 6 and 8, Run with DIPP and PPA Substrates.

(Table I). At pH 8, there is complete loss of affinity of PPA to the xerogel. It is evident from these studies that the surface near the receptor site can play a major role in the host-guest interactions. Investigation into ways to tailor these surfaces to improve affinities as well as tune the receptor to various conditions are forthcoming.

Binding Constants. Further evaluation of the molecularly imprinted xerogels was assessed through experimental binding isotherms obtained in water at pH 5. Table II shows binding constants (K_a) and binding capacities (guanidine:substrate) for blank, 4% guanidine randomly functionalized, and 4% guanidine-PPA imprinted xerogels with phosphate (PO_4) and PPA as guest substrates. Blank xerogel was found to have negligible affinity for phosphate and PPA. On the other hand, randomly functionalized silica exhibited good binding constants for phosphate (600 M^{-1}) and PPA (1100 M^{-1}) in water. Furthermore, molecular imprinting with PPA yields an enhancement of > 2 in binding constants for both phosphate (~1500 M^{-1}) and PPA (2600 M^{-1}) compared to randomly functionalized gel. These results show that it is possible to generate refined receptor sites in sol-gel materials through a surface imprinting technique. Additional effects, such as hydrophobicity, also appear to play a role in binding of substrates to the xerogels. For example, PPA, which has a hydrophobic phenyl substituent, exhibits about a two-fold higher K_a value than the highly water soluble phosphate in both imprinted and non-imprinted functionalized materials. As has been observed with Langmuir monolayers, interfacial phenomena may promote stronger partitioning of the hydrophobic substrate to the gel bound receptor.

Table II. Binding Data of Imprinted and Non-imprinted Xerogels

% guanidine	guanidine: template	substrate	K_a (M-1)[a]	guanidine: substrate
0	none	PO_4	0	
0	none	PPA	0	
4	none	PO_4	600	1.6:1
4	none	PPA	1100[a]	2.0:1
4	2:1 PPA	PO_4	1400	2.5:1
			1600[a]	2.8:1
4	2:1 PPA	PPA	2600[a]	2.8:1

a) Data obtained by ^{31}P NMR.

Conclusions

We have shown through these initial studies that molecular imprinting in sol-gel materials can be successfully accomplished with a simple surface imprinting procedure. This method does not explicitly provide size and shape imprinting in the matrix, which is a feature of molecular imprinting. However, the method is chosen to be compatible to the sol-gel matrix, provide quantitative surface functionalization while maintaining high surface area and large pore size, and allow rapid equilibration of the receptor sites to solution. Affinities measured in aqueous solutions of phosphate and phosphonate substrates to these imprinted xerogels are one to two orders of magnitude higher than analogous small molecule receptors including pre-organized bis- and tris-guanidine receptors (8,9). The guanidine-phosphonate pairing on the gel surface shows evidence of contributions from both ionic and hydrogen bonding interactions in aqueous solution. However, the silica matrix exhibits increasing influence over the efficiency of the receptor site with rising pH due to build up of surface charge. It is anticipated that by tailoring the matrix surface the specificity of the receptors will improve and result in highly efficient molecular recognition materials for aqueous systems.

Acknowledgments

We would like to thank Dr. Robert Lagasse and John Schroeder for the BET surface area analyses and their helpful discussions. We also thank the Technical Support Working Group of DoD's Office of Special Technology for financial support. Sandia is a multiprogram laboratory operated by Sandia Corporation, a Lockheed Martin company, for the United States Department of Energy under Contract DE-AC04-94AL85000.

Literature cited

1) Potter, B. V. L.; Lampe, D. *Angew. Chem. Int. Ed. Engl.* **1995**, *34*, 1933.
2) Klarlund, J. K.; Guilherme, A.; Holik, J. J.; Virbasius, J. V.; Chawla, A.; Czech, M. P. *Science* **1997**, *275*, 1927.
3) Kauffmann-Zeh, A.; Thomas, G. M. H.; Ball, A.; Prosser, S.; Cunningham, E.; Cockcroft, S.; Hsuan, J. J. *Science* **1995**, *268*, 1188.
4) *Chemical Warfare Agents: Toxicology and Treatment* Marrs, T. C.; Maynard, R. L.; Sidell, F. R., Eds.; John Wiley & Sons: New York, 1996.
5) Hirst, S. C.; Tecilla, P.; Geib, S. J.; Fan, E.; Hamilton, A. D. *Israel J. Chem.* **1992**, *32*, 105.
6) Czarnik, A. W. *SPIE - Fiber Optic Medical and Fluorescent Sensors and Applications* **1992**, *1648*, 164.
7) Deslongchamps, G.; Galán, A.; Mendoza, J. d.; Rebek, J. J. *Angew. Chem., Int. Ed. Engl.* **1992**, *31*, 61.
8) Dietrich, B.; Fyles, D. L.; Fyles, T. M.; Lehn, J.-M. *Helv. Chim. Acta* **1979**, *62*, 2763.
9) Dietrich, B.; Fyles, T. M.; Lehn, J.-M.; Pease, L. G.; Fyles, D. L. *JCS Chem. Comm.* **1978**, 934.
10) Sasaki, D. Y.; Kurihara, K.; Kunitake, T. *J. Am. Chem. Soc.* **1991**, *113*, 9685.
11) Blokzijl, W.; Engberts, J. B. F. N. *Angew. Chem. Int. Ed. Engl.* **1993**, *32*, 1545.
12) Schwabe, J. W. R. *Curr. Opin. Struct. Biol.* **1997**, *7*, 126.
13) For a comprehensive review, see: Wulff, G. *Angew. Chem. Int. Ed. Engl.* **1995**, *34*, 1812.
14) Wulff, G.; Grobe-Einsler, R.; Vesper, W.; Sarhan, A. *Makromol. Chem.* **1977**, *178*, 2817.
15) Sellergren, B.; Lepistö, M.; Mosbach, K. *J. Am. Chem. Soc.* **1988**, *110*, 5853.
16) Ramström, O.; Nicholls, I. A.; Mosbach, K. *Tetrahedron: Asymmetry* **1994**, *5*, 649.
17) Shea, K. J.; Spivak, D. A.; Sellergren, B. *J. Am. Chem. Soc.* **1993**, *115*, 3368.
18) Andersson, L. I. *Anal. Chem.* **1996**, *68*, 111.
19) Andersson, L. I.; Müller, R.; Vlatakis, G.; Mosbach, K. *Proc. Natl. Acad. Sci. U.S.A.* **1995**, *92*, 4788.
20) Calnan, B. J.; Tidor, B.; Biancalana, S.; Hudson, D.; Frankel, A. D. *Science* **1991**, *252*, 1167.
21) Cotton, F. A.; Day, V. W.; Hazen Jr., E. E.; Larsen, S.; Wong, S. T. K. *J. Am. Chem. Soc.* **1974**, *96*, 4471.

22) Raman, N. K.; Anderson, M. T.; Brinker, C. J. *Chem. Mater.* **1996**, *8*, 1682.

23) Morihara, K.; Kurihara, S.; Suzuki, J. *Bull. Chem. Soc. Jpn.* **1988**, *61*, 3991.

24) Wulff, G.; Heide, B.; Helfmeier, G. *J. Am. Chem. Soc.* **1986**, *108*, 1089.

25) Heilmann, J.; Maier, W. F. *Angew. Chem. Int. Ed. Engl.* **1994**, *33*, 471.

26) Dickey, F. H. *Proc. Natl. Acad. Sci. USA* **1949**, *35*, 227.

27) Bernatowicz, M. S.; Xu, Y.; Matsueda, G. R. *J. Org. Chem.* 1992, *57*, 2497.

28) Brinker, C. J.; Keefer, K. D.; Schaefer, D. W.; Ashley, C. S. *J. Non-Crystal. Solids* **1982**, *48*, 47.

29) *Sol-Gel Science: The Physics and Chemistry of Sol-Gel Processing* Brinker, C. J.; Scherer, G. W., Eds.; Academic Press, Inc.: San Diego, 1990.

30) Prakash, S. S.; Brinker, C. J.; Hurd, A. J.; Rao, S. M. *Nature* **1995**, *374*, 439.

Author Index

325

Subject Index

M

L

Bestsellers from ACS Books

The ACS Style Guide: A Manual for Authors and Editors (2nd Edition)
Edited by Janet S. Dodd
470 pp; clothbound ISBN 0–8412–3461–2; paperback ISBN 0–8412–3462–0

Writing the Laboratory Notebook
By Howard M. Kanare
145 pp; clothbound ISBN 0–8412–0906–5; paperback ISBN 0–8412–0933–2

Career Transitions for Chemists
By Dorothy P. Rodmann, Donald D. Bly, Frederick H. Owens, and Anne-Claire Anderson
240 pp; clothbound ISBN 0–8412–3052–8; paperback ISBN 0–8412–3038–2

Chemical Activities (student and teacher editions)
By Christie L. Borgford and Lee R. Summerlin
330 pp; spiralbound ISBN 0–8412–1417–4; teacher edition, ISBN 0–8412–1416–6

Chemical Demonstrations: A Sourcebook for Teachers, Volumes 1 and 2, Second Edition
Volume 1 by Lee R. Summerlin and James L. Ealy, Jr.
198 pp; spiralbound ISBN 0–8412–1481–6
Volume 2 by Lee R. Summerlin, Christie L. Borgford, and Julie B. Ealy
234 pp; spiralbound ISBN 0–8412–1535–9

The Internet: A Guide for Chemists
Edited by Steven M. Bachrach
360 pp; clothbound ISBN 0–8412–3223–7; paperback ISBN 0–8412–3224–5

Laboratory Waste Management: A Guidebook
ACS Task Force on Laboratory Waste Management
250 pp; clothbound ISBN 0–8412–2735–7; paperback ISBN 0–8412–2849–3

Reagent Chemicals, Eighth Edition
700 pp; clothbound ISBN 0–8412–2502–8

Good Laboratory Practice Standards: Applications for Field and Laboratory Studies
Edited by Willa Y. Garner, Maureen S. Barge, and James P. Ussary
571 pp; clothbound ISBN 0–8412–2192–8

For further information contact:
Order Department
Oxford University Press
2001 Evans Road
Cary, NC 27513
Phone: 1-800-445-9714 or 919-677-0977
Fax: 919-677-1303

Highlights from ACS Books

Desk Reference of Functional Polymers: Syntheses and Applications
Reza Arshady, Editor
832 pages, clothbound, ISBN 0–8412–3469–8

Chemical Engineering for Chemists
Richard G. Griskey
352 pages, clothbound, ISBN 0–8412–2215–0

Controlled Drug Delivery: Challenges and Strategies
Kinam Park, Editor
720 pages, clothbound, ISBN 0–8412–3470–1

Chemistry Today and Tomorrow: The Central, Useful, and Creative Science
Ronald Breslow
144 pages, paperbound, ISBN 0–8412–3460–4

Eilhard Mitscherlich: Prince of Prussian Chemistry
Hans-Werner Schutt
Co-published with the Chemical Heritage Foundation
256 pages, clothbound, ISBN 0–8412–3345–4

Chiral Separations: Applications and Technology
Satinder Ahuja, Editor
368 pages, clothbound, ISBN 0–8412–3407–8

Molecular Diversity and Combinatorial Chemistry: Libraries and Drug Discovery
Irwin M. Chaiken and Kim D. Janda, Editors
336 pages, clothbound, ISBN 0–8412–3450–7

A Lifetime of Synergy with Theory and Experiment
Andrew Streitwieser, Jr.
320 pages, clothbound, ISBN 0–8412–1836–6

Chemical Research Faculties, An International Directory
1,300 pages, clothbound, ISBN 0–8412–3301–2

For further information contact:
Order Department
Oxford University Press
2001 Evans Road
Cary, NC 27513
Phone: 1-800-445-9714 or 919-677-0977
Fax: 919-677-1303